THE MOVING FRONTIER

Ashgate Economic Geography Series

Series Editors:
Michael Taylor, Peter Nijkamp, and Tom Leinbach

Innovative and stimulating, this quality series enlivens the field of economic geography and regional development, providing key volumes for academic use across a variety of disciplines. Exploring a broad range of interrelated topics, the series enhances our understanding of the dynamics of modern economies in developed and developing countries, as well as the dynamics of transition economies. It embraces both cutting edge research monographs and strongly themed edited volumes, thus offering significant added value to the field and to the individual topics addressed.

Other titles in the series:

Network Strategies in Europe
Developing the Future for Transport and ICT
Edited by Maria Giaoutzi and Peter Nijkamp
ISBN: 978-0-7546-7330-9

Tourism and Regional Development
New Pathways
Edited by Maria Giaoutzi and Peter Nijkamp
ISBN: 978-0-7546-4746-1

Alternative Currency Movements as a Challenge to Globalisation?
A Case Study of Manchester's Local Currency Networks
Peter North
ISBN: 978-0-7546-4591-7

The New European Rurality
Strategies for Small Firms
Edited by Teresa de Noronha Vaz, Eleanor J. Morgan and Peter Nijkamp
ISBN: 978-0-7546-4536-8

Creativity and Space
Labour and the Restructuring of the German Advertising Industry
Joachim Thiel
ISBN: 978-0-7546-4328-9

The Moving Frontier

The Changing Geography of Production in Labour-Intensive Industries

Edited by

LOIS LABRIANIDIS
University of Macedonia, Greece

LONDON AND NEW YORK

First published 2008 by Ashgate Publishing

Reissued 2018 by Routledge
2 Park Square, Milton Park, Abingdon, Oxon, OX14 4RN
52 Vanderbilt Avenue, New York, NY 10017

Routledge is an imprint of the Taylor & Francis Group, an informa business

Publisher's Note
The publisher has gone to great lengths to ensure the quality of this reprint but points out that some imperfections in the original copies may be apparent.

Disclaimer
The publisher has made every effort to trace copyright holders and welcomes correspondence from those they have been unable to contact.

A Library of Congress record exists under LC control number:

ISBN 13: 978-0-367-14602-3 (hbk)
ISBN 13: 978-0-367-14612-2 (pbk)
ISBN 13: 978-0-429-05268-2 (ebk)

Contents

List of Figures

List of Tables

Notes on Contributors

Lois Labrianidis is Professor in the Department of Economic Sciences, University of Macedonia, Greece as well as founder and head of the Regional Development and Planning Research Unit (RDPRU – http://www.uom.gr/rdpru). He is an economic geographer (MA, Sussex; PhD, LSE) and has done research and published on topics such as industrial location, the spatial aspects of subcontracting, the economic implications of peripheral universities on their locality, rural entrepreneurship, foreign direct investment and economic consequences of immigration. His scholarly publications include four books: *The Spatial Aspects of Subcontracting Relations in Manufacturing Production*, Paratiritis; *Provincial Universities in Greece*, Paratiritis; (co-authored) *Albanian Immigrants in Thessaloniki*, Patakis; *Economic Geography*, Patakis; (editor) *The Future of Europe's Rural Periphery*, Ashgate; *Entrepreneurship in Rural Europe*, Patakis; and (co-editor) *Cities on the Verge*, Kritiki.

Alok Aggarwal is the Founder and Chairman of Evalueserve – a company that provides various high-value-added IT enabled services to North America and Europe from its offices in India and China. Prior to this position, he was the Director of Emerging Business Opportunities for IBM Research Division Worldwide. In 1997, he founded the IBM India Research Laboratory at the Indian Institute of Technology, Delhi, and grew it into a 60-member team (with 30 PhDs and 30 Masters in Electrical Engineering, Computer Science and Business Administration) by 2000.

Orna Berry has spent over 25 years in science and technology industries, as an academic researcher, entrepreneur, policy maker and most recently, venture capitalist. She chairs the Israel Venture Association, which represents the Israeli venture capital community. Israeli VCs have played a major role in making Israel an important global source of innovation. Dr Berry is also the outgoing Chief Scientist and Director of the Industrial R&D Administration of the Ministry of Industry and Trade of the Government of Israel. In this capacity she was responsible for implementing government policy regarding support and encouragement of industrial research and development. In 1993, she co-founded ORNET Data Communication Technologies Ltd.

Bolesław Domański is Professor of Geography and Director of the Institute of Geography and Spatial Management, Jagiellonian University in Krakow, Poland. He has recently been working on issues of local and regional development, foreign direct investment and activity of transnational corporations, restructuring of old industrial regions, urban and regional inequalities, post-socialist economic and social transformation. He is the author of five books including *Industrial Control over the Socialist Town: Benevolence or Exploitation?*, Greenwood, as well as many papers and chapters in various European journals and books.

Evgeni Evgeniev has completed his PhD in Political Economy at Central European University, Budapest (2006). He specialized in European Studies at the European College of Parma (2004) and served as visiting researcher at European University Institute, Florence (2005). He has been involved in teaching MA courses at Central European University with a focus on economic and political reforms of transition economies. He co-authored two strategy reports for the Bulgarian textile and clothing industry. He has published in the field of labour-intensive industries, global value chains, EU enlargement, international financial institutions, tourism policy and regulatory reforms. Currently, he works for the World Bank in Sofia.

Grahame Fallon is a Principal Lecturer in International Business at the University of Northampton, UK, as well as Postgraduate Programmes Manager and Deputy Director of the International Research Group in Northampton Business School. His teaching and research interests centre around foreign direct investment at the national and regional level in the EU, Eastern Europe and the former Soviet Union, together with economic transition issues and international political economy. He has also carried out research into international marketing strategies, small- and medium-sized businesses and ethnic businesses. His previous work has been published in a range of journals, including *Regional Studies*, *Journal of Small Business and Enterprise Development*, *European Business Review*, *Qualitative Research* and *The Journal of Contemporary European Studies*.

Robert Guzik is Adjunct Lecturer in the Department of Regional Development, Jagiellonian University in Krakow, Poland. He is an economic geographer (PhD – Jagiellonian University). He has participated in several international research projects and published on topics such as spatial accessibility, geography of innovations, regional development, automotive industry.

Krzysztof Gwosdz is an economic geographer in the Department of Regional Development of the Jagiellonian University in Krakow, Poland. His research is focused on issues of local and regional development and social and economic problems of urban areas and old industrial regions. His latest books deal with the long-term evolution of urban settlements within the Upper Silesian Industrial District and the effects of the Special Economic Zones programme in Poland. In recent years he has taken part in several international projects: *The Other Auschwitz – Economic Change and Dead Hand of History in Poland, Regional Development in Enlarged Europe.*

Margarita Ilieva is Associate Professor in the Department of Economic Geography, Institute of Geography, Bulgarian Academy of Sciences and Professor in the Institute of Geography, Kazimierz Wielki University in Bydgoszcz, Poland. She is an economic geographer (MS, Sofia University; PhD, Institute of Geography, BAS). She has done research into and published on topics such as agriculture development, changes of land use, rural areas changes, transformation of spatial systems, labour markets, border regions development and transborder cooperation studies. She is co-author of six books in Bulgaria (published by the Publishing House of the Bulgarian Academy of Sciences and by ForCom) and of one book published in Germany (Institute of Regional Geography, Leipzig). She has publications in journals and books in Bulgaria, Germany, Romania, Poland, Czech Republic, Hungary and Great Britain.

Christos Kalantaridis is Professor of Entrepreneurship and Regional Development and Director of the Centre for Enterprise and Innovation Research at Salford University. He has conducted research on the integration of formerly peripheral areas in the global network of production and distribution in Greece, the Ukraine and Russia. He has also explored rural entrepreneurship and contributed to the advancement of an institutionalist perspective in the field of entrepreneurial studies. He has published extensively in international, refereed journals, including: *Regional Studies*, *Environment and Planning A*, *Entrepreneurship and Regional Development* and *Journal of Economic Issues*.

Athanasios Kalogeresis is a senior researcher at the Regional Development and Policy Research Unit (RDPRU), University of Macedonia, Greece and an Adjunct Lecturer at the Department of Civil Engineering, University of Thessaly, Greece. An economist by education (PhD, University of Macedonia), he has participated in numerous research projects and published on topics such as rural innovation and entrepreneurship, international trade and FDI.

Martin Kenney is a Professor and Fellow at the Center for Entrepreneurship at UC, Davis, California. He has authored or edited five books and over 120 scholarly articles on the globalization of services, the history of venture capital, university-industry relations and the development of Silicon Valley. His two recently edited books *Understanding Silicon Valley* and *Locating Global Advantage* (with Richard Florida) were published by Stanford, where he is the editor of a book series on innovation and globalization. He has consulted for or presented to various US and international organizations. His research is currently supported by the NSF, the Sloan Foundation and the Kauffman Foundation.

Spartak Keremidchiev is Senior Researcher in Theory of the Firm Department, Institute of Economics, Bulgarian Academy of Sciences. He received his MSc from Sofia Economics University and PhD from the Institute of Economics. He has published in the field of privatization, corporate governance, enterprise restructuring and industrial democracy. His scholarly publications include approximately 40 articles, and he has co-authored four books: (1999), *Privatization and Economic Performance in Central and Eastern Europe; Lessons to be Learnt from Western Europe*, Elgar; (1999), *Corporate Governance of Privatized Enterprises*, Sofia, (in Bulgarian); (1998), *The Competitiveness of Countries in Transition During their Economic Depression and Recovery*, Ljubljana; and (1996), *Comparative Analysis of Structural Reforms in Transition Economies* (in Bulgarian).

Artemios Kourtesis is a researcher and member of the Regional Development and Policy Research Unit (RDPRU). He holds a PhD in Economics, University of Macedonia, Greece and an MSc in Management Sciences, University of Southampton, UK. His academic interests are in the areas of systems of innovation and regional policies.

Stefanie Ann Lenway is Dean of the College of Business Administration at the University of Illinois at Chicago. She has recently been working on a cross-industry and multi-disciplinary research project that looks at how global knowledge networks promote the international competitiveness of companies that participate in these networks. Her book *Managing New Industry Creation*, Stanford University Press (2001), co-authored with Tom Murtha and Jeffery Hart, focuses on the necessity for companies involved in the commercialization of liquid crystal displays to participate in global strategic alliances and knowledge networks. This research also found that US companies that relied on US strategic partners failed to understand the global competitive pressures in the industry and were forced to exit.

Rünno Lumiste is researcher at Tallinn University of Technology. He is a specialist in innovation management, innovation policy and industrial economics, and has participated in European Union Framework Programme projects. Previously he worked in the public sector (Ministry of Economic Affairs of Estonia) and non-profit sector (Confederation of Employers and Industry of Estonia). He has published in several journals and books; (1997), *Development of Innovation System for Estonian Industry*; (2000), *Product Innovation and Innovation Systems*; *Innovation in Estonian Companies 1998–2002*, as well as chapters on industry for *Estonian Encyclopaedia*.

Tina Mangieri is an Assistant Professor in the Department of Geography at Texas A&M University. An economic and cultural geographer (PhD, University of North Carolina at Chapel Hill), she has conducted research on globalization, development, and conceptualizations of identities, particularly as these processes contribute to (re)imagining and (re)materializing linkages within the Global South. She is currently pursuing these interests through a study of consumption spaces in Africa and Arabia, an interrogation of apparel manufacturing and (export) commodity fetishisms, and by focusing on the work of Islam, gender and fashion in the production of South-South spatialities.

Grzegorz Micek is Research Fellow in the Institute of Geography and Spatial Management, Jagiellonian University, Poland. He is an economic geographer (MSc, PhD, Jagiellonian University) and has conducted research and published on the following topics: factors and mechanisms behind spatial agglomeration of IT companies, geography of knowledge-intensive and business services, industrial clusters, spin-off and spill-over effects, academic entrepreneurship and foreign direct investment.

Carlo Pietrobelli is Professor of International Economics in the Law School of the University of Rome III. He has held positions in several Italian universities and has served as a consultant to international organizations such as the European Commission, the World Bank, the Inter-American Development Bank, UNIDO, UNCTAD, ECLAC and the OECD. He sits on the Board of the Master's Programme in Development Economics at the University of Rome, 'Tor Vergata'. He has worked extensively on industry, technology and trade in developing countries. His recent publications include (2002), *Failing to Compete: Technology Development and Technology Systems in Africa* (with S. Lall), Elgar; *The Global Challenge to Industrial Districts: SMEs in Italy and Taiwan* (with P. Guerrieri and S. Iammarino), Elgar; (1998), *Industry, Competitiveness and Technological Capabilities in Chile. A New Tiger from Latin America?* Macmillan.

Poli Roukova is a Researcher in the Department of Economic Geography, Institute of Geography, Bulgarian Academy of Sciences. She is an economic geographer (MS, Sofia University). She has done research on topics such as global commodity chains and regional development impact, socio-economic issues of transition, regional labour markets and cross-border regional studies. She has authored and co-authored 27 articles published in Bulgaria, the UK and Greece. She is an author of three chapters and a co-author of two chapters in three scientific monographs (two of which were published by the Publishing House of the Bulgarian Academy of Sciences and by ForCom, and one by Campus Verlag, Germany).

Federica Saliola has been at the World Bank, Financial and Private Sector Development – Enterprise Analysis Unit, since 2007, after working for the World Bank for two years. Her prior experience at the World Bank includes being a consultant for the Development Economics Research group and for the Middle East and North Africa region. She holds a PhD in Economics from the University of Rome III, and her research interests are Private Sector Development, Global Value Chains and firms' productivity.

Valerie Taylor is the Royce E. Wisenbaker Professor and Department Head of the Department of Computer Science at Texas A&M University. From 1991–2002 Dr

Taylor was a member of the faculty in the Electrical and Computer Engineering Department at Northwestern University. Her research interests are in the area of high performance computing, with particular emphasis on the performance of parallel and distributed applications and mesh partitioning for distributed systems. She has authored or co-authored over 90 papers in these areas.

Ivaylo Vassilev is a Research Fellow in the Department of Sociology in University of Aberdeen. He is an economic sociologist (MA CEU, Warsaw, PhD Lancaster University) and his research interests include issues of globalisation, governance, business strategy, transformation in Eastern Europe, and social theory, particularly around issues of trust and informality.

Preface

This book consists of three parts. The third part is based on invited papers, and I would like to thank all the authors involved for their invaluable contribution to widening the scope of this book thereby giving it a more international, rather than a simply European, perspective.

The other two parts of the book are primarily the outcome of a research project that was financed by the European Commission in the context of the Specific Targeted Research Project (STREP) of the Sixth EU Framework Programme for Research and Technological Development (FP6) (http://econlab.uom.gr/~move/index.php?lang=en). I would therefore like to take the opportunity here to thank the EC for providing us with the opportunity to realize this project. I am also grateful to the General Secretariat for Research and Development (GGET) for giving our national team a financial reward, acknowledging the importance of having our research financed by the European Commission.

Furthermore, I would like to thank my colleagues from the various national teams as well as those individuals who participated in this project; their names are listed with thanks in the appropriate chapters.

Most of all I want to express my gratitude to all those who contributed their valuable time in replying to our questionnaires (that is, 756 questionnaires to enterprises involved in some form of delocalization), or whose long conversations gave us valuable insights (that is, more than 120 key informants), whose names are impossible to mention.

The entire endeavour was quite an enriching experience for all of us, since we had to work collectively to produce this outcome. This partnership has a rather long history at this point. It began its interaction in its present form in 2002, when a proposal was drafted and submitted to the EU. However, most of us already knew one another and had already collaborated on a bi-lateral basis for several years before that. In a sense this book is a collaborative work throughout. That is, though individual teams are responsible for each particular chapter, it constitutes a collective authorship in the sense that there were extensive discussions over every minor detail of this book for more than three years. Needless to say, this is particularly true of the introductory as well as the concluding chapters.

At times it brought us up against the limits of our potential for agreement, as individuals and as national teams, and opened up some productive, and I must stress productive, tensions. I believe that we all learned not only through the actual research but also by working with other research teams from different countries as well as from different scientific disciplines, and I must admit that this was a very interesting and rewarding experience as well, though at times it proved to be quite difficult for all of us and particularly for the coordinating team.

I would like to thank the scientific officer of the EU Mr Marc Goffart for his continuing interest and helpful comments, as well as for his willingness to find solutions to the plethora of bureaucratic problems that have arisen during the course of this work so that we could proceed with our project without distractions.

Furthermore, I am grateful to all the people cited below for reviewing particular chapters of the book: Dr *Zografia Bika*, External Associate, The Arkleton Institute, University of Aberdeen, UK; *Trevor Buck*, Professor of International Business, Loughborough University Business School, UK; *Mark Cook*, Principal Lecturer, International Business, University of Wolverhampton, UK; Prof. Dr *Andreas Freytag*, Lehrstuhl Wirtschaftspolitik, Friedrich-Schiller Universität Jena, Germany; *Jane Hardy*, Research Leader, Economics Group at the University of Hertfordshire, UK; Prof. *Grigoris Kafkalas*, Department of Urban and Regional Planning and Development, Aristotle University of Thessaloniki, Greece; *Frank McDonald*, Professor of International Business, Bradford University School of Management, Chair of the Executive Committee of Academy of International Business, UK and Ireland; Dr *Alison Stenning*, Senior Lecturer, School of Geography, Centre for Urban and Regional Development Policies (CURDS), University of Newcastle, UK and *Roger Strange,* Senior Lecturer, Department of Management, King's College London, UK.

Thanks are also due to our colleagues who participated in the conference that we organized in Krakow on the 12–14 April 2007 and who helped us with their suggestions. These individuals include Prof. *Robert Begg*, Geography and Regional Planning, Indiana University of Pennsylvania, Indiana, USA; Prof. *Mick Dunford*, Economic Geography, University of Sussex, UK; Prof. *Gary Gereffi*, Professor of Sociology and Director of the Center on Globalization, Governance and Competitiveness at Duke University, Durham, NC, USA; Prof. *Costis Hadjimichalis*, Dept. of Geography, Harokopio University, Greece; Prof. *Antigone Lyberaki*, Economics, Dept. Economics and Regional Development, Panteion University, Greece; Prof. *John Pickles*, Globalization, Modernity, Geographies of Social Change, Dept. of Geography, University of North Carolina, Chapel Hill, Chapel Hill, NC, USA and Prof. *Adrian Smith*., Dept. of Geography, Queen Mary University of London, UK.

Finally, I would like to express my gratitude to my PhD students and close friends Aspa Kyriaki and Theodossis Sykas for generously providing their support by managing the questionnaires throughout the greater part of this project, as well as Nikos Vogiatzis for his meticulous technical work that shaped these papers into a book.

Lois Labrianidis
Thessaloniki, December 2007

List of Abbreviations

AGOA	African Growth and Opportunity Act
BPO	Business Process Outsourcing
CEE	Central and Eastern Europe
CEEC	Central and Eastern European Country
DC	Developed Country
EC	European Commission
EMS	Electronics Manufacturing Service
EPZ	Export Processing Zones
EU	European Union
EUR	Euro
FDI	Foreign Direct Investment
FYROM	Former Yugoslav Republic of Macedonia
GCC	Global Commodity Chain
GDP	Gross Domestic Product
GPN	Global Production Networks
GVC	Global Value Chain
IB	International Business
ICT	Information and Communications Technologies
ILO	International Labour Organization
IMF	International Monetary Fund
ISO	International Organization for Standardization
IT	Information Technology
JV	Joint Venture
KWNS	Keynesian Welfare National State
LDC	Less Developed Country
LII	Labour Intensive Industries
MFA	Multi-Fibre Agreement
NACE	Nomenclature of Economic Activities (Classification of Economic Activities in the European Community)
NAFTA	North American Free Trade Agreement
NGO	Non Governmental Organization
ODM	Original Design Manufacturers
OECD	Organization for Economic Cooperation and Development
OEM	Original Equipment Manufacturer
OPT	Outward Processing Trade
R&D	Research and Development

SEZ	Special Economic Zones
SMEs	Small and Medium Enterprises
SWPR	Schumpeterian Workfare Post–National Regime
TFP	Total Factor Productivity
TNC	Transnational Company
UK	United Kingdom
UNCTAD	United Nations Conference on Trade and Development
WTO	World Trade Organization
WWII	World War II

Chapter 1
Introduction

Lois Labrianidis

The Changing Geography of Production in a Globalized World

During the 60 years between the end of World War II (WWII) and today, the global map of production of goods and services has changed significantly. Although three-quarters of global manufacturing still takes place in the Developed Countries (DCs), the share of the Less Developed Countries (LDCs) has risen considerably: from 5 per cent in 1953 (Dicken 1998) to almost 24 per cent in 2001 (UNCTAD Globstat 2001).

The situation within the various groups of countries has also changed significantly. As far as LDCs are concerned, almost all the change in performance was due to the rapid increases in a relatively small number of countries in Southeast Asia. Led by the four Asian 'tiger' economies of South Korea, Taiwan, Singapore and Hong Kong, which were later followed by a number of other countries, the share of the region in world manufacturing value-added more than tripled during the last two decades. Among the countries of the region, it was China that displayed the most spectacular performance, recently overtaking Germany as the third most significant manufacturing producer in the world. In contrast, the LDCs of America lost ground, while Africa and the LDCs of West Asia and Europe have been marginal in the global map of production.

Within the DCs, there have also been considerable changes. North America emerged as the dominant region, with more than 30 per cent of the world manufacturing value-added in 2001, outperforming Western Europe during the last decade. The performance of the 'other' DCs seems to reflect the slump of the Japanese economy during the 1990s.

However, the regions where the most impressive (although very different in direction) changes took place are the Central Eastern European Countries (CEECs) and China. The former appear to have experienced a dramatic crisis, currently accounting for less than 3 per cent of the world manufacturing value-added (down from almost 20 per cent in 1980), while China is emerging as the fastest growing economy.

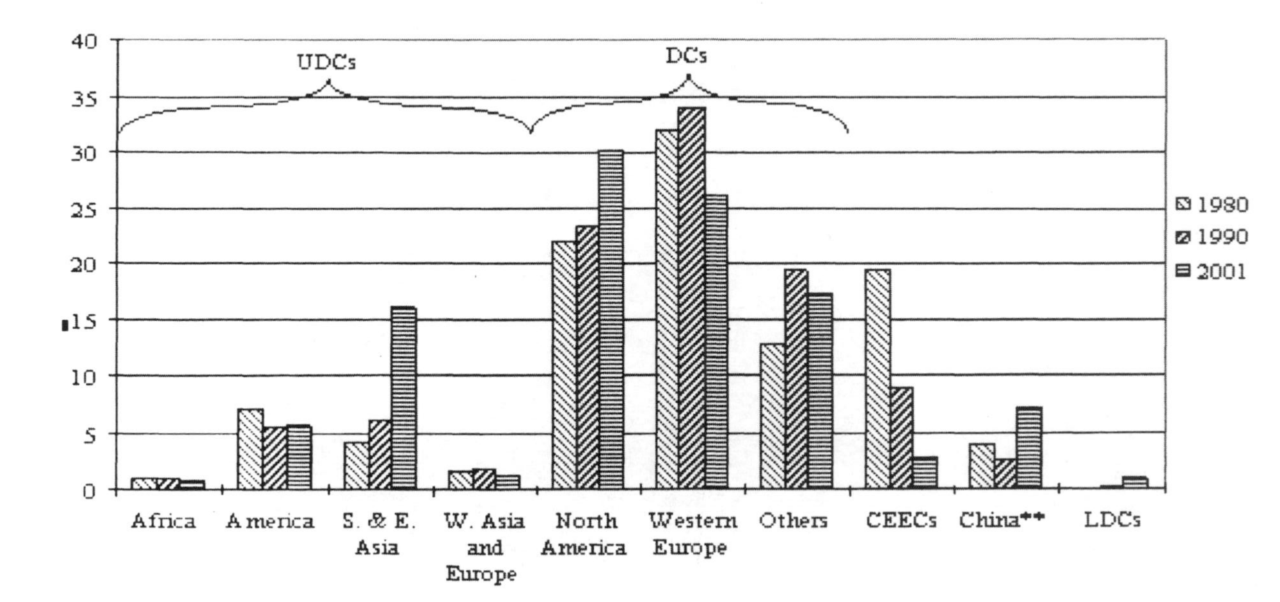

Figure 1.1 Distribution of world manufacturing value-added, at current prices, by region

Note: ** The data shown for the year 1980 corresponds to 1981 data at constant 1980 prices.

Source: UNCTAD Globstat (http://globstat.unctad.org/html/index.html).

Another significant feature of the world production map is that a very small number of countries produce a significant part of the global output. In 2000, the 15 most significant producers contributed 81.8 per cent of the global manufacturing value-added. As Dicken (1998, 27) noted, the 'manufacturing tail' of the world economy is very long indeed, even though concentration at the top has been reduced slightly during the last few years (according to Dicken the share of the 15 most important producers in 1994 was 85.8 per cent).

Table 1.1 The 15 most significant producer countries, 2003

Country	Industry value-added (in US $ m)	Percentage of world total
United States	2,192	26.8
Japan	1,540	16.3
China	738	8.4
Germany	504	5.0
UK	357	3.7
France	281	3.7
Italy	278	3.0
Canada	226	2.2
Korea, Rep.	216	2.4
Spain	168	2.4
Mexico	142	1.6
Brazil	133	1.7
India	131	1.9
Saudi Arabia	109	0.5
Russian Federation	105	0.8
Subtotal	7,122	80.3
World	9,135	

Source: World Bank, WDI database.

A simple illustration of the extent of the inequalities is the fact that the manufacturing value-added of Russia (the last country in our top-15 list) is more than the total of the 80 countries found at the bottom of the table!

What is Delocalization?

Delocalization is a term referring to the spatial restructuring of industry at a national, regional or global scale. According to Feenstra (1998), delocalization was originally conceived as another variant of the long list of terms referring to the splitting of a production process, including, but not limited to, disintegration, internationalization, intra-mediate trade, intra-product specialization, kaleidoscope of comparative advantage, multistage production, outsourcing, slicing up the value chain, splintering, subcontracting and vertical specialization.

This book adopts a wide definition of delocalization in order to include: Foreign Direct Investment (FDI); outsourcing; subcontracting; firms that traditionally bought the intermediate product (that is, never produced it in-house and therefore never stopped producing it) and are now outsourcing it; and horizontal FDI, which is very often not considered a component of delocalization, since it involves the movement of production abroad. Moreover, while in the literature the emphasis is on secondary data, large transnational companies (TNCs), chains or networks viewed from the perspective of the respective lead firm, our analysis has attempted to include even small firms that are not usually considered and might be very important for development.

To be more precise, our approach assigned the following definitions to the above terms:

- *FDI:* is an investment involving a long-term relationship and reflecting a lasting interest and control by a resident entity in one economy (foreign direct investor or parent enterprise) in an enterprise resident in an economy other than that of the foreign direct investor (FDI enterprise or affiliate enterprise or foreign affiliate) (UNCTAD 2004, 345).
- *Subcontracting:* is defined as the manufacture of goods by one firm (the subcontractor) for another (the lead firm) based on the specifications of the latter. Often there can be several layers of firms or intermediaries mediating the relationship between the actual production workers and the end product market. The lead firms normally exercise considerable control over their subcontractors in terms of price, quality and timing of the products they supply.
- *Outsourcing:* is the delegation of tasks or jobs from internal production to an external entity (such as a subcontractor). Most recently, it has come to mean the elimination of native staff to staff overseas (offshore outsourcing), where salaries are markedly lower. This is despite the fact that the majority of outsourcing that occurs today still occurs within country boundaries.
- *Offshoring*: can be defined as relocation of business processes (including production/manufacturing) to an overseas, lower-cost location.
- *Offshore outsourcing:* is the practice of hiring an external organization to perform some or all business functions in a country other than the one where the product will be sold or consumed.

Before analysing delocalization, let us first try to find our way through the 'forest' of terms and attempt to produce a taxonomy of a firm organization (Table 1.2). Let us suppose that we have a domestic firm producing a single product by using two intermediate goods (1 and 2 respectively, the former being labour- and the latter knowledge-intensive), domestic capital and domestic labour.

Let us start with a vertically integrated domestic firm. In this case, both intermediate goods are produced in-house, using domestic capital and labour. Under this configuration, the firm can only be linked to the world market through its final good exports or imports of raw materials.

The first possible deviation occurs when the firm realizes that, for a number of reasons, a foreign market can be better served by producing the product there rather than exporting it. This implies a duplication of the production process, as additional plants are established to supply different locations: a horizontal FDI.

Table 1.2 A summary of definitions of IB organization types

	Internalized or externalized production	
Location of production	**Internalized**	**Externalized**
Home country	Production kept in-house at home (vertically integrated domestic firm)	Outsourcing (production outsourced to third-party firm for example, a subcontractor)
Foreign country	Vertical or horizontal FDI	International outsourcing
For service industries offshoring	Intra-firm (captive) offshoring	Outsourced offshoring

Note: Categories in gray cells comprize the delocalization group.
Source: Compiled by the authors.

On the other hand, when the firm realizes that, for example, intermediate good 1 can be produced more efficiently in a less developed (and lower labour cost) country (LDC), it may decide to set up a plant there to produce it. Although such movements have traditionally been described as vertical FDI, quite recently the alternative term intra-firm or 'captive' offshoring has entered the International Business (IB) vocabulary. The differences between the two terms are almost nonexistent, although the latter tends to apply more to service industries, while the former to manufacturing industries.

Alternatively, the firm may decide that it is in its best interest to focus on its core competences, which one may assume are better employed in the production of intermediate good 2. This could imply that the firm wishes to stop producing intermediate good 1. Assuming that the latter is still essential to the firm, it would probably outsource it either to a local firm in a foreign country or to an affiliate of another TNC.

The inherent difficulties in defining the terms are now more or less evident. One question that arises is whether firms that traditionally bought the intermediate product (that is, never produced it in-house and therefore never stopped producing it) are outsourcing.[1] Another issue is related to subcontracting: is it merely a special case of outsourcing or a completely distinct category?

Delocalization is, therefore, a term that refers to the spatial restructuring of industry on a national, regional or global scale. Its primary elements are FDI and outsourcing,[2] although it also refers to all other types of cross-border business interactions. Traditionally, the direction of the movement was from the more developed countries (DCs) to the LDCs, although this is rapidly changing. In other words, delocalization is a term that is 'broader' than other terms, which, at least in mainstream international economics, are identified with and limited to the trade of intermediate products (for example, H. Egger and P. Egger 2003; Görg and Hanley 2003).[3] One last question that remains is how to treat horizontal FDI, which is very often not considered a component of delocalization. We feel that it should be included inasmuch as it involves the movement of production abroad. Nonetheless, it will not occupy a prominent position in our analysis, since, with the exception of the electronics sector, it is of minor importance to our sample.

FDI

Although there have been significant fluctuations, FDI has grown enormously during the last 35 years and especially since 1985. More specifically, during the period 1970–2003, the average rate of change of FDI inflows was 14.7 per cent, growing significantly faster than exports (11.2 per cent), and even faster than output (9.3 per cent – Figure 1.2). In fact, FDI growth was so spectacular that after three decades of growth, in 2000 (its peak year), it was more than 100 times higher[4] than its 1970 level, with more than 80 per cent of that growth occurring during the last decade.

With regard to the distribution of FDI, the main feature of the participation of LDCs in both inward and outward stocks appears to have been the considerable fluctuations within the last quarter of the century. Furthermore, although not easily visible, both figures display a long-term upwards trend, implying a strengthening of the position of LDCs in the overall distribution of FDI (Figure 1.3).

1 Gilley and Rasheed (2000) argue that abstention of producing a good in-house should also be considered outsourcing if the internalization of the good or service outsourced was within the acquiring firm's managerial and/or financial capabilities.

2 We understand outsourcing as a general term embracing subcontracting and offshoring. The reason is that the former is a subcategory of outsourcing, while offshoring is simply an alternative term, usually preferred by the services industries. Most importantly, the theoretical explanations of all terms would be very similar.

3 We should note that there have been efforts to analyse the two phenomena in a unifying way. For example, in his analysis the implications of 'international fragmentation', Kohler (2004) treats vertical FDI and subcontracting as the two components of fragmentation.

4 The figures for trade and output were 29 and 18 respectively.

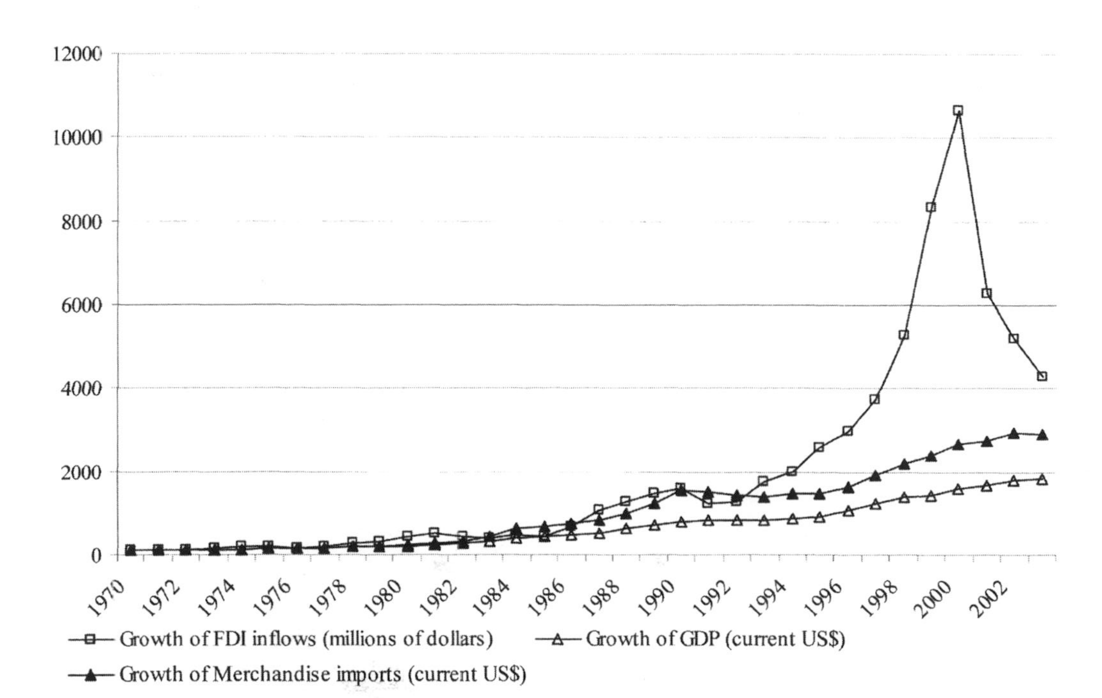

Figure 1.2 Growth and output (GDP) and merchandize imports and FDI inflows (1970 = 100)
Source: World Bank, WDI database (http://devdata.worldbank.org/dataonline/) for imports and output, and UNCTAD FDI database (http://stats.unctad.org) for FDI inflows.

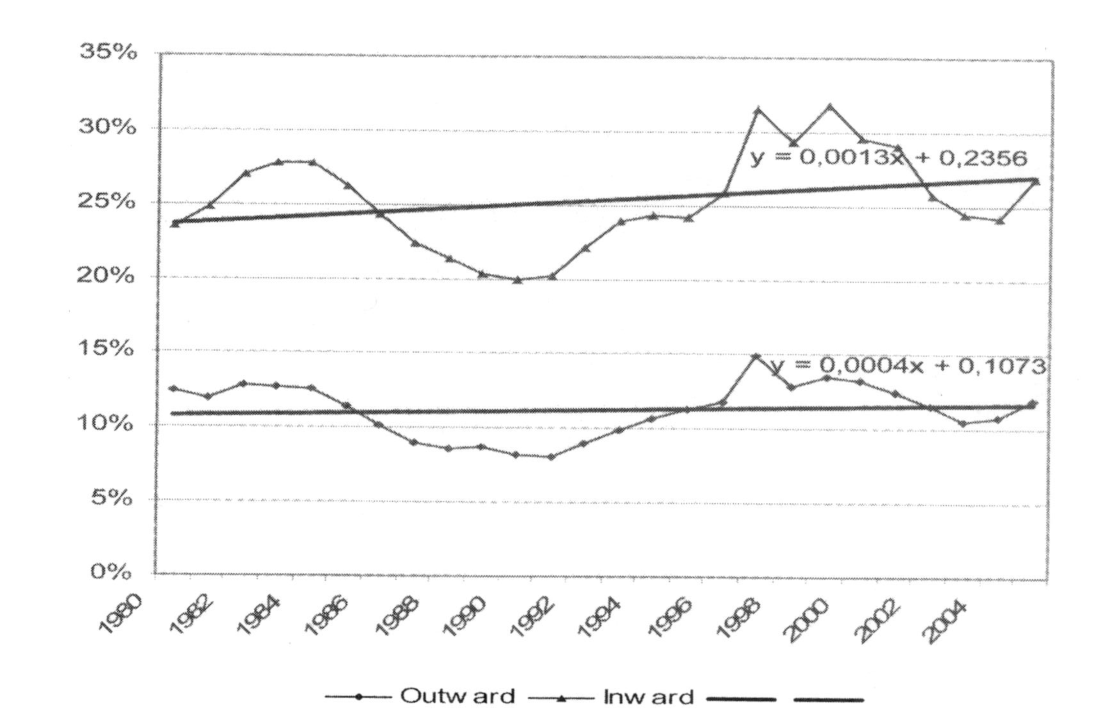

Figure 1.3 Inward and outward FDI stocks of LDCs as a percentage of the total
Source: UNCTAD FDI database (http://stats.unctad.org).

In spite of this trend, TNCs are still concentrated in the DCs. Low labour costs alone are not sufficient for a country to attract FDI. There are other more important factors, including, for example, physical and non-material infrastructure, socio-economic stability and human capital. In 2001, the inward FDI stock of the DCs amounted to $4,545 per capita, while the respective figure for LDCs was only $436. Furthermore, the stock of outward FDI of the DCs was $5,951 per capita, while the equivalent figure for LDCs was only $168. For the last 30 years, ten countries have accounted for approximately 85 per cent of outward investment stocks; during this period there were of course major changes in the importance of individual countries, the most prominent being the decline of the importance of the USA (Table 1.3).

Table 1.3 Outward investment cumulative stocks by country: Main players (per cent)

		1967	1973	1980	1990	2000	2003
1	USA	50.4	48.0	41.3	25.0	21.3	25.2
2	UK	14.1	7.5	15.4	13.3	14.8	13.8
3	France	5.3	4.2	4.7	7.0	7.1	7.8
4	Germany	2.7	5.6	8.3	8.6	7.7	7.6
5	Netherlands	9.8	7.5	8.1	6.2	5.1	4.7
6	Belgium/Luxemburg			1.2	2.4	6.3	4.1
7	Switzerland	2.2	3.4	4.1	3.8	3.7	4.2
8	Japan	1.3	4.9	3.8	11.7	4.6	4.1
9	Canada	3.3	3.7	4.6	4.9	3.7	3.8
10	Italy	1.9	1.5	1.4	3.3	3.0	2.9
	Subtotal	91.0	86.3	92.9	86.4	78.7	78.2
	Others	9.0	13.7	7.1	13.6	21.3	21.8
	Total	100.0	100.0	100.0	100.0	100.0	100.0

Source: Dunning (1993, 17), years 1967, 1973 and UNCTAD FDI database years 1980–2003.

Outsourcing

The problems related to the definition of outsourcing, briefly analysed above, are reflected in the difficulties in measuring it. In all types of efforts, outsourcing is measured in terms of trade flows or as a share of total trade flows. Intra-firm trade (trade between parents and affiliates in a TNC network) is one of the measures often used to capture outsourcing. As in the case of FDI, most of the empirical evidence comes from the USA, Japan and Sweden (Bonturi and Fukasaku 1993; Andersson and Fredriksson 2000; Slaughter 2000). In these three countries, intra-firm trade accounts for 38–45 per cent (in the case of the USA), 30 per cent (Japan) and 50 per cent (Sweden) of the total trade of the parent companies.

The second, and much more popular, measure is the various approaches of intermediate inputs. One such approach is that of Feenstra and Hanson (1999) and Hummels et al. (2001), who tried to estimate the level of 'Vertical Specialisation' as a measure of the imported inputs in products that are then exported. Their measure is a subset of all intermediate inputs, as some of these are not embodied in export products. Studying ten countries from the Organization for Economic Cooperation and Development (OECD) and three LDCs (Korea, Taiwan and Mexico), they observed increasing shares of vertical specialization trade in the majority of countries. Furthermore, they argue that about 30 per cent of the growth of exports of these countries is attributed to vertical specialization.

Why Labour-intensive Sectors?

Innovation is something that does not concern high-tech sectors only, while learning can (and in fact should) occur in all economic sectors (Lundvall and Borrás 1997). Regional innovation systems can consist of high-tech, science-based industries as well as of clusters of more traditional, medium or even low-tech industries. In fact, a great number of regions in continental Europe depend on industries with synthetic knowledge bases (for example, Germany, Nordic countries) where learning through interacting is more important for innovation processes than intense R&D and scientific knowledge breakthroughs. What seems to be important, however, is the ability of industries and regions to develop a strong learning capacity that will enable them to absorb and exploit new knowledge created both internally (within a given system) and externally.

Innovation in low-tech industries has attracted the interest of a number of scholars. This is – among others – due to the fact that in the OECD countries, high-tech industries, as defined by the OECD, account for only 3 per cent of value added, rising to 8.5 per cent if medium high-tech industries such as motor vehicles are included (OECD). Maskell (1996) claims that countries without specialization in high technology industries are not left in the backwaters of economic development. On the contrary, evidence suggests that a number of developed countries of Europe (that is, Austria, Belgium, Luxemburg, Denmark, Finland, the Netherlands, Norway, Sweden and Switzerland) specialize on medium- and even low-tech industries (Maskell 1996, 6). In fact, according to the author, it is primarily large countries (for example the USA, the UK and Japan) that can afford to invest in leading edge technologies, given the risks and costs involved, as well as the fact that high-tech sectors are usually highly subsidized through government subventions or public procurement (for example, in the defence industry). Another point Maskell makes is that the limited size of the national knowledge base influences the range of industries in which small countries might successfully specialize. Restrictions of size have gradually channeled the process of specialization towards industries with rather stable demands and low price elasticity. These industries are often medium- or low-tech, but can, nevertheless, yield high profits. In the long run no small country can cope with committing a larger part of its resources to industries with huge ups and downs.

In the late 1970s it was already apparent that there was, on the one hand, a shift of low-skill, low-wage work in industries such as clothing and electronics assembly to the world periphery and, on the other hand, an intensification of advanced design and production activities in core countries, a phenomenon which has been coined by Fröbel et al. (1980) the 'new international division of labor'. This term, though useful, as Scott (2006, 1520) argues, 'tends to impose an unduly schematic rigidity on the ways in which we think about the economic geography of contemporary globalization' and he suggests the notion 'of a worldwide mosaic of regional economies at various levels of development and economic dynamism and with various forms of economic interaction linking them together'. Hence we can acknowledge a) that there is much more than core-periphery b) there are many counterexamples such as: low-wage sweatshop industries in DCs; significant numbers of industrial agglomerations undergoing upgrading in LDCs; and dynamic new technology poles in countries such as Brazil, China and India.

Aim, Methods and Structure of the Book

Given the sweeping changes described above, the main aim of this book is to analyse the growing complexity of international business and, more importantly, how it is affected by the profound changes taking place during the last two decades in both the business environment and business conduct. To do so, the book will try to provide answers to four very important questions: Why and how firms delocalize, what the implications of delocalization are and whether there is room for policies that would answer to market or government failures related to the phenomenon.

The book is divided into three parts. The first two parts are based on the EU FP6 project 'MOVE' (Labrianidis et al. 2007). As with the vast majority of EU projects, MOVE was a multinational effort by six partners from five countries; two 'older' EU members, that is, the United Kingdom (UK) and Greece, two recent ones, that is, Poland and Estonia, and one from the very last 'wave' of accession, that is, Bulgaria. The focus of the project was on four sectors: clothing, footwear, electronics and software.

The issues raised in the first two parts of this book are based on an empirical study of the delocalization in four industries in five European countries, but they are not confined to these countries only. The issues raised have a global audience because people everywhere are trying to understand them. Many of the same themes that are examined in these two parts of the book (in the four industries in the countries presented) are also major issues in North America (Canada, the United States, Mexico), Central America and Asia; and these issues have global repercussions as well.

A total of 756 extensive interviews were conducted, whose distribution by country and sector can be seen in Table 1.4. Although our initial intent was to conduct a stratified sampling based jointly on the sector (that is, each national survey would contain equal numbers of randomly chosen firms by sector) and on whether the firm was in any way involved in delocalization, it was quickly realized that our aim could not be achieved. The main reason was that in some cases (for example, footwear in Greece, Estonia and Poland, and clothing in the UK), not only were we unable to randomly select from within our strata, no matter which source of information we used, but also we realized that these specific strata had been exhausted. As is evident in Table 1.4, all partners had trouble satisfying the quotas in all sectors. This is particularly evident in clothing and footwear, with the former being overrepresented and the latter underrepresented. Apart from the general picture, each country faced its own problems (with Bulgaria facing the least).

The instrument used in the survey was an extensive semi-structured questionnaire consisting of eight sections, four of which probed into delocalization (addressed to TNCs and other firms involved in outsourcing and insourcing, while at the same time examining the implications of delocalization).

There were also 120 Key Informant interviews (Bulgaria 29, Estonia 18, Greece 26, Poland 30 and the UK 17) with business associations, experts (academics, researchers in trade associations and trade unions, consultants among others), confederations of employers in all four industries, politicians, trade union leaders in all four industries, and so on.

Part 1 of the book is concerned with the more theoretical aspects, focusing on the experience of European countries, and consisting of four chapters.

Chapter 2 on 'Delocalization and Development in Europe: Conceptual Issues and Empirical Findings' analyses two questions: why and how do labour-intensive firms delocalize? It argues that explanations based only on production or consumption are partial. It is true that a significant part of the explanation concerning the variables affecting the decision to delocalize is inherently microeconomic and, therefore, production-based. However, to complete the picture and also analyse the organization of international production, one must look beyond production. On a conceptual level, there appears to have been surprisingly little cross-fertilization between the production-side theories (mainly the theories of the TNC) and those theories that claim to capture the whole array of activities, from production to consumption. On the one hand, the former seem to fall surprisingly short of grasping the complex realities of internationalization caused by the fixation with the hierarchy-market dilemma (therefore failing to deal with the incredible variety of the in-between types of organizations), as well as the minor role accorded to geography. On the other hand, the chain or network approaches – in spite of their stated intentions – largely ignore the firm as perhaps the single most important actor in all variations of modern capitalism (there is a 'black-box' attitude towards the firm). The chapter explores some possible linkages between the two schools of thought, which could be beneficial for both of them, despite the fact that their intellectual foundations may be completely different.

Table 1.4 Country sector cross tabulation

			Bulgaria	Estonia	Greece	Poland	UK	Total
Branch (V1b)	Software	Count	52	51	20	50	17	190
		R (%)	26	25.5	25	24.9	22.7	25.1
		C (%)	27.4	26.8	10.5	26.3	8.9	100
	Electronics	Count	44	78	21	25	24	192
		R (%)	22	39	26.3	12.4	32	25.4
		C (%)	22.9	40.6	10.9	13	12.5	100
	Clothing	Count	60	60	31	92	12	255
		R (%)	30	30	38.8	45.8	16	33.7
		C (%)	23.5	23.5	12.2	36.1	4.7	100
	Footwear	Count	44	11	8	34	22	119
		R (%)	22	5.5	10	16.9	29.3	15.7
		C (%)	37	9.2	6.7	28.6	18.5	100
Total	Count		200	200	80	201	75	756
	R (%)		100	100	100	100	100	100
	C (%)		26.5	26.5	10.6	26.6	9.9	100

Source: Enterprise Survey.

The conceptual framework proposed for analysing the delocalization of labour-intensive firms has four main analytical dimensions that are intersected by wider categories, that is, the firm, with its own unique set of resources and competitive advantages; the sector, with its given technologies and markets; the 'environment' (local/regional/national), with its unique institutions, civil society, history and policies; and the global 'environment', with its unique institutions, governance and power relations.

Chapter 3 on 'Patterns of Enterprise Strategies in Labour-Intensive Industries: The Case of Five EU Countries' enhances our understanding of enterprise strategies in labour-intensive sectors. Conceptually, strategies are viewed as multidimensional and considered to be influenced by a number of factors at work, while methodologically they are viewed through a move beyond the 'ideal type' models. The emerging picture is one of considerable diversity in enterprise strategies. Enterprises may opt for different strategies not only when they operate in different segments of the market and in different national contexts or in the same segment of the market but in a different national context, but even when they operate in the same segment of the market and in the same national context.

Chapter 4 on 'Social Consequences of Delocalization in Labour-Intensive Industries: The Experience of Old and New Members of the EU' argues that the social effects of delocalization are more limited than often claimed. They are mainly observed on the local scale, to a lesser extent on the regional and are almost negligible on the national scale. There are intermediating factors (for example, social and economic features of the locality/region/national labour regulations) that influence the intensity of the impact, rendering it either strong or weak. The net employment effects of delocalization within the EU are rather positive, at least in the mid-term, in the sense that more jobs remain within Europe rather than move to other parts of the world and that it lowers unemployment in the new member states far more than it increases unemployment in the DCs. The social characteristics of delocalization can hardly be interpreted as 'a race to the bottom' in terms of wages and employment conditions in the labour-intensive activities in the EU. This fact may be the effect of the regulated environment of the EU.

The public debate on the social consequences inflicted by the delocalization of labour-intensive industries is clouded by common misinterpretations. The analysis conducted in five countries shows that these effects are mainly observed on a local scale and to a lesser extent on the regional level. Widespread emphasis on job losses ignores the fact that this decline usually has no direct impact on unemployment levels. The balance of negative versus positive effects is place-dependent and determined by the role of the sector/employer in the labour market and by the overall performance of the regional/local economy. The main problem is not delocalization itself, but the 'weaknesses' of certain regions and localities. The net employment effects of delocalization within the European Union (EU) are rather positive. Delocalization facilitates lower unemployment in the new member states to a greater extent than it contributes to higher joblessness in the developed areas, where there are more alternative employment opportunities. A substantial part of manufacturing jobs and related improvements in skills and capabilities has moved to peripheral regions of Central and Eastern Europe (CEE) and to underprivileged social groups.

Chapter 5 on 'Between Policy Regimes and Value Chains in the Restructuring of Labour-Intensive Industries' argues that the notion of governance has come into prominence in the context of global, economic, social and political restructuring. One of the key changes introduced is that coordination is no longer exclusively in the hands of the states. Broad social and economic processes are becoming increasingly embedded into much more complex institutional arrangements that are organized around diverse spatial scales (sub-national, national, supra-national) and different networks. These changes raise questions some of which are central to understanding the process of delocalization of labour-intensive industries. Rather than juxtaposing different perspectives, the aim in this chapter is to discuss governance as a dynamic and multi-level process, where actors, with their motivations and time horizons, as well as objects of governance are constantly being created and reshaped. Thus, the chapter argues that while delocalization constitutes a key economic conundrum as well as a political and social concern, delocalization as such is not an appropriate object of governance, given the reduced powers of the state to influence processes within their own territories. Importantly, however, states are also acquiring new powers of coordinating, or steering, and thus have the ability to influence other levels of governance (for example, sub-national and supra-national).

The chapter studies the interrelationships of a state-centred (territorialized) and industry-centred (networked/deterritorialized) perspectives on governance. While focused on the same set of key players (states, global and local governing bodies, TNCs, Non-Governmental Organizations [NGOs], business associations, trade unions) it argues that the two perspectives offer different insights into the significance of these players for the coordination of the relations in the four industries studied.

Part 2 of the book is concerned with the analysis of the delocalization phenomenon in four European industries, consisting of four theoretically informed empirical chapters.

Chapter 6 is entitled 'The Impact of Internationalization on the Clothing Industry', and by drawing on the results of extensive fieldwork in the five European countries, it examines the impact of processes of global integration upon inter-organizational relationships and enterprise strategy in the clothing industry. The findings reported in the chapter suggest that international opportunities can be best exploited initially by early engagement with low commitment strategies, followed later by significant foreign investment and joint venture (JV) creation, and finally by an emphasis on buyer/supplier relationships. A gradual shift appears to be occurring from publicly- to privately-driven forms of governance, reinforcing the importance of such relationships.

Chapter 7 on the 'The Impact of Delocalization on the European Electronics Industry' describes major patterns of delocalization of European and global electronics industries. This is a timely topic, since two new factories are established and 500 new jobs are created every week in CEE, while at the same time one or two factories in Western Europe are closed. The chapter investigates the supply chains and the geographical patterns of the electronics industry in Europe. Major forms of delocalization in the electronics industry are foreign trade, subcontracting and acquisition/mergers of firms. The location and establishment of new factories involved in the electronics industry is influenced both by public sector policies and private sector demand. Public sector policies influencing the electronics industry are tariff rules, direct support by governments, intellectual property protection rules, environmental legislation, national education and science policy and general economic policy. The private sector influences the electronics sector via the growing purchasing power in Eastern Europe and the search for new clients in Eastern Europe by transnational companies. Investments of TNCs also act as initiators of the creation of local supply networks. Electronics production is mainly influenced by the cost of the input factors and the ability to create new products. This ability is determined by the education level and the entrepreneurship of engineers and managers. The chapter also investigates the social consequences of delocalization and possible public policies.

Chapter 8, entitled 'Footwear Industry: Delocalization and Europeanization', argues that the 1990s and 2000s have been marked by the enlargement of the EU and the further liberalization of international trade. The dynamic process of the relocation of the European footwear industry has introduced significant diversification in the different forms of delocalization, as well as in the networks connecting firms and regions that are embedded in different local, historical, political, economic and social environments. This chapter aims at contributing to a better understanding of the recent delocalization trends in the European footwear industry. The main research objectives are to identify the industry-specific and country-specific factors and the effects of delocalization. The focus is concentrated on how national production networks are integrated into European ones, on outlining the causes and the effects of this process, and on what opportunities and constraints the existing relationships create in terms of international competitiveness of footwear firms.

Chapter 9 concerning 'The Impact of Delocalization on the European Software Industry' analyses the internationalization of the European software industry in the context of subcontracting and FDI. Forms of delocalization and their extent in the European software industry, the reasons behind delocalization from the perspectives of both host and home countries and the prospects of further delocalization to locations outside Europe are the issues discussed in this chapter. The software sector is one of the most rapidly growing ones in OECD countries, with strong increases in value added, employment and Research and Development (R&D) investment. Rapid growth, especially in CEECs and some non-OECD countries such as India, deserves recognition as a new wave of globalization in global Information and Communications Technologies (ICT). Despite the dynamic development of offshoring activities, there has been no job loss in DCs. Global expansion of Information Technology (IT) firms is driven firstly by the need for market access and growth, secondly by economies of scale and costs savings and finally by access to skills and technology. The success of many CEE IT companies is attributed to the quality of human capital, as well as the companies' flexibility and level of expertise rather than to their low cost. Further delocalization of IT sector activities to India or to other low-cost countries is not perceived as a danger to the European software industry.

Part 3 presents the delocalization experience outside Europe, which significantly broadens the geographical breadth of the book. It comprises of three chapters, two of them concerning sectors that are also studied in the second part of the book (Chapter 10 on software and Chapter 12 on the garment trade) while the third (Chapter 11) concerns clothing as well as other six sectors.

These three contributions that were outside the MOVE project do not directly present a single theme. In fact, one could argue that they share nothing, conceptually, methodologically or empirically. Yet they serve a very important purpose, which is none other than to broaden our understanding of a phenomenon which appears to affect all countries and sectors directly or indirectly in a kaleidoscope of ways.

When it comes to the diversity of the modes of delocalization, the types of actors and countries involved, one would be hard pressed to come up with a better case than that of the software industry presented in Chapter 10 entitled 'Corporate Strategies for Software Globalization' analysed by Aggarwal et al. Specifically, they argue that the reasons for offshoring vary by firm and particular recipient nation (for example, in the case of the elite R&D laboratories, the desire to tap into the most talented individuals, wherever they might be in the world, is clearly the foremost motivation), and often decisions are made for a complicated amalgam of reasons, something which is true even when one compares relatively similar firms. However, along with diversity, which has been the foremost theme of this book, there are some rather clear common trends. For example, firms are becoming increasingly willing to entrust core activities to their offshore subsidiaries, and it is likely that more sophisticated work will be relocated during the coming decade. Whereas some believed that a certain size was necessary prior to offshoring, the case studies of startups showed that this is not true. The labour-cost arbitrage factor is and will remain significant and all executives, in large and small firms, are considering the most economical footprint for their IT operations. In certain cases, the real reason for offshoring might be simply that competitors have already done it or the board of directors is demanding an offshoring initiative to save money. There can be little doubt that offshoring is still small in comparison to how large it is likely to become.

Chapter 11 entitled 'Newly Emerging Paradigms in the World Economy: Global Buyers, Value Chain Governance and Local Suppliers' Performance in Thailand' completes the puzzle to approach diversity exceptionally well. The focus of the chapter is on the patterns of governance arising in value chains led by Global Buyers and their impact on suppliers' performance with specific reference to the Thai manufacturing industry. One of their main findings is that the relationships TNCs have with their suppliers are multifold, as they become engaged in their suppliers' process or product R&D and send their experts to work to disseminate and diffuse new technologies more often than other buyers do. In contrast, firms which are part of domestic value chains and those that sell to their global buyers demonstrate that these firms follow modes of governance that imply only involvement in defining design and product characteristics. Moreover, their estimates show that more intense buyers' involvement with local suppliers – not only in the definition of product characteristics, design and quality, but also in technology dissemination and R&D – is associated with higher productivity.

The fluid, emerging and largely underexplored workings of the global software industry, analysed by Aggarwal et al. (Chapter 10), contrast strikingly to the even more underexplored workings of three subsectors of the garment market in Kenya (that is, African print cloth, export-oriented mass-produced clothing and second-hand clothing) studied by Mangieri in Chapter 12 entitled 'African Cloth, Export Production, and Second-hand Clothing in Kenya'. Here, local and regional developments come to the fore, allowing us a more distant view of the omnipotent role of the global processes.

In this context, Mangieri's contribution stresses the importance of issues of South-South trade (that is, trade within the LDCs) and the conflict between the different LDCs (for example, since the end of the Multi-Fibre Agreement [MFA] there has been devastation in Kenya's garment manufacturing that Kenyans call the 'Chinese tsunami'), which is underrepresented in the literature on global apparel. Attention to these contexts is needed in order to provide conceptual alternatives to normative understandings of globalization as fundamentally limited to Western practices and experiences. Furthermore, through the specific analysis of how the garment industry in Kenya was affected by the preferential trade arrangements with the EU (the Cotonou Convention) and the USA (the African Growth and Opportnity Act [AGOA]), the chapter points out how the global regulatory environment affects developments of particular manufacturing sectors. Mangieri's contribution, through the specific analysis of development in the garment industry in Kenya, reveals that offshoring influences not only the particular sector or subsector in which it is directly involved but the other subsectors as well. Finally, at a more general level, the chapter is a very interesting exposition of the influence of global forces on the local or regional orientations of countries whose international position is still affected by 'old world' policies, as exemplified by the gradual strengthening of trade relations and networks in the Indian Ocean.

Chapter 13 advances the conclusions based on the synthesis of the preceding chapters.

References

Andersson, T. and Fredriksson, T. (2000), 'Distinction between Intermediate and Finished Products in Intra-firm Trade', *International Journal of Industrial Organization* 18:5, 773–93.

Bonturi, M. and Fukasaku, K. (1993), 'Globalisation and Intra-firm Trade: An Empirical Note', *OECD Economic Studies* 20, 145–59.

Dicken, P. (1998), *Global Shift: Transforming the World Economy*, 3rd edition (London: Paul Chapman Publishing Ltd).

Egger, H. and Egger, P. (2003), 'Outsourcing and Skill Specific Employment in a Small Economy: Austria after the Fall of the Iron Curtain', *Oxford Economic Papers* 55, 625–43.

Feenstra, R.C. (1998), 'Integration of Trade and Disintegration of Production in the Global Economy', *Journal of Economic Perspectives* 12:4, 31–50.

Feenstra, R.C. and Hanson, G.H. (1999), 'The Impact of Outsourcing and High-technology Capital on Wages Estimates for the United States, 1979–1990', *Quarterly Journal of Economics* 114:3, 907.

Fröbel, F., Heinrichs, J. and Kreye, O. (1980), *The New International Division of Labour* (Cambridge: Cambridge University Press).

Gilley, M.K. and Rasheed, A. (2000), 'Making More by Doing Less: An Analysis of Outsourcing and its Effects on Firm Performance', *Journal of Management* 26:4, 764–90.

Görg, H. and Hanley, A. (2003), 'International Outsourcing and Productivity: Evidence from Plant Level Data', The University of Nottingham Research Paper Series: *Globalisation, Productivity and Technology*, Research Paper 2003/20.

Hummels, D., Ishii, J. and Yi, K. (2001), 'The Nature and Growth of Vertical Specialization in World Trade', *Journal of International Economics* 54, 75–96.

Kohler, W. (2004), 'Aspects of International Fragmentation', *Review of International Economics* 12:5, 793–816.

Labrianidis, L., Domanski, B., Kalantaridis, C., Kilvits, K. and Roukova, P. (2007), 'The Moving Frontier: The Changing Geography of Production in Labour Intensive Industries: Final Report', Financed by European Commission 6th Framework Programme, MOVE Project [website] (updated 20 November 2007) <http://econlab.uom.gr/~move/index.php?lang=en>, accessed 5 June 2008.

Lundvall, B. and Borrás, S. (1997), 'The Globalising Learning Economy: Implications for Innovation Policy', Brussels: DG XII (published online December 1997) <ftp://ftp.cordis.lu/pub/tser/docs/globeco.doc>, accessed 12 January 1998.

Maskell, P. (1996), 'Localised Low-tech Learning in the Furniture Industry', *Danish Research Unit for Industrial Dynamics*, Working Paper 96–11.

Scott, A.J. (2006), 'The Changing Global Geography of Low-Technology, Labor-Intensive Industry: Clothing, Footwear, and Furniture', *World Development* 34:9, 1517–36.

Slaughter, M.J. (2000), 'Production Transfer within Multinational Enterprises and American Wages', *Journal of International Economics* 50:2, 449–72.

UNCTAD Globstat (2001), 'Production of Manufactures' [website] <http://globstat.unctad.org/html/index.html>, accessed 13 February 2007.

UNCTAD (2004), *World Investment Report 2004: The Shift towards Services* (New York and Geneva: United Nations).

PART 1
Thematic Analysis of the European Experience

Chapter 2
Delocalization and Development in Europe: Conceptual Issues and Empirical Findings

Athanasios Kalogeresis, Lois Labrianidis

Introduction

The beginning of the twenty-first century could easily be considered as one of the most exciting times for the social scientist. Consumers, firms – ranging from micro- enterprises to transnational behemoths – localities, regions and countries, to mention only a few of the actors involved, are increasingly being affected by forces that are inherently global in nature, while, on the other hand, the role of geography is increasingly becoming more pronounced.

At the same time, the global map of the production of goods and services has been changing, with LDCs currently producing almost a quarter of the global value-added goods, compared to only 5 per cent in 1953. This change is mostly due to a small minority of LDCs which have joined the ranks of the world's significant producers.

The main aim of this chapter is to look into the inner working of the organization of production and its implications for development. On a conceptual level, there appears to have been surprisingly little cross-fertilisation between the production-side theories (mainly the theories of the TNC) and those theories that claim to capture the whole array of activities, from production to consumption. On the one hand, the former theories seem to fall surprisingly short of grasping the complex realities of internationalization caused by the fixation with the hierarchy-market dilemma (thereby failing to deal with the incredible variety of the in-between types of organization), as well as the minor role accorded to geography. On the other hand, the chain or network approaches – in spite of their stated intentions – largely ignore the firm as perhaps the single most important actor in all variations of modern capitalism. It is our intention, therefore, to explore some possible linkages between the two schools of thought.

On an empirical level, even a casual look at the literature reveals a clear preference for analyses which are either based on secondary data (exclusively concerned with the issue of FDI or outsourcing, especially when the latter is identified with flows of intermediate products) or case studies about large firms which are, for example, creating their own Global Production Networks (GPNs – for example, see Coe et al. [2004] about BMW's GPN) or organising extensive Global Commodity Chains (GCCs – as in Gereffi and Mayer's work on Gap

[2004, 22]). In turn, our empirical analysis will be based on an extensive survey database, which, although not really suitable for inferences, may give valuable insights into those frequently taken-for-granted 'small' players (that is, second- or third-tier subcontractors or small affiliates), who may be less fascinating than the 'big players' such as the central or lead firms, or the large TNCs, but are, however, central to both the creation and capturing of value, and therefore to development.

In fact, our analysis is based on the ostensible 'black box' of the firm, since we are interested in the impacts of its ownership and therefore decision making structure (including how the firm interacts with other agents, that is, market, hierarchy or network[1] and why it decides to become involved in international production); its strategic orientation and embeddedness on how it creates, enhances and captures value; and lastly, its power within the chain or network (seen not only as the control exerted on others, but also the freedom to operate independently, which is usually essential in functional upgrading).

The structure of this chapter is the following: the first section contains an exposition of our understanding of how firms delocalize, a framework which is taken forward in the second section, where our conceptual framework of the impacts of delocalization on development is discussed. The third section is an empirical assessment. Lastly, in the fourth section, the concluding points of the chapter are drawn.

Conceptual Issues

The organization of international business

The constantly changing new geography of production, particularly (although by no means exclusively) in labour intensive industries (LII) is characterized by a multiplicity of ways of integrating firms and regions into global networks of production and distribution.

What emerges as a key question in this context is how formerly localized enterprises globalize. This first key question (the second will be discussed below) has two distinct, although closely interrelated, dimensions which have to be addressed in conjunction in order to get a clear picture. The first dimension is that of the individual firm, which we consider the basic building block of the economy (the 'why' of the main object). It is interesting to note that, although both the GCC and the GPN approaches of explaining international production (on which our analysis will also be based) claim to consider the firm to be very important (according to Henderson et al. 2002, firms constitute one of the conceptual dimensions of the

1 Even if we assumed that the market was of minor importance (which by the way, we do not), it is still a mode of managing transactions that cannot be ignored, as has been the case in the majority of the relevant literature. In a sense, GCC and GPN theorists have gone to the extreme of providing oversocialized accounts of the organization of economic life.

GPN), not only has this variability been consistently treated as exogenous, but also the empirical evidence appears to be surprisingly scant. This weakness may be due to the fact that, as Hess and Yeung (2006) argue, most of the empirical studies are interview-based qualitative ones, attempting to follow mechanisms and processes. In fact, in the majority of network or chain approaches, there seems to be a 'black-box' attitude towards the firm that is enhanced by the reluctance to pose the most basic questions (which have been central in the international business literature) revolving around the decision to delocalize. Specifically, questions such as 'why do some firms decide to invest abroad while others prefer not to?' or 'why is FDI more preferable to subcontracting?' necessitate the use of tools or theories that are usually considered alien – at best – to the network or chain approaches (the theory of the firm, internalization, Dunning's [1993] eclectic approach, and so on).

The second dimension concerns the broader picture of how the economy is organized from production to consumption. Before going into any assumptions or theorizations, we should note that our main analytical instruments will be theories falling into the wider group of chain or network theories, mainly the GCC and the GPN. The former has been the most successful paradigm in a number of – perhaps contradictory – ways: firstly, it was the culmination and therefore the most elaborate of a number of chain conceptualisations of the economy; secondly, it was the most successful effort to link production and consumption in a coherent framework; thirdly, in an era of globalization, or at least excitement about it, the approach provided an insight into truly global phenomena (such as the GCCs). The final proof of the theory's success is that it has received significant criticism, mainly from the GPN proponents, who surprisingly acknowledge GCC as perhaps the most significant precursor.

GCCs, according to Gereffi and Korzeniewicz (1994, 2), are sets of inter-organizational networks, clustered around one commodity or product, linking households, enterprises and states to one another within the world economy. A 'commodity chain' traces the entire trajectory of a product from its conception and design, through production, retailing and final consumption. GCCs are the network of labour and production processes whose end result is a finished commodity. These networks are situationally specific, socially constructed and locally integrated, underscoring the social embeddedness of economic organization. Initially, Gereffi (1994, 45) distinguished two types of GCCs governance: a) *Producer-driven commodity chains*, where TNCs or other large industrial enterprises play the central role in controlling the production system, mainly in capital and technology intensive industries (for example, automobiles, computers, aircraft and electrical machinery); and b) *Buyer-driven commodity chains*, where large retailers, brand-name merchandizers and trading companies play a central role, predominantly in labour-intensive and consumer-good orientated industries (for example, clothing, footwear, toys, consumer electronics, housewares and hand-crafted items). More recently (Gereffi et al. 2005), it was acknowledged that the governance of value chains depends on three factors (the complexity of transactions, the degree to which knowledge can be codified and the capabilities of suppliers vis-à-vis the requirements of the transaction), giving rise to five different governance structures. Not unexpectedly, the two extreme positions are occupied by market and hierarchy, while between these (moving from market towards hierarchy) we find the modular, the relational and the captive types of governance, each one characterized by different combinations of the three factors and an increasing degree of explicit coordination and power asymmetry going from market to hierarchy.

Among the various criticisms that GCC theory has received (for example, Leslie and Reimer 1999; Henderson et al. 2002; Smith et al. 2002), the most relevant appears to be that the theory is overly preoccupied with flows and 'systems', while individual nodes more often remain at the periphery of the analysis. In this context, the focus on specific sectors implies a neglect of the history and social context of the various nodes of the chain. The point here is that history and social relations impose a path dependency on the chains (for example, the impact of the former communist state regimes on the incorporation of CEE firms, regions and countries in chains – Henderson et al. 2002). Therefore, it would appear that firms within the GCC framework appear as largely disembedded from their local or national, social and institutional context. Hence, firms appear to have little or no autonomy to develop independent strategies (Henderson et al. 2002).[2] In a similar context, Smith et al. (2002) argue that the region is remarkably downplayed, while, in contrast, the nation is the crucial barrier and divide.

GPN proponents argue that their approach deals successfully with the majority of criticisms. According to Henderson et al. (2002, 445), a production network is a 'nexus of interconnected functions and operations through which goods and services are produced, distributed and consumed'. These networks integrate firms and national or regional economies in ways that have enormous implications for their well-being. The interaction of firm-centred networks with the socio-political contexts in which they are embedded is a very complex, often bi-directional process, also because the former can potentially be very mobile, while the latter are territorially specific.

It would appear that our approach shares more features with the GPN rather than the GCC theory. Assuming that this is the case, and we will not argue the contrary, this is to a great extent symptomatic of our data, which by being very diverse allows for the inclusion of more external factors, more akin to the GPN approach.

Our conceptual framework for analysing the delocalization of labour intensive firms has four main analytical dimensions that are intersected by wider categories.

2 Although Gereffi (1994) claims that GCCs have a territoriality in the sense that the various activities, nodes and flows are geographically situated.

Analytical dimensions

Dimension 1: The firm with its own unique set of resources and competitive advantages The main approaches in explaining why firms internationalize revolve around three main themes: the firm's ownership advantages, the internalization decision and the role of resources (internal or external to the firm). The two first dimensions, along with location, are the three elements of Dunning's (1993) extremely influential Ownership-Location-Internalization (OLI) or eclectic paradigm of the TNC. The third (that is, the resource-based approach) is, in fact, not widely considered a mainstream explanation of why firms expand abroad. However, following Kay (2000) and Pitelis (2000), we believe that a treatment of resources as a separate factor is essential in understanding the growth of firms. On the other hand, the treatment of location as an analytical category can be better implemented.[3] In turn, the framework presented here has the potential to bring into the analysis a much wider array of exogenous factors, central to the analysis of internationalization.

Firms trying to operate in a foreign market are faced with a number of barriers (Hymer 1974). Those firms that manage to compensate for these disadvantages of 'foreignness' need to possess certain competitive advantages, such as (Graham 2002, 37): a) 'preferential' access to cheaper factors of production compared to competition; b) a production function of lower cost; c) access to better (cheaper or more extensive) networks of distribution, and d) product differentiation.

However, the possession of such advantages can only represent a firm's competitive advantage vis-à-vis its competitors at home or abroad. Whether these will be exploited by the firm itself or leased to some foreign firm is an issue discussed by the internalization theory. Hence, ownership advantages are a necessary, but by no means a sufficient, condition.

The same is true for the firm's resources. According to Penrose (1959), productive resources are not general and unspecified categories to which all firms have access. Therefore certain resources, and especially the services that they can offer, are particularly important to each firm, since they constitute the base of firm differentiation. In fact, even if two firms have exactly the same resources, it is almost impossible for the way they combine their services to be identical, so inevitably they will be led to producing different products. Therefore the growth of firms depends on factors that are slightly or not at all predictable. On the one hand, we have the resources of the firm that change constantly, while the accumulated experience determines novel ways of combining them in a by and large turbulent external environment. On the other hand, we have the very important role of

3 This is not to say that Dunning missed something. Location advantages include a great variety of factors such as (Dunning 1993, 81): the spatial distribution of natural resources and markets, the prices and the quality of inflows and the productivity, the investment incentives, or disincentives, the particular social characteristics of the receiving region, the economies of concentration of R&D and so on.

entrepreneurship. Without the 'psychological predisposition' for discovering opportunities, which requires considerable effort, along with the engagement of certain resources of the firm, it is almost impossible to achieve change (with the characteristics of innovation).

We should note that the resource-based theory cannot constitute a complete theory of the TNC. According to Kay (2000), the resource-based theory is useful for the analysis of the direction of expansion (that is, expansion at home or abroad). For the analysis of the mode (that is, whether the firm will advance in the direction it has selected alone or with collaborators), the use of the internalization theory is required.

In fact, what we are looking for here is an explanation regarding the decision to make or buy. Almost all answers to that question can be traced to Coase's seminal 1937 article on the boundaries of the firm. According to Coase, outside the firm, it is price movements that direct production, which is coordinated through a series of exchange transactions in the market. Within a firm, these market transactions are eliminated and the entrepreneur-coordinator who directs production replaces the complicated market structure with exchange transactions. It is clear that these are alternative methods of coordinating production.

In other words, the firm internalizes the operations of the market to the extent that the cost of this internalization is lower than the cost of using the market mechanisms. Therefore, the decision about whether to make or buy is a trade-off between the cost of running a large and less specialized organization (similar to those described above) and the costs involved in finding partners and incomplete contracting (Grossman and Helpman 2002).

There have been a few other efforts to explain how firms grow and how they internationalize, but none has been more influential than Dunning's (1993 – particularly regarding the internationalization issue) eclectic approach. This was also the theory that paid more attention to the exogenous factors affecting (mainly) FDI.

Dimension 2: The sector with its given technologies and markets Sectors are central in the determination of the possibilities to upgrade because firms from the same sector will tend to share two characteristics that we consider central in the decision to delocalize, that is, technology and market orientation. As Henderson et al. (2002) further argue, firms in the same sectors will tend to create similar (in terms of organization and governance and institutional framework) networks, and share common 'languages' and communication structures.

Dimension 3: The environment with its unique institutions, civil society, history and policies The environment (local, regional, national or beyond) has its own unique historically shaped institutions – including among others the local or national governments, labour unions and business associations – that constantly affect and are affected by the civil society and the prevailing norms and attitudes. How do the characteristics of the environment influence the decisions to delocalize?

Somewhat paradoxically, in a period of increased globalization there has been a strong revival of academic (and policy makers') interest in the role of regions as loci of innovation and economic activities. Undoubtedly, success stories of industrial districts and regions across the globe (for example, 'Third Italy' in Europe, Silicon Valley in the USA) in the late 1980s–early 1990s have contributed significantly in this direction. As Coenen et al. (2004) mention, researchers in economic geography and innovation (Porter 1990; Saxenian 1994; Asheim 1996) argued that processes of localized learning played a crucial role in fostering innovation within territorial agglomerations.

At about the same time, Soete and Freeman (1987) and Lundvall (1988, 1992) used the term 'innovation system' in order to describe the complex nature of innovation process involving intense and multiple interactions between various actors (firms, employees, research organizations, universities) within a – usually nationally defined – institutional framework. The National Innovation Systems (NIS) approach emphasized the importance of interactive learning and the role of nation-based institutions in explaining the difference in innovation performance and economic growth across various countries (Coenen et al. 2004). Building on both these approaches, Cooke (1992) developed the concept of Regional Innovation Systems (RIS), stressing the fact that regions are in several cases geographical (and administrative) units that play an important coordinating, economic and institutional role. Later, Braczyk et al. (1998) provide us with a more detailed definition of an RIS. As cited in Coenen et al. (2004, 2), in order to have an RIS in place, two subsystems must be systematically engaged in interactive learning: a) the regional production structure (or knowledge exploitation subsystem) consisting primarily of firms, especially when there are cluster tendencies, and b) the regional supportive infrastructure (or knowledge generation subsystem), consisting of private and public research labs, universities, technology transfer agencies and so on. In addition, the role of informal institutions (trust, norms, routines) is emphasized as the main factor facilitating communication and interactive learning within a region. Much along the same lines, Lundvall and Borrás (1997), also mention: 'the region is increasingly the level at which innovation is produced through regional networks of innovators, local clusters and the cross-fertilizing effects of research institutions' (Lundvall and Borrás 1997, 39).

In an opposite direction (away from the regional and even national scale), one has to take into consideration the changing environments running parallel to the trend towards globalization and trade liberalization, in the formation of trading blocks and regional agreements, among which, *EU further integration and recent enlargement* holds a prominent position. During the 1980s and 1990s, the EU made unprecedented progress towards greater integration, making economies of scale and agglomeration more relevant, thus altering the geography of production. Moreover, the larger size of the market and the dynamic effects this may create (in terms of productivity growth) could strengthen the degree of integration of the EU – or parts of it – into the global economy (Baldwin 1992). Indeed, there is empirical evidence suggesting that changes in governance structures have spurred the re-organization of operations of TNCs located in the EU to a much greater extent than in the case of affiliates based outside the area (Dunning 1996). The recent enlargement (completed in 2007 with Bulgaria and Romania) introduced significant quantitative and qualitative changes, as well as new challenges and opportunities, for both the new members and the EU as a whole. These two processes have already put in motion a potentially much more drastic series of changes in the structure and hierarchy of European economies. Based on rather different theoretical backgrounds and policy assumptions, two diverging scenarios seem to emerge, that is, convergence of European countries and regions versus divergence. Nevertheless, the historical evolution of the EU has shown that both scenarios are possible and have in fact taken place (Fagerberg et al. 1997).

Dimension 4: The global environment with its unique institutions, governance and power relations Globalization has come to prominence over the past two decades, following changes in the 'real world', namely improvements in ICT that have facilitated global exchange, changes in the institutional framework governing world trade and production, as well as the pursuit of a liberal economic policy agenda fostering integration of product and capital markets. The opening-up of international markets resulted in intensified (and growing) international competition and forced enterprises to adopt an international perspective. Even businesses focusing primarily, or even exclusively, on their domestic markets must become internationally competitive to ensure long-term survival and growth (Karagozoglu and Lindell 1998).

This need is not limited to individual firms, but also encompasses sectors, regions and countries. According to Dicken (2000, 287), like firms, states also engage in *price competition,* in their attempts to capture a share of the market for mobile investment, and in *product differentiation*, by creating particular images of themselves (that is, the strategic nature of their location, the attractiveness of the business environment, the quality of the labour force and so on). Nevertheless, it must be pointed out that globalization has so far been an uneven and asymmetric process with differentiated results, across both sectors of the economy and regions (Lundvall and Borrás 1997; Petit and Soete 1999).

On the other hand, individual strategies, which shape and are shaped by the decision to delocalize, the role and position of the firms within their networks, all greatly affect how much and in what ways value is created, enhanced and captured by the firm, its labour force, the region and the country in which it is located.

Delocalization and growth

The creation, enhancement and appropriation of value bring us to the second key question, which is how delocalization affects growth. In analysing how our framework may help us better understand the impacts of delocalization on development, we follow Coe et al. (2004), who claim that endogenous factors are inadequate to generate growth in an era when competition is increasingly global. According to them, development (which is identified with value creation, enhancement and capture) is a product of the interplay between three large groups of variables.

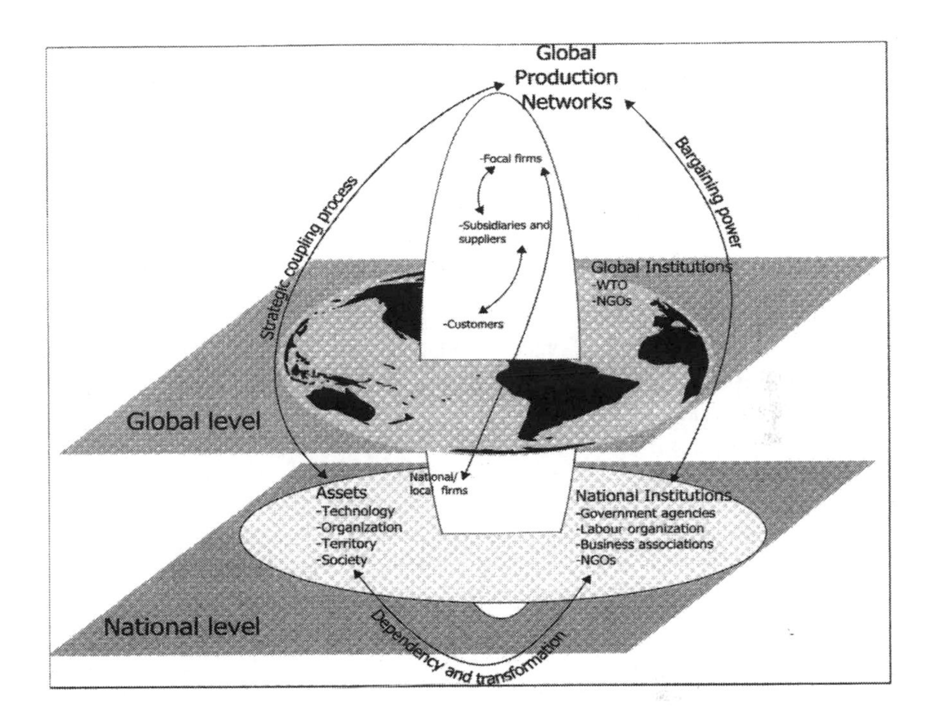

Figure 2.1 A conceptual framework for analysing the impacts of delocalization

Source: Based on Coe et al. 2004.

Each country or region is endowed with (or has created) a set of assets, some of which are ubiquitous, while others are more or less exclusive to the country. Although technology, organization and territory appear in the framework since they are considered the 'holy trinity' of (regional) development (Storper 1997, 26), these three elements seem to imply the conditions of a whole range of resources including labour, capital, infrastructure and others.

Firms lie at the other apex of the triangle. These may be the focal firms, their subsidiaries or suppliers or their customers. For the needs of our analysis, this has to be expanded to firms that are not part of GPNs or GCCs. Firms operating within bi-national or regional networks, which are most usually considered as quite irrelevant to not only the more mainstream IB or neoclassical approaches, but also to the more 'socialized' network approaches, need to be seriously considered, as they often support or outright substitute GPNs in their role of growth enhancement accorded to them by Coe et al. (2004). Specifically, they argue that a region's (and therefore a nation's) assets (in fact the economies of scale and/or scope implied by the specific assets) can bring growth inasmuch as they 'complement the strategic needs of trans-local actors situated within GPNs' (2004, 471). This strategic coupling (of the firms' of GPNs' strategic goals and the regions' economies) is therefore of primary importance to development.

Nevertheless, we should try not to ignore the possibilities of a de-coupling process that can be equally detrimental to development. This possibility points to the importance of embeddedness to the sustainability of development. Here we view embeddedness as defined and understood by Liu and Dicken (2006, 1232), who 'use the term "embed" in a very precise way, in order to capture the extent to which an activity becomes fixed in a particular place and, as such, contributes to local/national economic development through both direct and indirect spinoffs (including backward linkages with local suppliers)'. It is clear then that even local firms can become dissembedded.

Finally, we have the institutions at all levels of governance (national, regional or local). During the last decades, there has been a growing devolution of political and economic power from the central state to local and regional institutions. This has resulted in a multitude of configurations of institutions, affecting both the regional or national sets of assets and the bargaining power vis-à-vis the TNCs. In this context, the case of China is rather unique. Describing the Chinese automobile industry, Liu and Dicken (2006) show the enormous bargaining power of a large and centralized country like China that is able to play off TNCs against one another. At the same time, automobile TNCs improvized quite a few ways to avoid bargaining with the central government and instead tried to play off Chinese provinces (eager to attract FDI) against one another.

The result of this interplay between countries or regions with their assets and institutions and firms trying to take advantage of the assets is development, which is conceptualized as value that is created, enhanced and captured. Assuming that value is created in the country, the big question then is how to enhance this value and, more importantly, how to capture it.

What Coe et al. (2004) did not explicitly consider is the role of the global environment, with its evermore pervasive institutions, which are increasingly capable of greatly influencing national policies and, therefore, the immediate environment of firms and networks at all territorial levels. However, the global environment is not only about the infamous Washington Consensus. Along with the institutionally induced collapse of trade barriers, the world is also getting smaller in more tangible ways, mainly through the increasing efficiency of transport and more importantly ICT.

Finally, transcending the two levels of analysis (the global and the national/local) as well as the three tangible dimensions parallel to the two levels, that is, the firms', the assets' and the institutions' flows, the central object of this chapter is none other than the value created in the production and consumption of goods. The territorial embeddedness of the firms as well as the power they possess within their respective networks more or less defines how much of this value will be captured by the regions.

It is clear that these factors combine in a rather individualistic way, consequently rendering prescriptions extremely difficult. This fact, nevertheless, does not render the approach completely chaotic. Indeed, there are very clear directions towards development in the intangible categories, which are always mediated by the pursuit of the firms' goals and the balance of bargaining power.

What, then, are the possibilities for upgrading created by delocalization? We will follow Kaplinsky (1998) in both the definition of 'upgrade' (although in the following sections we will make use of the more systematic and widely used definition of Humphrey and Schmitz (2002), which is none other than the ability 'to appropriate a greater share of the returns accruing from the whole production cycle', as well as the types of economic rent, which in the Schumpeterian tradition stem mainly from innovation and, to a significantly lesser extent, from scarcity. In this context, apart from resource rents, that is, those that are based on the availability of otherwise scarce resources (the only such resources found within European territories that spring to mind are the North Sea oil reserves), a country (or a sector or a firm) may benefit from the existence of policy rents that may be local, national, regional or global. An example of the last type was the MFA that, as will be shown later, was central in the development of the Greek clothing sector. Closely related to the resource and policy rents are infrastructure, human resources and finance rents, since all of these are, to varying extents, exogenous to the firm. Of course, this does not mean that these resources are freely available to all firms. Each firm is constantly forced to operate within given production frontiers, with scarce resources. In this context, not all firms in a country that has heavily invested in education can employ post-docs for all the available positions.

Instead, in our times of trade liberalization and integration, it would appear that the rents that are internal to the firm – such as technological, organizational, relational, product and marketing rents – are far more significant. Regarding technology, it appears that with the shortening of the life cycles of most current technologies, what is more important is the creation and appropriation of new technology and its rents. However, what is paramount in any analysis of economic rents is the realization of the transient nature of economic rents, making the ability to identify and pursue constantly new sources of rents perhaps the most significant of economic rents.

Having more or less established value, what are the roles of power and embeddedness? In the context of our analysis, they both have considerable implications for whether value created in a country or region is actually captured locally (development *in* a region versus development *of* a region). The role of embeddedness is rather straightforward. In a sense, territorial embeddedness refers to implanting a firm into deeply rooted social and economic relations with which it becomes interwoven. In practical terms, a firm is territorially embedded if it draws resources (for example, labour or intermediate products) from local sources which possess qualities that are hard to replicate. The more embedded a firm is, the more the value that it creates will be captured by the region in which it operates. In a similar manner, value creation and capture of a firm are also conditioned by the power they possess within a network. An interesting, although not necessarily typical, and quite extreme representation of the role of power is provided by Sacchetti and Sugden (2003, 674). They define a network as a number of nodes and links amongst actors, where each one dynamically aims at improving its position within the network, as relating the distribution of resources amongst actors to the structure of actors' interdependencies. Therefore, according to Sacchetti and Sugden (2003, 675), we can understand networks either as reciprocal dependence based on the complementarity of resources and shared objectives and on the agreement not to act against the interests of others in the network, or *not* as a reciprocal, preferential and mutually supportive locus of production (for example, relation of prime contractor to subcontractors).

Therefore, networks entail an idea of governance in production in which power becomes a crucial determinant of the nature of relationships between actors, with or without the presence of market relations. According to Sacchetti and Sugden (2003, 670), this is different from the 'market model' (where all actors have equal power), as it does not confine the presence of power asymmetries to exceptional circumstances – that is, a market failure – but embodies power as a constituent element of network relationships.

In this context, as Sacchetti and Sugden (2003, 671) argue, networks may be viewed as having a centre, the big firm (the star) managing the actors of its 'constellation' (the planets), which are partially controlled and partially autonomous, without a centre, where – in order to obtain reciprocal advantages – relationships among participants are mutual. Hence, one of the most central characteristics of networks is the distribution of power, which may give rise to two quite distinct governance structures, that is, networks of direction and networks of mutual dependence,[4] with very different implications for the roles of the various members of the network. In the TNC literature, the issue of the relationships between parents and affiliates has attracted considerable attention (for example, Bartlett and Goshal 1989; Birkinshaw and Hood 1998). Expanding this logic to encompass all other relations (contractual or not) could be particularly interesting in our context.

4 Tracing the actual source of dependence is a very interesting and relevant issue. In this context, considering the embeddedness of the firms to be the source of dependence, instead of purely economic (that is, transaction) reasons will have different implications for the network, as well as for the firms comprising it (Uzzi 1997).

Empirical Findings

The first aim of this section is to analyse how the basic dimensions of our analytical framework affect the decision to delocalize (specifically, 'why' and 'how' firms delocalize). The second is to examine the impacts of delocalization on development through the use of some stylized examples.

Explanatory power of the analytical dimensions

The sector Not unexpectedly, the specificities of the sectors appear to influence significantly not only the reasons for delocalization, but also the form it takes. In fact, it would appear that the different competences (or put in a different way, the technology) characterizing each sector will lead to different types of delocalization. In this context, the software and electronics sectors place considerably more value on knowledge than do the clothing and footwear sectors, where the generic category of skills is more significant. This was evident in the huge differences of the mean share of workforce with tertiary education: 83.6 per cent of the personnel of the software sector had tertiary education, while the shares of the other sectors were considerably smaller (electronics 34.2 per cent, clothing 11.5 per cent and footwear 8.7 per cent).

Apart from the level of education, firms in the clothing and footwear (and some segments of the electronics) sectors express their concern about the fact that they actually face an ageing problem, since their employees keep getting older, while at the same time young people do not seem to be willing to work in these sectors:

> People working in the shoe trade are getting older and older. (British shoe firm)

> Younger people don't want to work in the shoe industry now. (British shoe firm)

> What is missing are young people willing to do this work. They rather seek employment in other spheres and not in the footwear industry. (Bulgarian shoe firm)

> There is scarcity of young qualified personnel willing to get employment in the footwear industry. (Bulgarian shoe firm)

On the other hand, the software sector does not appear to be experiencing the same problem; on the contrary, there are many young people who wish to be employed in the sector after finishing their studies (sometimes there is some kind of cooperation between firms and local universities). To some extent, this cooperation is also the case for some firms in the electronics sector.[5]

5 This is due to the great diversity of the electronics sector comprised of an extremely wide array of firms with different technology and knowledge intensities, from simple assembly to product development and R&D.

> We cooperate with universities such as The Silesian University of Technology, Warsaw University of Technology, Warsaw University. Thanks to that the company recruits workers from these universities. (Polish software firm)

> We have young people with technological experience. (Polish software firm)

> I am really happy to be looking for young graduates and PhDs. We have employed two new graduates and two new post-docs and two people from industry, so there is a balance. (British software firm)

This phenomenon was further mirrored in the differences in the perceived sources of competitive advantage of the sectors. Hence, although R&D was important to both the software and electronics sectors, design and marketing was much more important to the former, as was inputs supply to the latter. Similarly, the ability to produce cheaply (labour-intensive products) was most important to both the clothing and the footwear sectors. Nevertheless, the latter appear to place more emphasis on skills (one of the most elusive characteristics of our survey).

The reasons for delocalizing reflect the different competences of the sectors. Hence, knowledge being the most significant strength of the software sectors, the lack of specifically skilled labour becomes the most important reason to outsource. In a similar manner, the higher capital and R&D intensity of the electronics sector makes the lack of the appropriate technology or equipment the most significant reason to outsource. High labour costs in the home country were of paramount importance to the clothing sector and rather important to the footwear sector, where access to natural resources was the primary motive.

Regarding the motives of FDI, low-cost unskilled labour was of primary importance only for the clothing sector, while market-related factors (market size, growth and per capita income) dominated the other three sectors. Given the predominantly vertical structure of FDI in clothing, the importance of low cost unskilled labour is not surprising. However, footwear in particular and to a lesser extent electronics seem to behave rather unexpectedly. We will try to understand why in the next section.

The country Not unlike the sectors, the countries of our sample share some features while they are completely different in others. Hence, although Greece[6] is (in terms of Gross Domestic Product [GDP] per capita) conventionally considered

6 Greece is a Mediterranean country with a strong eastern European orientation. Although its relative position within the EU was improved by the recent enlargement, it is still characterized by average levels of economic prosperity, reliance on agriculture, considerable degree of concentration of economic activity in Greater Athens and Thessaloniki, and the presence of some of the most peripheral locations – both at the national and European level. The degree of integration of the Greek economy in the European and global networks of production and distribution increased considerably during the post-1974 period. It became

a DC, as is the UK,[7] one could argue that this is the only similarity between the two countries. Their economic and socio-political history, their geography and their industry structure (to mention just a few parameters) clearly differentiate them. The importance of national idiosyncrasies could, nevertheless, hardly be more pronounced than in the case of the former socialist-state countries. Despite a period of 45 years of more-or-less similar historical experiences, the three former socialist countries in our sample have followed very different pathways since the beginning of the transition process.[8] Features such as the size of the country, its progress in the transition process (despite the fact that all three countries are currently EU members), its geography, as well as its pre-WWII socio-political situation and therefore position in the international status quo all combine to create some very unique trajectories. Therefore Poland, being the largest CEE country (except the Ukraine), has been defined by rapid advances in the process of post-socialist transformation and below average levels of economic prosperity. Increasing integration had diverse effects on LII, offering a multitude of opportunities for clothing, textiles and electronics, but posing considerable threats to agriculture. Bulgaria, on the other hand, advanced more hesitantly towards reform and has only recently managed to join the EU. Bulgaria is characterized by very distinct historical trajectories defined by its position in the fault-line between Orthodox Russia and the Muslim Ottoman Empire. The prevalence of an idiosyncratic form of socialism perpetuated the specific characteristics of the Bulgarian socio-economic structure and its marginal integration in the global economy. However, during the last decade, the country has gone a long way in reforming its socioeconomic structures and becoming an EU member. Finally, Estonia, a country with strong Nordic ties, went through a rapid transition process in the 1990s, accompanied by the restructuring of the economy and rapid growth of the service sector. Not unexpectedly, the three countries have developed quite distinct orientations of economic relations, with the role of geography being very pronounced. Germany (in the case of Poland), Finland and Sweden (in the case of Estonia) and Greece and Italy (in the Bulgarian case) have been the main trade and FDI partners.

---

a member of the EEC during the second wave of enlargement in the 1980s. LII account for a very large part of the total economic activity and display varying degrees of dynamism.

7 The UK was the first industrial nation and probably an early exemplar of a global economy. It is also an advanced industrialized economy that maintained a global orientation for the past two hundred years or so. However, during the post-war era, it had to change the earlier patterns of international trade and production factor flows and acquire a stronger European orientation. It became a member of the EEC during the first wave of expansion in 1977. LII in the UK have been undergoing significant restructuring accompanied by significant employment decline for the best part of the twentieth century.

8 This does not imply that the recent history is unimportant. In fact, of the five countries in our sample, the ones which at some time in their recent history shared most characteristics would be Greece and Bulgaria before WWII. The 45 years that followed set the two countries on completely diverging pathways.

Apparently, the common socialist history of the three CEE countries is evident in the similar competences of their firms. Specifically, skills are considered a very important aspect of human capital in all three countries, while experience and knowledge are important in Greece and the UK. The three CEE countries also seem to affect the competitive advantages of their firms in broadly similar ways, since the majority claimed that, prior to delocalization, they were competitive in skill and labour-intensive products. The backwardness of the Greek firms is evident in their contrast with their UK counterparts. In fact, the British firms were the only ones in our sample to define design as their most significant competitive advantage, while Greek firms admitted to depending on the production of labour-intensive products.

The differences in the attitude of the firms towards relations are slightly more complex to comprehend. Specifically, the UK and Estonian firms stand out as those operating in a more supportive environment, while Bulgaria and Poland are at the other end of the spectrum, with significantly more firms claiming to have no relations with other firms, institutions, and the central or local government.

The impact of the country is perhaps best depicted in the geographical orientation of the delocalizing firms. Although the findings are consistent in all types of delocalization, for reasons of simplicity (and wealth of information, since the relative sample is considerably more sizeable) we will concentrate on outsourcing. Figure 2.2 summarizes the geographical orientation of outsourcing from the UK and Greece. Three interesting and rather unique points characterize the geography of British outsourcing. Firstly, there is the fact that China and India were clearly the most important 'host' countries for the British firms, accounting for almost 20 per cent and 13 per cent respectively. Secondly, British firms were the ones with the greatest diversity in terms of the location of their subcontractors. In fact, they reported doing business with firms located in twice as many countries as the other four counties of our sample. Finally, there does not seem to be any geographical concentration of the countries receiving orders from British firms, which appear to be the only really 'global' in our sample. This 'global' argument is further reinforced by the fact that the vast majority of British firms outsourced to firms in more than one country (50 per cent to at least three countries and only 27.8 per cent to only one country).

On the other hand, the most important feature of Greek outsourcing is the concentration in the Balkan region, with a single country, Bulgaria, accounting for almost a third of the subcontractors. Adding the other Balkan countries, the figure easily exceeds 50 per cent. However, the most interesting feature of Greek outsourcing is its relatively small depth, with most firms (56.8 per cent) outsourcing to only one country, thus implying a lock-in of Greek firms to a single country. Furthermore, there are cases where Greek companies outsource from companies located in the Balkans, which are however of Greek interests.

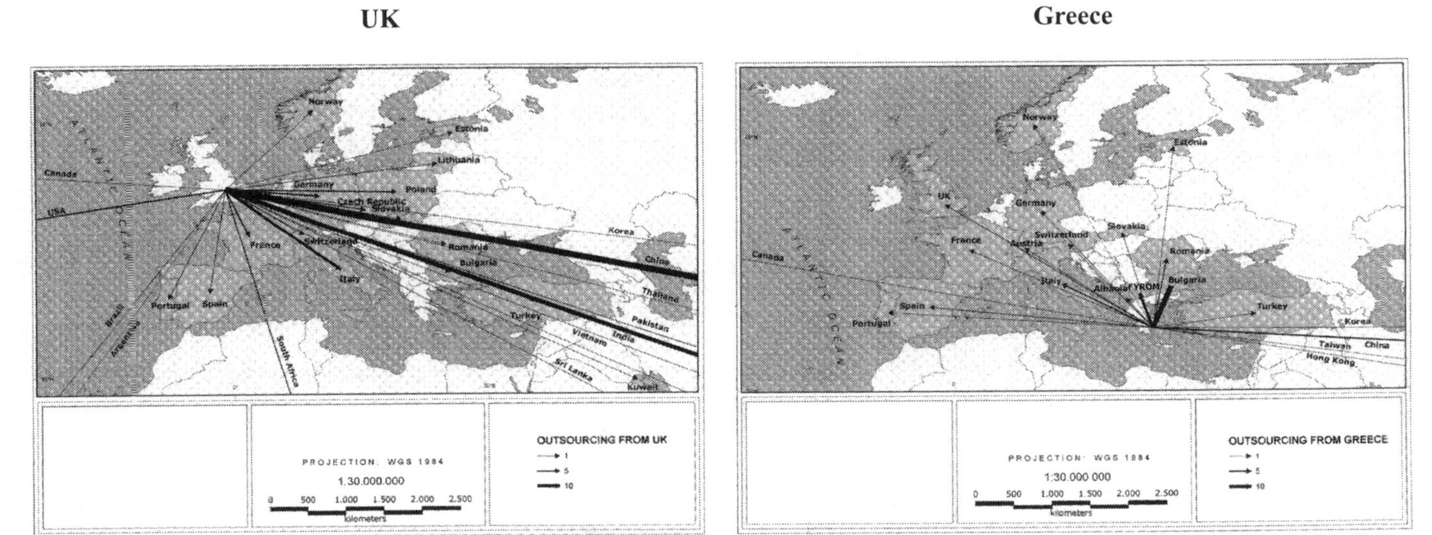

Figure 2.2 Geographical orientation of outsourcing from the UK and Greece
Source: Enterprise Survey.

Polish firms most resemble the British ones in many respects. Firstly, China is the most important recipient, while there is a spread of countries that is equally wide in terms of geography, although the number of countries is considerably smaller. Interestingly, Polish firms outsourced to firms located in both LDCs (for example, China, India and a few other SE Asian countries) and DCs (Italy, the UK, USA, Germany and so on). The same is also true for the Estonian firms; however, this is where similarities end. Estonian firms were significantly less 'global' than their Polish counterparts – in fact among the 24 countries, Estonian firms assigned contracts to only four non-European ones (China, India, the USA and Taiwan), while there was an obvious regional focus. Finland alone is home to almost a quarter of the suppliers (mainly through outsourcing) of Estonian firms (many of which are owned by foreigners), while the Baltic and Nordic regions together account for more than 60 per cent of the total.

The firm As we have argued in the previous sections, the firm is perhaps the single most significant source of variation in the decision to delocalize, as well as in the forms delocalization takes. What this point implies is that firms that are in most (observable) ways identical are expected to behave in quite different ways. Given the variability in strategies among the firms of our sample (Kalantaridis et al. 2008), the main aim of this section is to look into the importance of the competences (resources) of the firms and transaction costs involved when doing international business to explain the different outcomes.

As will become evident, the nature of our sample has some adverse implications for the expected outcomes. This small section will begin with a simple finding from our sample which is counterintuitive to the basic argumentation of the resource-based theory that large firms are more likely to become TNCs. The mean size of the firms owning at least one subsidiary abroad in our sample was 509 employees, while the respective figure for the firms involved in outsourcing was 154. So far, there is nothing counterintuitive. However, remove two outliers from the TNC group (Siemens UK and Logica CMS – two giants of 21,000 and 6,000 employees respectively) and the image changes to 229 and 154 employees. Even if we assume that these figures point in the direction of the resource-based view, explaining Figure 2.3 requires some departure from the conventional understanding of resources.

What is troubling about the figure is that an annoying 30 per cent of the 95 TNCs of our sample employ less than 50 people and also that 60 per cent of both groups is accounted for by firms with less than 100 employees.

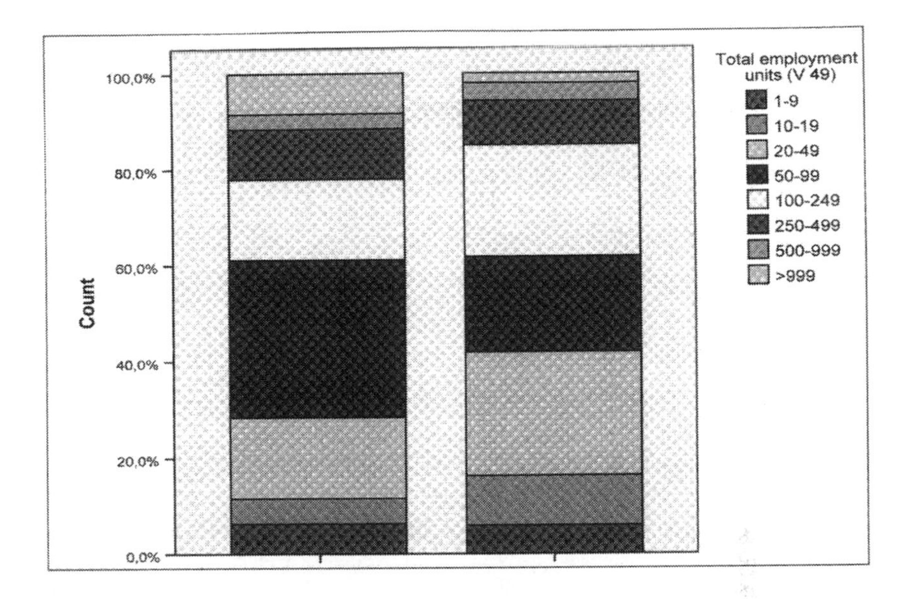

Figure 2.3 Size Composition of the TNC (left) and outsourcing (right) groups

Source: Enterprise Survey.

We believe that the answer could lie in the special nature of some resources available to some firms combined with the resource requirements of some forms of FDI. The former is exemplified by the commuting entrepreneurs of the Greek-Bulgarian borders. Nineteen out of the 22 (85 per cent) Greek clothing firms of our sample (only two of which had more than 100 employees) have invested in Southern Bulgaria, the vast majority being within commuting distance. Can these projects be called FDI? We prefer to use the term 'local delocalization', since the small distance appears to alter the characteristics of the decision to delocalize in many ways. Apparently, according to the entrepreneurs, the penalty of foreignness is considerably smaller in those regions. Not only do they not have to employ expensive administrative staff, since it is usually the entrepreneur or a member of the family performing these tasks, perhaps more importantly there is an intangible resource available to all incumbent or potential Greek ventures in the region, which is none other than the extensive community of Greek entrepreneurs and their support mechanisms.

The ethnic entrepreneurs in the UK clothing sector operate in a similar way. Although the distance is considerably longer (for example, to Bangladesh or Pakistan), the most considerable barriers are outrightly demolished. With no (or very small) culture, language and religious barriers, these firms face practically no foreignness penalties.

On the other hand, for the majority of the Greek clothing firms, the establishment of a factory is very cheap, as it is very often equivalent to the cost of transporting the existing equipment no more than 100 km away.

Therefore, the decision to delocalize is very much an issue of resources. The choice between the different types of delocalization was more of a mixed bag. Not unexpectedly, the factor most often mentioned was market failures.

Delocalization and growth

Upgrading The focus of this subsection is the individual firm and it is approached through the use of an extremely rich qualitative dataset, gathered during semi-structured interviews. What we are interested in is to highlight the diversity of experiences of firms in the same sector and country, with regard to upgrading. This is not as straightforward as it may initially seem, since very different views on what constitutes upgrading appear to exist. Unfortunately, the data from the UK and Estonian firms was scarce and is, therefore, not reported.

Looking for a unifying framework for upgrading, we turned to Humphrey and Schmitz (2002), who identified four distinct types of upgrading, namely: a) *process:* transforming inputs into outputs more efficiently by reorganizing the production system or introducing superior technology, b) *product:* moving into more sophisticated product lines (which can be defined in terms of increased unit values), c) *functional:* acquiring new functions (or abandoning existing functions) to increase the overall skill content of activities, and d) *inter-sectoral:* involving the movement of firms into new productive activities.

Although not explicitly mentioned, these four types implicitly represent a procession from less to more desirable types of upgrade. However, as Martin Bell (quoted in Humphrey and Schmitz 2002, 13) noted while criticizing Gereffi's (1999) idea of a virtuous circle of upgrading of East Asian clothing manufacturers from assembly to undertaking the entire production process, to own design, to sale to domestic and foreign markets,[9] the movement towards the final stages of upgrading is by no means guaranteed.

Table 2.1 is a summary of the various responses we received regarding the changes in processes after the delocalization of the firm interviewed. It is quite straightforward that all three countries present very different pictures. Even Bulgaria and Poland (the two CEE and overwhelmingly host countries) cannot be grouped together. The majority of Greek respondents (54.7 per cent) were unaffected by delocalization, while 28 per cent of the respondents (a figure corresponding to more than 60 per cent of the firms that claimed to have upgraded) went through process upgrade. The vast majority of software firms was unaffected by delocalization, while those that claimed to have upgraded, managed it mostly because of the experience they had acquired through the years, which enriched their know-how and rendered them capable of widening the range of the products offered.

9 We should note that these 'stages' are not seen as equivalent to those described by Humphrey and Schmitz (2002). The possibility of discontinuity is, however, very similar.

Table 2.1 Changes in processes after delocalization (percentage of firms that responded by sector and by country)

		Bulgaria					Greece				Poland				
		Software	Electronics	Clothing	Footwear	Total	Software	Electronics	Clothing	Total	Software	Electronics	Clothing	Footwear	Total
	N	36	30	45	37	148	17	13	23	53	31	22	84	23	160
Process	Equipment – technologies	5.6	16.7	11.1		8.1		23.1	13	11.3		9.1	2.4		2.5
	Greater capacity – higher turnover – employment	2.8				0.7			8.7	3.8					
	QMS	2.8			2.7	1.4									
	Quality	2.8	3.3			1.4			21.7	9.4					
	Quick response – flexibility				2.7	0.7			8.7	3.8					
Product	More complex prod – more activities – whole products – more activities		13.3	40	43.2	25.7					16.1	13.6	3.6	13	8.8
	Wider range	2.8	3.3			1.4					9.7				1.9
	R&D-design – innovativeness	5.6	6.7	11.1		6.1		15.4	8.7	7.5	6.5	18.2	10.7	17.4	11.9
	Know how – experience – models from sketches	27.8	30	4.4	13.5	17.6	17.6			5.7	3.2	9.1	7.1	4.3	6.3
	Own product										12.9		4.8		5
	Relations										3.2		2.4		1.9

Table 2.1 continued

		Bulgaria					Greece				Poland				
		Software	Electronics	Clothing	Footwear	Total	Software	Electronics	Clothing	Total	Software	Electronics	Clothing	Footwear	Total
	N	36	30	45	37	148	17	13	23	53	31	22	84	23	160
Functional	Distribution network – own brand – marketing – niche markets	5.6		6.7	2.7	4.1			4.3	1.9	6.5	4.5	7.1	17.4	8.1
	Diversification from manufacturing to services										12.9		2.4		3.8
	Diversification from subcontractor to lead firm										3.2		2.4		1.9
	Move up subcontracting layer												2.4	13	3.1
	Other							7.7		1.9	3.2		2.4		1.9
Down Grading	Diversification from first to second layer											4.5	1.2		1.3
	Narrower range of activities												1.2	4.3	1.3
	Seize own production/only subcontracting											9.1	1.2		1.9
	Seize R&D			2.2		0.7							1.2	4.3	1.3
	Unchanged	44.4	26.7	24.4	35.1	32.4	82.4	53.8	34.8	54.7	22.6	31.8	47.6	26.1	37.5
	Total	100	100	100	100	100	100	100	100	100	100	100	100	100	100

Source: Enterprise Survey.

While this trend is not the case with electronics, since those that were unaffected were fewer, while the upgrade paths were slightly more diverse, it was interesting to note that three companies attributed this upgrading in terms of technology either to the demands of their customersor to the competition they face. In both cases, this upgrading is a result of factors that are exogenous to the company itself.

Clothing was the most diverse (in terms of upgrade strategies) Greek sector, since for the 12 companies stating that they had been upgraded, there were no less than ten different upgrade strategies. Of particular interest here are the impacts on the successful domestic brand names, that is, the firms organizing their own national and lately international (although still by no means global) production networks. Only one firm in this 'elite' group attributed its upgrade not to delocalization but to the general strategy it has followed. On the contrary, other brand-name companiesexplicitly attribute their upgrading to their decision to delocalize part of their production activities abroad. More specifically, these companies state that if it weren't for delocalization, they wouldn't have achieved their objectives.

Finally it is quite surprising that no company expresses a downgrading of its position in the supply chain, especially given the general picture of the sector in Greece (and the quantitative findings from the field work).

On the contrary, considerably more Bulgarian and Polish firms appear to have benefited from delocalization, although in quite different ways. Concerning the former, the vast majority of respondents went through some type of product upgrade. Bulgarian firms appear to correspond more closely to the virtuous circle described above. Specifically, clothing and footwear firms (the two sectors that became involved in the delocalization processes earlier) were considerably underrepresented in process upgrading, which most of the respondents appear to have gone through. On the other hand, it is the more high-tech industries (electronics and software) that appear to be going through process upgrading, even though product upgrading is far more important.

However, Polish firms clearly stand out, as a significant minority (almost 18 per cent) is undergoing functional upgrade, while at the other end of the spectrum only four firms went through process upgrade.

This upgrading could be either proactive, in the cases where it is consciously pursued by the firm, or reactive, when it is imposed by a client. Nevertheless, in both cases, it is evident that a learning process derives from the subcontracting activities, the results of which are evident in the firm's functioning.

We consider upgrade to be proactive when a firm involved in insourcing as subcontractor mentions that its relation with the foreign client sets off an evolutionary course for the firm itself. This course occurs when the firm actually takes advantage of insourcing, in order to improve its own position in the value-added chain, as part of a development strategy. On the other hand, we have reactive upgrading, when a firm involved in internationalization reports an improvement either of its production process or its products (quality), but purely as a result of the client's demands.

Polish companies appear to be involved, to a wider extent than Bulgaria, more in proactive upgrading, in the sense that they express that subcontracting resulted in:

- Learning a particular know-how from the client, which is then embodied in their own production, either for the domestic or foreign markets (sometimes even under their own brand).
- Developing their own brand/product for the domestic market. In some cases, this also means that they also start to assign subcontracting to third parties, while in others that they stop acting as subcontractors themselves.
- Undertaking a wider range of activities:

> We turned into a subcontractor mainly for German companies. Over the time we have tried to develop our own brand names through taking over design activity and development of distribution channels for our products. Now, we outsource part of our manufacturing activity to subcontractors in Poland, China, Indonesia and India. (Polish clothing firm)

Bulgarian companies, on the other hand, seem to be more involved in reactive upgrading, since they mostly talk about upgrading of the equipment and the technologies used for the production, complying with various health and safety regulations that improve the working conditions within the firm.

> We go up – the customer requires more, that lead to adding a value in the product. We use always the new technologies. (Bulgarian software firm)

At this point, it should be made clear that this classification describes the general tendency of the companies in these two countries and does not mean that the opposite is ruled out (Bulgaria undergoing reactive upgrading and Poland proactive).

A further interesting point is that the attitude of the Bulgarian firms towards upgrading appears similar to that reported by many Greek firms, which, however, never managed to create their own competences and remained locked-up in subcontracting arrangements.

Undoubtedly, it is very difficult to compare the relative success of upgrade strategies, when the whole sample is comprised of relatively successful firms. However, we feel that a general assessment is possible. The general rationale behind this was succinctly expressed by the manager of a Bulgarian clothing firm in the following: '*Quality is standard (You cannot sell anything without it)*'. This simple phrase points directly to the transient nature of economic rent. If we assume that a few decades ago quality was scarce, then it is normal to expect that firms managing to produce high-quality products would reap the benefits of producing a scarce product.

In other words, the value of any specific upgrade strategy lies, to a considerable extent, in its uniqueness. Hence, this is the first distinction of upgrade strategies, that is, the common and the unique upgrade strategies. A second possible distinction could be between strategies that are more or less imposed on the firm, as opposed to strategies decided by the firms. In a sense, these two distinctions are the two sides of the same coin. For example, in the quotation of the Bulgarian manager, a subcontractor may not be directly forced to adopt a new technology by his lead firm. However, the knowledge that this technology may become widespread, in which case the firm will be outpriced, is a substantial indirect pressure. So, technology rents, as they were discussed here, may be a suboptimal upgrade strategy. In fact, with many very different manifestations (for example, purchase of new machinery, production of more complicated part of products or even whole products, increased automation and most often higher quality) simple technological upgrade was the only response for the majority of the firms. On the other hand, R&D (equivalent to the production of new technology or knowledge) or design was pursued by relatively fewer firms, which were again based in the three countries. However, it is the less frequent responses that are more interesting. Product or marketing rents were pursued by a small number of Greek and Polish firms and a couple of Bulgarian firms. Polish firms were the only ones to attribute their upgrade to organizational innovations, as well as a functional upgrade out of subcontracting.

Equally interesting is the relatively poor performance of Greek firms, particularly in relation to the absence of all types of organizational or relational innovations. Perhaps one needs to look more carefully into the specific historical context of the Greek economy, something that, to some extent, is accomplished in the next section.

A Case Study of the Greek-Bulgarian Clothing Sector 'Connection': The Role of Proximity and the Creation of Transnational Clusters

The main aim of this case study is the analysis of the northwards movement of a Greek clothing cluster and its transformation into what appears to be an international cluster in northern Greece-southern Balkans. Similar phenomena have also been noticed in other countries (for example, the clothing industry moved from Japan to China in the 1990s – Yamamura et al. 2003). During the last few years, clothing in Northern Greece has been developed in what could be called 'triangular manufacturing' (Labrianidis and Kalantaridis 2004). Recently, one of the triangle's apexes, namely the textile-clothing industry cluster in Northern Greece, is gradually shifting more to the north, crossing the borders of the country to include parts of southern Bulgaria and to some extent parts of southern Albania and Former Yugoslav Republic of Macedonia (FYROM). Therefore, the central questions are the following: what are the characteristics of this '*transnational cluster*', can it provide some competitive advantage to the companies involved and finally, what type of policies can support it?

As a result of the decentralising strategies pursued by the DCs during the 1960s and mainly from Germany, the textile-clothing industry developed in Greece because of the country's relatively low labour cost, as well as some privileges it enjoyed in its trade relationships with the EU (Simmons and Kalantaridis 1995, 290). Consequently, the sector was developed on the basis of undertaking subcontracting by companies from the DCs, something that has continued to a considerable extent to the present. This can largely explain the important role of clothing in Greek exports (mainly towards the EU and especially Germany). In fact despite its problems the sector is still highly export-orientated.

The clothing industry has been one of the most important sectors of the Greek economy. Since the mid-1980s, it has undergone crucial crises. However, despite the drastic reduction in the number of enterprises, the clothing industry still remains one of the predominant sectors of Greek manufacturing. In 2005 (NSSG 2008a, b) it contributed to a large extent to the country's manufacturing production (2.9 per cent), employment (4.8 per cent) and exports (9.3 per cent).

With the gradual abolition of all possible support measures available to the Greek clothing industry, it became apparent that the only alternative option the industry had in order to recover its waning international competitiveness was the relocation of part or the entire production to CEECs and mainly in the Balkans, a policy which could lead to considerable reductions of average per unit production costs. Therefore, in many cases, Greek clothing manufacturing enterprises created a triangular manufacturing arrangement. Specifically, they reach into the Balkans assigning a (second level) subcontracting part of – or the whole of – their production, for which, in turn, they had already been assigned a (first level) subcontracting from a company in a DC. In certain cases, the production is carried out in hired plants, using imported equipment from the Greek plants.

The ability of the lead firms to control the market renders them capable of maintaining their position at the top of the subcontracting chain. In this way, they achieve the lowest possible cost, while at the same time, by relying on the mediatory role of the subcontractors located in Greece (first level subcontractors), reduce the risk factor. On the other hand, companies in Greece maintain an intense interest in assigning subcontracting to the Balkan countries (second level subcontractors), since it constitutes a means for staying competitive.

Since the beginning of the 1990s, when Greek companies started assigning third party subcontracting, a gradual transfer of more complex production parts to the Balkan countries is observed. Initially, the seaming process was the main one to be relocated – which is also the most common example of the labour-intensive part of the production – but even more operational parts of these enterprises are now done this way.

Up until today, the experience of *triangular manufacturing* has been mixed. For example, in the case of Bulgaria, average wages in the southern provinces have increased substantially, while unemployment rates are among the lowest in the country. Conversely, most of the new jobs have been low-skilled, which could hamper the upgrading of the local human capital. Furthermore, the integration of the Balkans in the European and global production and distribution networks is characterized by a high degree of dependence and vulnerability. As the Greek experience points out, redefining the position of the Balkan countries in the international division of labour is a difficult, and not necessarily guaranteed mission, since up to the present, Greece has not managed to improve substantially its position in the international division of labour in the textile and clothing industry (Labrianidis 2000).

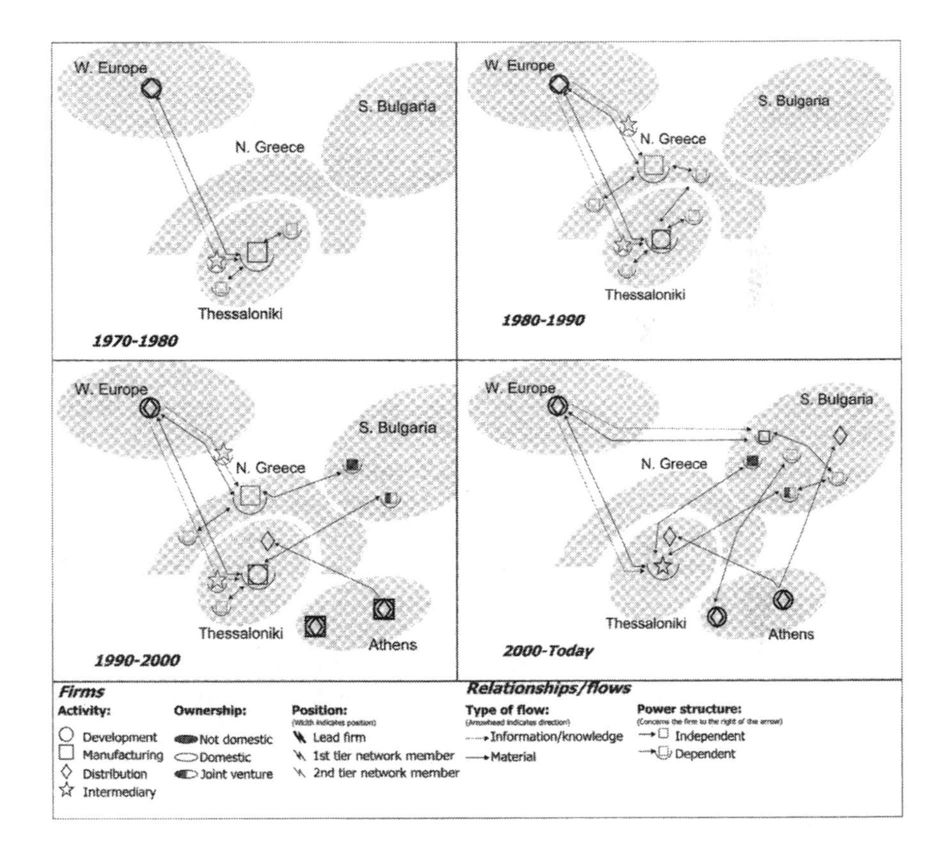

Figure 2.4 The evolution of the clothing sector in Greece (1970 – today)
Source: Compiled by the authors from the literature review.

Developing a relationship with a new contractor usually takes a long time, and thus assignors may prefer to follow their established patterns wherever they decide to shift to a new country and to keep on negotiating the conditions of their relations rather than look for new suppliers. Moreover, dependence and asymmetrical power relations may co-exist with mutual confidence (Kalantaridis et al. 2008). These two factors can explain, at least in part, why Greek firms continue to form one of the apexes of the German-Greek-Bulgarian clothing manufacturing triangle.

Figure 2.4 describes the various stages in the transformation of the sector, viewed from the Greek perspective (Labrianidis, 1996, 2000; Kalogeresis and Labrianidis 2007), while Figure 2.5 tells the story from the Bulgarian perspective. The former describes four quite distinct stages, which, although highly stylized, reflect the reality of the vast majority of the sector. During the first stage (1970–1980) orders to Greek subcontractors started coming from Western European (mainly Germany) and to a lesser extent American lead firms. These were normally executed within the company, although in some cases, there were also second level subcontractors within the same city (for example, Thessaloniki), so as to lower labour costs.

The change that takes place during the second stage (1980–1990/91) comes from the Greek subcontractors, since they gradually develop quite extensive subcontracting networks of second level subcontractors within the same city (for example, Thessaloniki) and mainly in the surrounding villages. Although a considerable part of the less labour-intensive tasks are still performed in-house by the first level subcontractors, the simpler tasks are assigned to second level subcontractors, who, in their turn often develop their own – smaller – networks of third level subcontractors, often homeworkers.

The third stage (1990/91–2000) is the 'triangular' manufacturing stage. Lead firms, for example in Germany or the USA, continue placing their orders to their established Greek partners. However, soon after the collapse of the socialist-state regimes, the latter are beginning to transfer parts of their activities to second level subcontractors in the southern parts of the neighbouring countries, particularly Bulgaria, and FYROM. By the end of this period, most Greek clothing firms are mainly responsible for the organization of the network. It is interesting to note that for variable periods (ranging from several months to a few years) the lead firms were uninformed of this development. This gave first movers a considerable advantage, since they were paid Greek prices for Eastern European labour costs. The extent to which this behaviour gave rise to the fourth stage (2000–present), when more direct links between Germany and Bulgaria are being created, thereby circumventing one of the apexes (that is, Greece), is discussed in the Bulgarian case below.

We feel that there are two central points of this story. The first is none other than the immensely important role of proximity in the creation of the cluster. In fact, the Greek-Bulgarian borders highlight the case that intense delocalization in the border areas can provide a fertile ground for companies that otherwise would not be able to go international (Labrianidis 2001). On the other hand, the very fact that even very small companies can easily go international can create a shock, at least in the short term, to the local economy of the home country.

The development of local delocalization around the Greek – Bulgarian borders was mainly based on two factors. On the one hand geographic proximity in the sense of *commuting distance* was crucial for the delocalization of the SMEs located mainly close to the borders areas. This allowed the entrepreneurs themselves to commute daily to their subsidiary in southern parts of Bulgaria. There are businesspeople and highly paid technicians that have been doing this cross-border commuting every single weekday for years, spending almost five hours a day on the road.[10]

On the other hand, historic ties, cultural affinity, common religion (Greeks and Bulgarians being Orthodox Christians) mixed marriages and other kinship relationships, the existence of Greek students as well as of political refugees from the Greek Civil War who lived in Bulgaria were very important in guiding those intending to invest, especially during the early 1990s (Labrianidis 1996; Kamaras 2001).

Delocalization 'comes' from Greece, but the lead firms and the main market are German. From the early 1990s until the present, it has been very important both for Greece and Bulgaria. For the Greek border areas it meant opportunities, even for very small-sized companies, to go international so as to take advantage of new markets and low-cost labour. However, in the short term, it led to increasing bankruptcies and unemployment.

For Bulgaria it was also important, since Greece is one of the country's most significant trade partners, following Italy and Germany. For Bulgarian border regions Greek delocalization is crucial, because it contributed in solving the problem of unemployment, which was particularly high in the first years of change. Nowadays, it is still important, because working in Greek clothing firms or Bulgarian firms undertaking subcontracting from Greece is the only job option for women in many settlements of the border areas.

This is not only a matter of physical proximity, since there are instances of Greek firms located in Athens and flying to Sofia. Thus, this difference might be attributed primarily to *cultural* reasons (Greeks do not want to leave their home, or perhaps they 'look down' on the particular societies where they have delocalized their business).

The second point is related to the development implications of the evolution of the cluster in Northern Greece, and particularly Thessaloniki. In Figure 2.4 it becomes evident that the clothing sector is in a sharp recession phase. In addition to what was already mentioned about the role of the city as an apex of the triangle, it is interesting to note that, as if the increasing establishment of direct links of their original (foreign) buyers with Bulgaria was not enough, local producers never really managed to gain a foothold in the local, not to mention national, market, which during the last few years is increasingly dominated by firms originating from Athens. In other words, functional upgrading for the cluster's firms was a real rarity. How did this come to happen? This is certainly not a case of obstacles imposed by the buyers, as Schmitz and Knorringa (1999) argue is often the case in the footwear industry. What seems to be the answer to this question is the fact that Greek subcontractors became locked into what was, at least initially, a very rewarding arrangement. Almost overnight these firms broke into foreign markets,

10 A similar case might be USA firms delocalizing to Mexico. All the American staff – engineers, technicians, managers and so on – want to reside in the USA and commute to work in Mexico every day.

eschewing the considerable sunk cost mentioned by Roberts and Tybout (1995) as necessary in order to become exporters. In fact, most of these firms never even searched for new clients. In a sense, the obstacles to upgrading were imposed by the firms themselves and their unfortunate short-termist view of success. As one interviewee said:

> ... it is often pointed out to us that in order to be a successful exporter one has to be good in design and marketing. Surprise. surprise; we are good exporters. Why then should we make all this risky investment in design and marketing?

When it comes to the Bulgarian side, although, as will become evident below, the history of the sector was completely different from the Greek one; it seems to face very similar problems, since it is now entering its critical phase.

The manufacturing of clothing in Bulgaria began in the 1960s, while the following two decades (1970s and 1980s) saw the establishment of large state enterprises. Each enterprise had many workshops located in areas where a free female labour force was available. These areas specialized in mining and metallurgy industries and tobacco growing. The clothing industry had tight production linkages with the domestic textile industry. Not surprisingly, the bulk of exports were directed to the CEECs, but also to West Germany, the UK, some North African countries and Near East countries.

The early reform period (1990–1996/97) was dramatic in two ways. The first was – very – negative and was none other than the collapse of the Council for Mutual Economic Assistance (CMEA, also known as COMECON) markets, which resulted in a sharp decline of production volumes, closures of workshops, decreases in employment in large plants and wide-scale restructuring of ownership. The second development was positive and was linked with the establishment of new, small, clothing firms working for the domestic market and/or under subcontracting for small Greek and Turkish entrepreneurs. FDI in SMEs was pursued by Greek firms in southwestern Bulgaria.

During the last decade (depicted as the bottom frame of Figure 2.5), the situation appears to have changed dramatically. Specifically, the number of new medium and large firms has considerably increased, along with employment in the sector. Production volume, most of which is being exported under subcontracting (more than 90 per cent), has also increased. This has not been the case with FDI, which is mainly pursued by Greek firms, followed by other EU and Turkish firms. Germany is the most significant market, either through direct subcontracting, which is based on relations extant from the 1980s or indirectly through Greece. Other markets are the UK, Italy, France, and Spain. A small part of the export is directed to the American market, mainly through Turkey. No information for outsourcing by Bulgarian clothing firms is available, but if there are any cases, they are few in number. Among Bulgarian firms, subcontracting is the most widespread practice. Some large firms have their own brands and products, but they are sold in the domestic market.

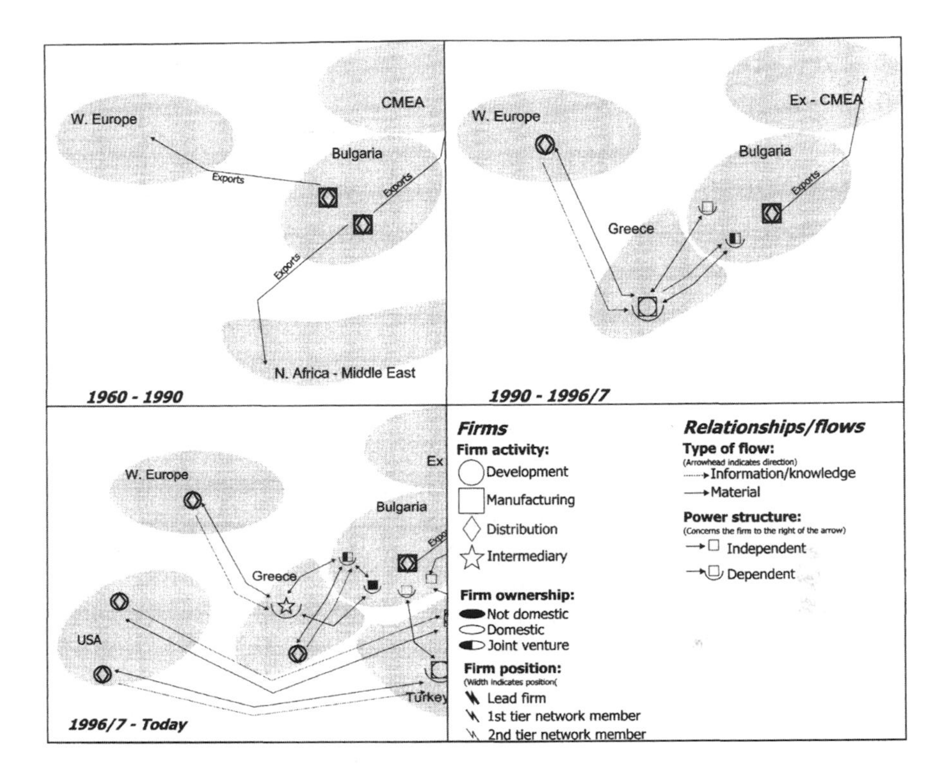

Figure 2.5 The evolution of the clothing sector in Bulgaria (1960 – today)
Source: Compiled by the authors from the literature review.

Conclusion

Our effort is obviously prolusory in the sense that more has to be said about the role of the nature of networks, as well as the role of embeddedness, both of which were merely touched upon.

A brief reading of the findings presented could give a rather chaotic picture. The main reason for this impression, we believe, is that there are inherent difficulties in bringing together exclusively micro- with relatively macro-approaches in order to understand why and how firms delocalize. Our findings seem to be at odds with all conventional wisdom. At the level of the firm, globalization and regionalization seem to have blurred the distinctions between FDI and outsourcing, creating circumstances and specific loci in which the two are equivalent at least in terms of the organizational stress they impose on firms. This does not mean that resources (or competitive advantages) and market failures are becoming irrelevant in the decision to delocalize, but merely that changes in the external environment may affect the nature of resources.

At the level of the sector, there appears to be considerably more variation than what is assumed by the conventional understanding of the technological and market orientations of sectors. Sectors are definitely different in some respects, but not so different in others, and while this will depend on the various definitions and understandings of sectors, it more or less prohibits the creation of any hierarchy of 'sector desirability'. Hence, there may be a higher technology content in the electronics sectors, than even the software sector, but the crucial questions must always be related to the implications for value creation, enhancement and capture. According to Kaplinsky (1998), technological rents are one of no less than nine distinct types of rent, none of which is in any way superior to the others. In fact, since all of the rent types (as rent itself) are dynamic and transient in nature, there are no easy recipes to development. Hence, it may be more important to be able to capture value than to simply create it, and perhaps the most significant determinant of value capture is functional upgrading, as it is always the focal or central firms that capture most of the value. Furthermore, it may in fact be easier for firms in sectors that are not technologically advanced to upgrade, often by creating local production networks and specializing for the internal market, as have some clothing and footwear firms in our sample.

In one of the few efforts to study such turns of inward-looking-upgrade strategies, Schmitz (2006) wonders whether upgrading depends on the choice between global versus national chains or captive versus even relationships. Our findings point to the importance of sector, and more importantly country. In the context of our study, the differences between Polish and Bulgarian subcontractors in the clothing sector were minimal. However, the former were considerably more active in the direction of upgrading than the latter. In fact, the behaviour of Bulgarian subcontractors in 2006 resembles the behaviour of their Greek counterparts ten years ago (Labrianidis 1996, 2008) who were forced to 'upgrade' in order to match the requirements of their customers. Are we, then, dealing with some kind of economic determinism?[11] Although we are not actually equipped to answer that, it would suffice to note that some Greek firms (contrary to national trend) managed to break out into creating their own branded product, although we have very little information about whether they are former subcontractors. In any case, and this is perhaps the main contribution of the chapter, firm behaviours are codetermined by a vast array of factors, of which we have studied only a small fraction.

In conclusion, the decision to delocalize is obviously affected by the form delocalization takes and vice-versa, and is obviously considerably more complex than what the theory of the firm or economic geography usually assumes. Ours was merely an exploratory effort, and we feel that more work needs to be done in this direction.

11 To succumb to a deterministic claim that firms from a given country will not upgrade would be equivalent to arguing that the firm is unimportant, actually negating our own argument about the significance of the firm.

References

Asheim, B.T. (1996), 'Industrial Districts as "Learning Regions": A Condition for Prosperity?', *European Planning Studies* 4:4, 379–400.

Baldwin, R. (1992), 'On the Importance of Joining the EC's Single Market: The Perspective of EFTA Members', *Rivista di Politica Economica* XII, 267–84.

Bartlett, C.A. and Ghoshal, S. (1998), *Managing Across Borders: The Transnational Solution*, 2nd edition (Boston, MA: Harvard Business School Press).

Birkinshaw, J. and Hood, N. (1998), 'Multinational Subsidiary Evolution: Capability and Charter Change in Foreign-owned Subsidiary Companies', *Academy of Management Review* 23, 773–95.

Braczyk, H.-J., Cooke, P. and Heidenreich, M. (eds) (1998), *Regional Innovation Systems: The Role of Governances in a Globalized World* (London: UCL Press).

Coase, R.H. (1937), 'The Nature of the Firm', *Economica* 4:16, 386–405.

Coe, N.M. et al. (2004), 'Globalizing Regional Development: A Global Production Networks Perspective', *Transactions of the Institute of British Geographers* 29:4, 468–84.

Coenen, L. et al. (2004), 'Nodes, Networks and Proximities: On the Knowledge Dynamics of the Medicon Valley Biotech Cluster', *European Planning Studies* 12:7, 1003–18.

Cooke, P. (1992), 'Regional Innovation Systems: Competitive Regulation in the New Europe', *Geoforum* 23, 365–82.

Dicken, P. (2000), 'Places and Flows: Situating International Investment', in G.L. Clark, M. Feldman and M. Gertler (eds), *The Oxford Handbook of Economic Geography* (Oxford: Oxford University Press), 275–91.

Dunning, J.H. (1993), *Multinational Enterprises and the Global Economy* (Wokingham: Addison-Wesley).

Dunning, J.H. (1996), 'The Geographical Sources of Competitiveness of Firms. Some Results of a New Survey', *Transnational Corporations* 5:3, 1–21.

Fagerberg, J., Verspagen, B. and Caniëls, M. (1997), 'Technology, Growth and Unemployment Across European Regions', *Regional Studies* 31:5, 457–66.

Gereffi, G. (1994) 'The Organization of Buyer-driven Global Commodity Chains: How US Retailers Shape Overseas Production Networks', in G. Gereffi and M.E. Korzeniewicz (eds), *Commodity Chains and Global Capitalism* (Westport, CT: Greenwood Press), 1–14.

Gereffi, G. (1999), 'International Trade and Industrial Upgrading in the Apparel Commodity Chain', *Journal of International Economics* 48:1, 37–70.

Gereffi, G. and Korzeniewicz, M. (1994), 'Introduction: Global Commodity Chains', in G. Gereffi and M.E. Korzeniewicz (eds), *Commodity Chains and Global Capitalism* (Westport, CT: Greenwood Press), 95–122.

Gereffi, G. and Mayer, F.W. (2004), 'The Demand for Global Governance Working Paper', *Terry Sanford Institute of Public Policy*, Duke SAN04-02, 1–36.

Gereffi, G., Humphrey, J. and Sturgeon, T. (2005), 'The Governance of Global Value Chains', *Review of International Political Economy* 12:1, 78–104.

Graham, E.M. (2002) 'The Contributions of Stephen Hymer: One View', *Contributions to Political Economy* 21, 27–41.

Grossman, G.M. and Helpman, E. (2002), 'Integration versus Outsourcing in Industry Equilibrium', *Quarterly Journal of Economics* 117, 85–120.

Henderson, J., Dicken, P., Hess, M., Coe, N. and Yeung, H.W. (2002), 'Global Production Networks and the Analysis of Economic Development', *Review of International Political Economy* 9:3, 436–64.

Hess, M. and Yeung, H.W. (2006), 'Whither Global Production Networks in Economic Geography?', *Environment and Planning A* 38:7, 1193–1204.

Humphrey, J. and Schmitz, H. (2002), 'How Does Insertion in Global Value Chains Affect Upgrading in Industrial Clusters?', *Regional Studies* 36:9, 1017–27.

Hymer, S. (1974), *A Study of Direct Foreign Investment* (Cambridge, MA: MIT Press).

Kalantaridis, C., Vassilev, I. and Fallon, G. (2008), 'Patterns of Enterprise Strategies in Labour-intensive Industries: The Case of Five EU Countries', in L. Labrianidis (ed.), *The Moving Frontier: The Changing Geography of Production in Labour Intensive Industries* (Aldershot: Ashgate).

Kalogeresis, T. and Labrianidis, L. (2007), 'Delocalisation of Labour Intensive Activities in a Globalized World', in P. Getimis and G. Kafkalas (eds), *Overcoming Fragmentation in Southeast Europe* (Aldershot: Ashgate), 121–126.

Kamaras, A. (2001), 'A Capitalist Diaspora', *Discussion Paper*, Hellenic Observatory, LSE.

Kaplinsky, R. (1998), *Globalisation, Industrialisation and Sustainable Growth: The Pursuit of the nth Rent* (Institute of Development Studies, University of Sussex, IDS Discussion Paper 365).

Karagozoglu, N. and Lindell, M. (1998), 'Internationalization of Small and Medium-sized Technology-based Firms: An Exploratory Study', *Journal of Small Business Management* 36:1, 44–58.

Kay, N. (2000), 'The Resource-Based Approach to Multinational Enterprise', in C.N. Pitelis and R. Sugden (eds), *The Nature of the Transnational Firm* (London: Routledge), 140–61.

Labrianidis, L. (1996), 'Subcontracting in Greek Manufacturing and the Opening of the Balkan Markets', *Cyprus Journal of Economics* 9:1, 29–45.

Labrianidis, L. (2000), 'Are Greek Companies that Invest in the Balkans in the '90s Transnational Companies?', in A. Mitsos and E. Mossialos (eds), *Contemporary Greece and Europe* (Aldershot: Ashgate), 457–82.

Labrianidis, L. (2001), 'Geographical Proximity Matters in the Orientation of FDI: Greek FDI in the CEECs' in G. Petrakos and S. Totev (eds), *Economic Co-operation in the Balkans* (Aldershot: Ashgate), 463–89.

Labrianidis, L. (2007), 'La relation ambiguë de la Grèce avec les Balkans: Afflux d'immigrés en Grèce et "fuite" des entreprises à haute intensité de main-d'œuvre, les deux visages de Janus', *Mesogeios* 32–33.

Labrianidis, L. (2008), 'Delocalisation of Labour Intensive Industries in a Globalized Economy', in Y. Caloghirou, L. Papayanakis and Y. Papadopoulos (eds), *Greek Manufacturing: Towards a Knowledge Intensive Economy* (Athens: Technical Chamber of Greece).

Labrianidis, L. and Kalantaridis. C. (2004), 'The Delocalisation of Production in Labour Intensive Industries: Instances of Triangular Manufacturing between Germany, Greece and FYROM', *European Planning Studies* 12:8, 1157–73.

Leslie, D. and Reimer, S. (1999), 'Spatializing Commodity Chains', *Progress in Human Geography* 23:3, 401–420.

Liu, W. and Dicken, P. (2006), 'Transnational Corporations and "Obligated Embeddedness": Foreign Direct Investment in China's Automobile Industry', *Environment and Planning A* 38, 1229–47.

Lundvall, B.-A. (1988), 'Innovation as an Interactive Process: From User-Producer Interaction to the National System of Innovation', in G. Dosi, C. Freeman, R. Nelson, G. Silverberg, and L. Soete (eds), *Technical Change and Economic Theory* (London and Washington: Pinter), 349–69.

Lundvall, B.-A. (1992), *National Systems of Innovation* (London: Pinter Publishers).

Lundvall, B.-A. and Borrás, S. (1997), *The Globalising Learning Economy: Implications for Innovation Policy* (Brussels: DG XII).

NSSG (National Statistical Service of Greece) (2008a), *Annual Industrial Survey*, <http://www.statistics.gr/table_menu_eng.asp?dt=0&sb=SIN_2&SSnid=%20-%20Manufacturing&Dnid=%20-%20Annual%20Industrial%20Survey>, accessed 4 June 2008.

NSSG (National Statistical Service of Greece) (2008b), *External Trade* (INTRASTAT-EXTRASTAT) <http://www.statistics.gr/table_menu_eng.asp?dt=0&sb=SFC_1&SSnid=External%20Trade&Dnid=%20-%20External%20Trade%20(INTRASTAT-EXTRASTAT)>, accessed 4 June 2008.

Penrose, E. (1995 [1959]), *The Theory of the Growth of the Firm* (Oxford: Oxford University Press).

Petit, P. and Soete, L. (1999), 'Globalization in Search of a Future', *International Social Science Journal* 51, 165–81.

Pitelis, C.N. (2000), 'A Theory of the (Growth of the) Transnational Firm: A Penrosian Perspective', *Contributions to Political Economy* 19, 71–89.

Porter, M.E. (1990), *The Competitive Advantage of Nations* (New York: The Free Press).

Roberts, M.J. and Tybout J.R. (1995), 'An Empirical Model of Sunk Costs and the Decision to Export', Policy Research Working Paper 1436, Policy Research Department, Finance and Private Sector Development Division, The World Bank.

Sacchetti, S. and Sugden, R. (2003), 'The Governance of Networks and Economic Power: The Nature and Impact of Subcontracting Relations', *Journal of Economic Surveys* 17:5, 669–91.

Saxenian, A. (1994), *Regional Advantage: Culture and Competition in Silicon Valley and Route 128* (Cambridge, MA: Harvard University Press).

Schmitz, H. (2006), 'Learning and Earning in Global Garment and Footwear Chains', *The European Journal of Development Research* 18:4, 546–71.

Schmitz, H. and Knorringa, P. (1999), 'Learning from Global Buyers', IDS Working Paper 100.

Simmons, C. and Kalantaridis C. (1995), 'Labour Regimes and the Domestic Domain', *Work-Employment and Society* 9:2, 287–308.

Smith, A., Rainnie, A., Dunford, M., Hardy, J., Hudson, R. and Sadler, D. (2002), 'Networks of Value, Commodities and Regions: Reworking Divisions of Labour in Macro-regional Economies', *Progress in Human Geography* 26:1, 41–63.

Soete, L. and Freeman, C. (1997), *The Economics of Industrial Innovation* (London: Pinter No. 824).

Storper, M. (1997), *The Regional World: Territorial Development in a Global Economy* (New York: The Guilford Press).

Uzzi, B. (1997), Social Structure and Competition in Interfirm Networks: The Paradox of Embeddedness, *Administrative Science Quarterly* 42:1, 35–67.

Yamamura, E., Sonobe, T. and Otsuka, K. (2003), 'Human Capital, Cluster Formation, and International Relocation: The Case of the Garment Industry in Japan, 1968–98', *Journal of Economic Geography* 3:1, 37–57.

Chapter 3
Patterns of Enterprise Strategies in Labour-Intensive Industries: The Case of Five EU Countries

Christos Kalantaridis, Ivaylo Vassilev, Grahame Fallon

Introduction

One of the noteworthy features of contemporary economic change involves the emergence of growth patterns resting squarely upon the global diffusion of production of labour-intensive industries (Scott 2006). This is facilitated by 1) advances in ICT; 2) the changing architecture of production through modularization, itself the result of the advent of digital technologies (Steinfeld 2004); and 3) the opening-up of national markets through changes in the global governance framework and the advance of neo-liberal views (Levitt 1995). This outward-oriented pattern of growth is of particular importance in the case of formerly planned economies of CEE and what was the Soviet Union. This is partly because their semi-detached position from the global marketplace for the best part of the post-war era – or even longer in the case of the Commonwealth of Independent States – offers a multitude of opportunities for growth. More importantly, however, the collapse of the old system was linked to the disintegration of old distribution channels and significant decline in levels of domestic demand that could offer alternatives for growth. Thus, there is now a large body of empirical evidence suggesting that growth based upon the global diffusion of production in labour-intensive industries is increasingly evident in formerly planned economies (Smallbone et al. 1996; Kalantaridis et al. 2003; Smith 2003; Pickles et al. 2006; Scott 2006).

During the past 20 years or so, research in this area – in formerly planned economies and beyond – has been heavily influenced by the GCC approach. This approach provided researchers with useful insights into the reconfiguration of industrial dynamics in increasingly integrated networks of production and distribution. These insights have been used to inform policy-making by transnational organizations such as the International Monetary Fund (IMF 2004), UNCTAD (2003; 2004), the World Bank (2004) and the Organization for Economic Cooperation and Development (OECD) (2004). As is always the case with approaches that dominate the research and policy agenda, GCCs have been the focus of intense scrutiny (Raikes et al. 2000; Henderson et al. 2002; Coe et al. 2004, Palpaceur et al. 2005). Among the plethora of critiques, Smith et al. (2002)

advanced the thesis that there is 'a tendency to neglect the dynamics and fluidity of organizational forms in GCC analysis ... [and] [t]here is consequently little detailed analysis of complexity in either intra- or inter-organizational relations' (Smith et al. 2002). Advocates of the GCC approach acknowledged the importance of the difficulties created by the relatively high level of abstraction of early works and provided a number of correctives (Gereffi and Mayer 2004; Bair 2005; Gereffi et al. 2005; Neidik and Gereffi 2006). This constitutes the point of departure for this chapter. We set out to explore the micro-dynamics of industrial change. The enterprise is the subject at the heart of our inquiry, whilst deciphering the strategies adopted by firms and their implications for external linkages and performance constitute key areas of our work.

The chapter is organized into three large sections. The first, and largest section, reviews the accumulated body of empirical evidence around adjustment strategies in labour-intensive industries. This is followed by a discussion of the results of our fieldwork research in Bulgaria, Estonia, Poland, Greece and the UK. Lastly, we offer some concluding remarks.

A Review of the Literature

A framework for exploring the literature

Previous empirical research into the successful adjustment strategies adopted by enterprises in LIIs has often adopted two, not mutually exclusive, viewpoints (see Figure 3.1): internal and external (Humphrey and Schmitz 2001; Schmitz 2006). The former refers to the study of dimensions that can be controlled and directed as they lie within the organizational boundaries. Within this context, particular attention has been paid to products, processes and production. The second viewpoint explores the interrelationship between the firm (and its strategy) and its environment, focusing particularly upon relationships with other firms and organizations. Particular emphasis is placed here upon the organization of different types of relationships over space and through time. A third dimension involves patterns of integration (ranging from market exchange to hierarchical linkages through the creation of subsidiaries). We would like to stress here that the boundaries between these two viewpoints are at best blurred. For example, a decision to externalize part or the whole of the production process is closely interlinked with decisions about the nature of the relationships to be established as a result. Lastly, there is the nature of emerging relationships.

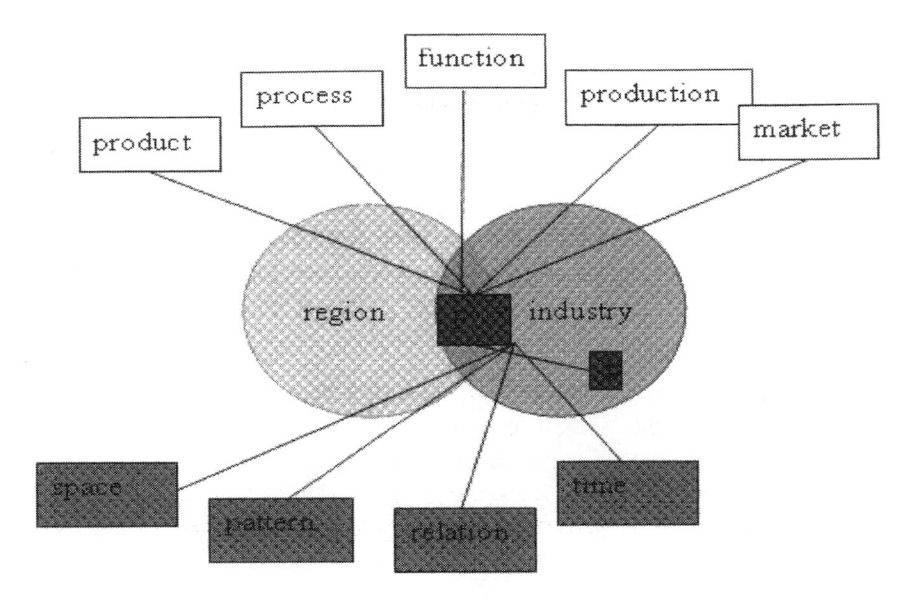

Figure 3.1 Analytical framework enterprise strategy
Source: Compiled by the authors.

Enterprise strategy

There is widespread agreement among researchers as well as policy-makers that product innovation is one of the main means of enhancing long-term enterprise competitiveness in LII (Dunford 2002; Gereffi and Memedovic 2003; Schmitz 2003). It enables the enterprise to move away from cut-throat price competition towards market segments where price (and as a result production costs) is a secondary consideration. However, the manifestations of product innovation may vary considerably between LII. In the clothing and footwear industries, one manifestation of product innovation is the introduction of new design. Although each 'generation' of new designs may have a relatively limited 'shelf-life', it may increase 'brand-awareness' (assuming of course that a new design is used in own-brand products). Another manifestation of product innovation in the clothing industry is the use of new man-made fibres and technical textiles. Technical textiles are particularly significant in that they offer opportunities to develop new kinds of product and are suited to new uses in, for example, the transport sector, furniture and furnishings and construction (Dunford 2002). Again however, product innovation has to be visible to the consumer, as the advantage derived from each generation of changes may be short-lived. The development of brand awareness is therefore a key consideration. Another manifestation of product innovation is the development of incrementally different products (Ran and Kuusela 1996), something that is particularly the case in the electronics and software industries. These industries also offer scope to launch radical innovations in the form of products that are totally different from those that preceded them. In both incremental and radical product innovation each 'generation' of changes may be more sustainable – at least in the short- to medium-term – as they may involve difficulty in replicability, either on the grounds of technological or intellectual property rights.

It is useful to note here that product innovation matters not only because of the short- or medium-term competitive advantage gained. Probably more importantly, product innovation matters because of the competences (in a wide sense of the term) that it affords the enterprise (Corso and Pavesi 2000). These competences are both tangible (for example, capital due to short-term profit rises, technology and skills) and intangible (in terms of knowledge, market recognition and so on). These competences are the real source of competitive advantage, if they are used in a process of continuous product innovation.

We must stress here that product innovation is an activity that involves considerable difficulty, and therefore may often be unsuccessful (in the sense of failing to achieve the objectives originally envisaged, rather than necessarily leading to bankruptcy or downsizing). In fact, for every successful product innovation, there may be literally tens of unsuccessful ones.

Process-related strategies revolve around two key considerations. The first involves matching demand with production and storage capacity, whilst the second involves technological advancement. The key concept regarding the former consideration is 'lean retailing'[1] (Wrigley and Lowe 2002). This aims at the optimization not only of intra-organizational but also of inter-organizational processes. The underlined aim is to optimize the flow of products and information along the entire value chain, starting at the point of sale with the collection of highly detailed data on customer demand (Von der Heydt 1999). This procedure results in faster reaction of supply to actual sales. Lean retailing is made possible through using more accurate sales forecasts and the adoption of electronic sales registers and bar-coding, the latter also providing valuable information about customer behaviour (that is, who buys what, when and where) that can be used for the development of new marketing strategies (Wortman 2003).

Technological change constitutes an element of process-focused strategy that has not often received sufficient attention in empirical literature on enterprise adjustment in labour-intensive industries. In some instances the implementation of new (either to the company or the industry) technology has been identified as a source of productivity gains and increased production capabilities (that is, better quality of production or enhanced capacities) (Kalantaridis 2000). Within this context, accessing new technologies is viewed as a source of enhanced competitiveness for the enterprises involved. In many other instances, technological change is viewed as an enabler in the processes of either product innovation or functional upgrading,

1 Another term often used in the literature is 'agile supply chain' (for a review see Christopher 2000).

which will be discussed in greater detail below. In this context, access to new technologies is a means of implementing enterprise strategies. In both instances, however, we have to be aware of the distinction between accessing and exploiting new technologies. The latter requires strong assimilation capacity and ability to utilize spill-over technology: Japan in the 1980s being the best such example (Watanabe et al. 2001).

Functional upgrading has emerged during the past ten years or so as a key element of enterprise adjustment in labour-intensive industries (IMF 2004; World Bank 2004). It is a strategy often identified with the GCC approach. The argument goes like this: European producers and distributors seek monopolistic rents through strategies centred on design, fashion and branding. Another strategy (deployed by producers both in Europe but also in Newly Industrialized Countries) is aiming to introduce changes in the distributive order. This is to be achieved through changing the weight attached to different functional roles in the value-added chain by concentrating, for example, on knowledge-intensive activities, marketing and logistics. Essentially producers seek to reposition themselves in the overall value-added chain, with a specialization on what are seen as core competences (Gereffi and Memedovic 2003).

Improving the position of an enterprise in the global networks of production and distribution involves organization learning. Thus, participation in these networks emerges (at least in the GCC literature) as an essential precondition that may initiate dynamic learning curves (Bair and Gereffi 2003). However, there are other obstacles in the process of functional upgrading, as higher-level roles are more demanding than lower-level ones. In order to overcome these obstacles, enterprises require physical and human capital as well as access to effective networks (invariably identified in the literature as social capital). Within this context, building and managing networks, where power is complexly constituted rather than simply 'possessed' by one of the partners (Tokatli 2007), emerges as an issue of at least equal importance as accessing financial, design and marketing resources. Some forms of upgrading may meet obstacles of different kinds: access to resources, restriction from partners[2] and so on; addressing obstacles such as restriction may require 'underground' risk-diversification (Bazan and Navas-Aleman 2001).

In achieving functional upgrading there are differing views regarding the origin[3] of resources. On the one hand, local cluster theory emphasizes that the knowledge needed for upgrading comes from within the cluster (Fujita et al. 1999; Audretsch 2003). On the other hand, Global Value Chain (GVC) theory emphasizes that the know-how comes from outside the cluster, in particular from the global buyers (Schmitz 2003).

2 Schmitz (1999) discusses a case of a footwear cluster in Brazil where few big local exporters, included in global chains, tried to prevent conflicts of interest with the lead firms in the value chains, and were instrumental in preventing a collective upgrading strategy.

3 This will be discussed in greater detail in the following sub-section.

Functional upgrading is invariably viewed in the literature as progressive change, whilst functional downgrading is viewed more or less by definition as inherently negative. In fact, there has been precious little research exploring in particular functional downgrading, even though we are aware that this is also a strategy that may be deployed by enterprises in labour-intensive industries. This is because the main approach adopted by the majority of scholars in the field focuses upon the long-term economic development of a spatial unit (locality, region or nation). This tension between enterprise on the one side and wider economic development on the other means that we currently possess precious few insights into the process of functional downgrading. This situation occurs despite the fact that there is ample evidence to suggest that functional downgrading is often used as a short- to medium-term strategy by enterprises in labour-intensive industries. For example, a number of enterprises in post-socialist regimes opted for functional downgrading during the early stages of reform, in order to safeguard survival and link into the global network of production and distribution (Kalantaridis et al. 2003). An example of an Italian company which opted for functional downgrading is presented in Rabellotti (2001).

Production constitutes the final element of the internal viewpoint. If product innovation revolves around the question: what to produce, and process about how to produce it, then production focuses squarely upon how much to produce and where to produce. It is interesting that there is precious little discussion in the literature about production levels per se. Instead, discussion about production focuses more on where it takes place, an issue that will be discussed in greater detail in the following sub-section.

Enterprise strategies and external linkages

The geography of production (identified as the 'space' dimension in Figure 3.1) constitutes the first aspect of the external viewpoint. A number of competing explanations have emerged regarding *spatial enterprise strategies*. These could be broadly clustered in two groupings: the first stresses the importance of locality, and enterprise embeddedness as a source of competitive advantage. In sharp contrast, the second approach focuses upon industrial dynamics, and thus views enterprise strategy emerging in a global but structured space. However, an issue common to both approaches is the importance of environmental influences in the process of enterprise strategy formation.

The 'locality' view falls within a broader shift in paradigm, supported by a voluminous body of empirical research regarding the role of spatial externalities on economic activity (Fujita et al. 1999; Dunford 2006; Yeung et al. 2006). This 'new learning' perceives individual ventures as structural elements of territorially defined networks, whereby emphasis is placed on the interaction between firms and the local milieu (Audretsch 2003). Within this context, geographical, industrial, organizational and institutional proximities are perceived to be instrumental in facilitating the emergence of shared patterns of behaviour and cognitive rules, which in turn underpin collective learning processes (Kirat and Lung 1999; Malberg and Maskell 2002). This shift in emphasis towards localized interacting agents rather than their behaviour in isolation, long accepted in regional science, has become more common in 'mainstream' economics (Anselin 2003; Karlsson and Dahlberg 2003). As a result, concepts such as location, spatial interaction and spatial externalities are increasingly common in theoretical formulations in a growing number of fields of study within economics. The empirical evidence that lends support to the new paradigm draws upon a growing number of celebrated cases of localized systems the world over (see Castells and Hall 1994; Cooke 1996; Ottati 1996). The 'new learning' advances the idea of local enterprise cooperation as a key element of economic development initiatives (DTI 2005). Such strategies build upon notions of participation and endogenous development and involve the exploitation of human, natural and economic resources that are specific to a geographically defined locality (Laschewski et al. 2002). Within this context, both policy-makers and academics have become concerned with the role that public agencies can play in enabling or even stimulating inter-organizational cooperation and networking (Huggins 2000). Two central assumptions underlie these local development initiatives. The first is the assumed existence, or the likely possibility of creating, relationships between local actors which themselves may engender mutual trust and shared learning (Curran et al. 2000). The second assumption is that economic activity is typically socially embedded, which is generally taken to imply local embeddedness (Jack and Anderson 2002).

The importance of the locality as a source of competitive advantage has been emphasized by authors coming from different traditions. There are, however, different views on exactly how the locality is important. Thus, some authors emphasize the linkages between local enterprises and institutions (Scott 2002) and those who stress the importance of extending those links to the meso-level and the global level (Messner 2002; Bair 2006). The importance of the region has prompted further distinctions between types of regions beyond the 'flexibly specialized' region, as discussed mainly in relation to Italian cases (Rama et al. 2003).

However, there is also a growing appreciation of the disadvantages associated with 'over-embeddedness' in a regional or local setting. These arguments are inspired by the work of economic sociologists who suggest that local embeddedness can also act as a constraint. Uzzi (1997) identifies three conditions that may turn embeddedness into a liability: the unforeseeable exit of a key player; the prevalence of institutional forces that rationalize markets; and overembeddedness, which is of greater importance in a rural context. Burt (1992) argues that overembeddedness can reduce the inflow of information into the local setting if there are few or no links to outside members who can contribute innovative ideas. He takes the argument further, suggesting that people who stand near structural holes 'are more familiar with alternative ways of thinking and behaving, which gives them more options to select from and synthesize' (Burt 2004). Whilst people connected across groups may be able to generate good ideas, locally embedded entrepreneurs may become 'ossified and out of step with the demands of their environment, ultimately leading to decline' (Uzzi 1997, 59).

One important issue discussed in the external view of enterprise strategy revolves around the choice of *patterns* (or modes) of integration. The literature on transaction costs provides us with an understanding of the full complement of options confronting the firm, ranging from spontaneous contracting in the marketplace to hierarchical control through internalization. Each pattern of integration possesses a number of advantages and disadvantages. Moreover, the accumulated empirical evidence suggests that whilst there are a multitude of patterns of integration, no obvious regularities emerge. It is impossible to sustain arguments of the type that in industry A, sub-contracting outconstitutes the main pattern of integration. Such arguments are not sustainable even within the same country, as a multitude of patterns of integration exist comfortably side by side. The specific characteristics of the enterprises involved and often the attributes of the entrepreneurs (for example, their attitudes to risk) are instrumental in defining this diversity.

Of course, it is worth pointing out some suggestive contributions to the exploration of this issue. Schiavone (2005) distinguishes between two types of enterprise strategies regarding patterns of integration. The first revolves around the creation of new business ventures in lower-wage countries and is termed entrepreneurial delocalization. The second involves simply changes in the supply chain without necessarily the externalization of part of the production process. This involves the creation of subsidiaries of the very same firm in foreign countries and is termed 'productive delocalization'. This distinction has significant implications for the nature and characteristics of the enterprise strategies adopted. In the latter case, there is a much greater degree of alignment of interests that may prevent independent action in lower-wage countries – though there are significant resource advantages concerned. In this case what is essentially argued is that there is a significant trade-off between risk and flexibility on the one hand and control and standards on the other.

Relying extensively on inter-organizational *relationships* raises questions of first, migration of responsibility and ways of controlling standards and, second, managing the diversity of contexts, locations and relationships. Thus, for example, the migration of bureaucracy leads to changes in negotiating, administrating and monitoring contracts[4] (MacKenzie 2002), as well as putting in place mechanisms for assessing the quality and the work of subcontractors before they are contracted and before they deliver the product (Assmann and Punter 2004). Further, the problem of co-ordination becomes a central strategic task (Abernathy et al. 2006), making it necessary for companies to develop distributed management execution systems (Huang 2002). Humphrey and Schmitz (2001) address the question of governance by asking how parameters are set and then enforced, and whether

4 Contrary to what is often believed, this process does not necessarily lead to the dismantling of hierarchies but on the contrary to their reproduction.

firms within the chain (for example, the lead firm) or external entities enforce these parameters. The empirical evidence, however, does not offer a single and straightforward answer to the question about what types of relationships and governance mechanisms employed within the process of restructuring work best, as they are always socially embedded and are simultaneously positioned within different, and often contradictory, discourses, structures of interests and priorities. For example, developing close relations or arms-length relations can have both advantages and disadvantages, being useful in certain cases yet harmful in others. The ability to manage the inbound logistics and to cooperate with other companies appears to be essential for the success of subcontractors, and one of the positive consequences of such relations is that they can lead to knowledge transfer (Deardorff and Djankov 2000). Thus, looking at Finnish manufacturing companies, Lehtinen (1999) argues that there is an increased significance of long-term and commitment-based supplier-customer relationships, while Lazzeretti et al. (2004) emphasize the importance of trust and informal credit for the industrial development of the Italian district of Prato.

Because developing a relationship with a new supplier usually takes a long time, companies may prefer to follow their established partners wherever they decide to move and keep on negotiating the conditions of their relations rather than looking for new suppliers. Discussing the clothing sector in the UK, Gibbon (2001) argues that there is a tendency to reduce the number of suppliers, while also increasing the expectations of the range of services and functions expected to be carried out. In contrast, long-term relations can also be harmful and partnerships between manufacturers and retailers may create binding ties on both sides, where buyers may be forced to buy things just because the producer has got the capacity to produce them (Gibbon 2001) and/or at non-competitive prices. Furthermore, the positive effects are never guaranteed. Indeed, there are structural constraints to the inter-organizational learning process, while buyers might also be concerned with future competition and thus would be cautious in transferring knowledge and technology to their partners.

The variety of the observed relationships can further be extended into studying and conceptualising different forms of networks. Ponte (2005) distinguishes four forms of co-ordination:[5] hierarchy, relational contracting (tighter forms, not easy to standardize, repeated interaction, understanding the mindset, 'captive' contractors), relational contracting (looser forms in which standardization is possible but needs some degree of customization, 'modular' contractors and contract manufacturing) and market contracting (homogeneous product, universally understood quality and so on).

5 Gereffi, Humphrey and Sturgeon (2005) link the shape of the network to the degree of complexity of transactions, the possibility to codify transactions and the capabilities in the supply-base, and come up with similar categories. Thus they distinguish between markets, modular value chains, relational value chains, captive value chains and hierarchies.

The diversity of possible outcomes from similar types of relationships,[6] given different contexts, leaves the question open as to what is a 'good' choice.[7] However, being unable to predict and to firmly establish relationships of the cause-effect type does not mean that strategic choices are made totally in the dark. On the contrary, here we argue that it is possible to identify significant mechanisms, analyse the wide diversity of ways in which processes can work and refine the existing distinctions, and that this is what practitioners often do in deciding on their strategies.

Time is the last fundamental element of enterprise strategies in labour-intensive industries, in that each firm has a history comprised of significant events that occurred at specific points in time. Acknowledgement of the time dimension is implicit in a number of studies emanating from different disciplinary settings (GCC, Actor Network Theory, incremental models, etc). Based on insights gained from these approaches, we would like to distinguish here between chronological time and time sequences. Chronological time is the same for all firms, that is, all firms in a given area operating in the 1990s experienced the same macro-environmental influences and passed through the same periods of economic growth and recession. Thus, chronological time patterns may emerge. Moreover, it is likely that the same enterprise may differ in its strategic decision-making over chronological time. Therefore, it is important to peg the firms' strategies against a relevant historical backdrop. Whilst chronological time is shared, time sequences are specific to each enterprise. They refer to the stages in the evolution of the firm and their implications for the resources, skills and attributes of the enterprise.

Enterprise Strategies

Patterns of enterprise strategy in the clothing industry

Drawing primarily upon the literature, a number of prototype strategies were identified (in a manner similar to the GCC). The prototype strategies for the clothing industry are presented in Table 3.1. We used this matrix as a means of clustering enterprise strategies: the variables used were those in the vertical axis of Table 3.1 (product, process, function, production, market) and the coding for each company reflected the horizontal axis (lock-in, hybrid, break-out). In order to process the data, we used a hierarchical cluster analysis, particularly the Ward method, a common clustering algorithm which has also been used effectively in

6 Buckley and Ghauri (2004) offer a comprehensive literature review on the links between ownership and location strategies.

7 Sacchetti and Sugden (2003) contrast the externalization activities of large TNC, which are concerned with flexibility, but also more control over governments, labour and subcontractors, and argue that different networks would have different effects on socio-economic development.

previous studies. This technique was performed separately for each country, so as to allow for context specificity. The only exception was the case of Greece and the UK, where the two were processed together due to the relatively small number of cases (31 and 12 respectively). The determination of the appropriate number of groups or types is a key, but arbitrary, decision in hierarchical cluster analysis. In our case, guidance was provided by the increase in within-cluster distances as groups were merged. Relatively large increases that signify the merging of less similar cases (Harrigan 1985; Carlyle 2001) were apparent at different solutions for each. Overall, 16 groupings were compiled.

Table 3.1 Overview of strategies

	Competence lock–in	Hybrid	Break-out competences
Product/ service	Not own product range so limited scope for action	New product or product design for some of the product range	New product design and brand development
Process	Technological change (invariably in production) in line with needs of parent Enterprise	Technological change in order to gain manufacturing competences (often knowledge transfer from one dimension [OPT] to the other)	All encompassing technological change including manufacturing and/or lean retailing
Function	Moving up or moving down the production chain but remaining within manufacturing	Moving up and/or moving down the production chain – often simultaneously in two different production dimensions	Moving up the production chain – often away from manufacturing towards distribution. Proximity to the consumer a key source of competitive edge
Production	Production competences remain at the heart of enterprise strategy		The importance of production competences and volume–production decline
Market	Serving in the main price-sensitive and to a lesser degree flexibility-focused segments of the market	Serving flexible-response focus plus one more of the other two (flexibility-focus or design-sensitive) segments of the market	Serving in the main design-focus segments of the market

Source: Compiled by the authors.

The analysis of the enterprise strategy demonstrates how enterprises operating in the same segment of the market but in different national contexts may opt for different strategies (for example, Poland 2, Estonia 3, and Bulgaria 3). This point can be taken further, as the analysis of strategy illustrates that even companies in the same country and same market segment may adopt significantly different strategies (for example, Poland 1 and 4, Poland 2 and 3, Estonia 1 and 2, Bulgaria 2, 3, and 4). Interestingly, the analysis suggests that companies that operate in different segments of the market, in different countries, may actually adopt the same strategy (for example, Poland 2 and Estonia 1, Greece 1 and Poland 1). Of course, there are also some similarities. Companies both in Greece and the UK fall in clusters 2 and 3; clusters Poland 2 and Bulgaria 4 are identical, as are clusters Poland 1 and Bulgaria 1.

The cluster analysis of the enterprise strategies allows us to identify a number of interesting patterns (see Figure 3.2). Competence lock-in strategies do not link exclusively to the price-sensitive segment of the market. This type of strategy also appears to be of importance in market segments where success is conditional upon flexible response. This is apparent in the case of Estonia 1 and Bulgaria 1 clusters. Interestingly, both clusters maintain a strong export orientation and enjoy foreign investment to a considerable degree. However, in both instances there is little evidence of functional upgrading, with most companies reporting no change. Competence lock-in strategies are, of course, apparent in the price-sensitive segment of the market: for example, Poland 2 and Bulgaria 4 as well as Greece 1 fall within this grouping. Another grouping (Estonia 3) appears to adopt a very similar strategy, except for the development of some design competences for some of the product range. A strong export orientation and significant foreign involvement (but for the Poland 2 grouping) are also apparent. Among these groupings functional upgrading is commonplace only in one grouping, namely Bulgaria 4.

Another interesting pattern emerges around those hybrid strategies, aiming to use competences used through engagement in global networks of production and distribution in order to enhance the enterprise's position in the domestic market. However, none of the emerging clusters reflect the ideal-type developed in Table 3.1. There are five groupings that appear to fall broadly within this pattern. Those in Poland (4) and Estonia (2) appear to have developed greater design competences and own-brand products than originally envisaged. Functional upgrading among these groupings is widespread. The Bulgarian case (3) has failed to introduce functional change. The remaining two, which differ somewhat because of the declining importance of production, are UK 1, and UK/Greece 1. Moving upwards in the production chain is common in the former grouping but less so in the latter. Interestingly, the evidence suggests that those groupings which developed a strong interest in the domestic market were, overall, more successful in moving up the production chain. In many instances, this was often achieved at the same time as downgrading – or at best no change – in the position of companies in GCCs.

Figure 3.2 Patterns of enterprise strategies
Source: Enterprise Survey.

There is only one grouping (UK/Greece 2) that resembles to a considerable degree the break-out competences category, comprising only nine companies that maintain a strong domestic focus.

Lastly, there are three outlier groupings. The first (Poland 3) is located in between hybrid and break-out strategies. These firms have developed their own brand and are increasingly moving away from production, linked with a move further up in the chain, but they still have not completed the implementation of advanced technologies in all aspects of production and distribution, and their focus remains on the price-sensitive segment of the market. The remaining two groupings provide us with useful illustrations of two processes of transition between competence lock-in and hybrid strategies (as can be seen in Figure 3.2).

Patterns of enterprise strategy in the footwear industry

In analysing footwear strategies we followed the same general approach as we did with clothing. However, the categories that were used to represent lock-in, hybrid and break-out competences were adapted to the specificities of the footwear sector. Thus, the prototype strategies for the footwear industry are presented in Table 3.2.

In order to process the data, hierarchical cluster analysis was performed separately for two groups of companies in our sample: those located in new member states, Bulgaria, Poland and Estonia, and those located in old member states, the UK and Greece. The cases were combined into only two groups due to the small number of cases for Estonia and Greece. While these broader groupings restrict our ability to derive country-specific conclusions we can, nevertheless, distinguish between strategies in countries that are differently positioned in the value chain.

Similar to our analysis for clothing, in footwear we also observe that companies operating in the same market segment may adopt different strategies (for example, 1 and 3; 2 and 4; 12 and 13; 11 and 14). Only companies that are located in new member states are targeting the price-sensitive segment of the market (1 and 3). Companies that are competing on design are present in both old and new member states (2, 4, 12, 13). Only two groups (1 and 2), both located in the new member states, demonstrated lock-in competences. Cluster 4 was closest to the ideal-type of hybrid strategy, while strategy groups 3, 12, 13 and 14 adopted different versions of hybrid strategies also addressing different market segments. Interestingly, companies located in different market segments adopted the same strategies (3 and 12). Only cluster 11 coincided with our ideal-type of break-out competences.

Country of origin is a significant factor for the type of strategies adopted in the footwear sector, although less so than in the case of the clothing sector (Table 3.3). Thus, companies in cluster 1 are almost exclusively located in Bulgaria, while the presence of Bulgarian companies in cluster 4 is very weak. In the grouping of old member states, Greek companies are almost exclusively located in cluster 12, with UK companies present in all four clusters. Bulgarian companies are the largest in the sample with the highest dependence on export markets also.

Table 3.2 Overview of strategies

	Competence lock-in	**Hybrid**	**Break-out competences**
Product/ service	Not own product range so virtually no action	New product or product design for some of the product range or movement to niche markets	New product design and brand development
Process	Technological change (invariably in production) in line with needs of parent enterprise	Technological change in order to gain manufacturing competences; this may include sharing of technological production secrets from parent company as well as knowledge transfer from one dimension (OPT) to the other	All encompassing technological change including manufacturing and/or lean retailing
Function	No movement within production chain remaining firmly within manufacturing	Moving up and/or moving down the production chain – often simultaneously in two different production dimensions. Often targeting simultaneously specific markets both at home and abroad	Moving up the production chain – often away from manufacturing towards distribution. Proximity to the consumer a key source of competitive edge
Production	Production competences remain at the heart of enterprise strategy	Moving away from volume production and/or the production of parts for footwear	The importance of production competences decline. Fully develop own retail network in the home and/or international markets
Markets	Competition occurs primarily on the basis of price and to a lesser degree quality	Competition occurs primarily on the basis of design mainly for the domestic market	Competition occurs mainly on the basis of design and quality, mainly for international markets

Source: Enterprise Survey.

Table 3.3 Country and strategy

	Country					Total
	1	2	3	4	5	
1 Count % Within Count	13 31	1 11	0 0	0 0	0 0	14 12.5
2 Count % Within Count	15 35.7	3 33.3	11 34.4	0 0	0 0	29 25.9
3 Count % Within Count	12 28.6	0 0	11 34.4	0 0	0 0	23 20.5
4 Count % Within Count	2 4..8	5 55.6	10 31.4	0 0	0 0	17 15.2
11 Count % Within Count	0 0	0 0	0 0	1 14.3	4 18.2	5 4.5
12 Count % Within Count	0 0	0 0	0 0	6 85.7	4 18.2	10 8.9
13 Count % Within Count	0 0	0 0	0 0	0 0	5 22.7	5 4.5
14 Count % Within Count	0 0	0 0	0 0	0 0	9 40.9	9 8
Total Count % Within Count	42 100	9 100	32 100	7 100	22 100	112 100

Source: Enterprise Survey.

The cluster analysis of the enterprise strategies allows us to identify a number of interesting patterns (Figure 3.3). There are only four clusters that fall into one of our ideal-types; these are clusters 1 and 2 for lock-in, cluster 4 for hybrid, and cluster 11 for break-out competences. Interestingly, competence lock-in strategies (1 and 2) are linked to both the price-sensitive and the design-focused segments of the market.

At the other end of the spectrum is cluster 11, which stands for companies that have break-out competences and compete mostly in international markets.

There is only one cluster (4) that represents companies oriented towards the domestic market and is likely to incorporate both companies that are aspiring to develop their own brand as well as companies that are focusing on narrow niche markets. Significantly, here they are all employing very similar strategies, which is in contrast to their counterparts in old member states. The remaining three clusters from new member states are oriented to the price-segment of the market, and for them selling on the national market is supplementing their export focus.

The remaining four clusters (3, 12, 13, and 14) developed different forms of hybrid strategies. Clusters 12, 13 and 14 are at different stages in terms of product development, with cluster 13 being the least advanced of the three. Cluster 3, located in the new member states, is the only cluster with a hybrid strategy that is targeting the price-sensitive segment of the market.

Similar to the situation in clothing, the evidence for footwear suggests that those groupings that developed a strong interest in the domestic market were, overall, more successful in moving up the production chain. In many instances moving upwards thus was often achieved at the same time as downgrading – or at best no change – in the position of companies in GCCs. Thus, advance in the main market is linked to functional retreat in the secondary market.

Figure 3.3 Patterns of enterprise strategies: Footwear
Source: Enterprise Survey.

Patterns of enterprise strategy in the software industry

Drawing primarily on the literature, a number of prototype strategies were identified (see Table 3.4). Not unexpectedly, these strategies differed significantly from those in the clothing and footwear industries. The specificities of the sector mean that there is a much greater emphasis on innovation, both in terms of products and processes, as well as intellectual property (IP). Moreover, the characteristics of the supply chain mean that low value-added stages are concentrated in the middle, with higher value-added at the two ends (design and distribution).

The evidence regarding the software industry shows a relatively lower degree of diversity than in the case of footwear and clothing (see Figure 3.4). Enterprises in the same segment of the market tend to adopt, to a greater degree than in the other two sectors, broadly similar strategies. Thus, we have a number of groupings across different countries falling within our prototype categories. This relatively lower impact of the national context is an interesting feature of the industry and may be attributed to the ease of flows across space.

Another interesting characteristic of the industry is the concentration of competence lock-in strategies exclusively in Eastern European countries, where they account for a significant percentage of the total. However, all of the enterprises surveyed do not remain focused exclusively on price-sensitive segments of the markets. Indeed, price competitiveness in the markets in which they operate is combined with innovativeness. Overall, 43 enterprises fall into this grouping – making up around 30 per cent of the total enterprises in the sector.

Interestingly, there is a considerable concentration of strategy groupings around the hybrid prototype strategy. However, there are some interesting differences between enterprises adopting hybrid strategies. A number of these enterprises fall squarely within the prototype strategy developed here, thus focusing on markets where competition occurs in terms of design, and to a lesser degree price and quality. Some 54 enterprises (38 per cent of the total) fall in this grouping and can be found in all the countries surveyed. However, hybrid strategies are also adopted by enterprises that try to focus on niche markets and provide solutions for major (often international) players. Groupings from Estonia and Poland fall in this broad category – accounting for some 40 enterprises. Lastly, there appears to be grouping UK/Greece 2, which is moving away from competence lock-in towards hybrid strategies. Altogether, 73 per cent of the enterprises surveyed adopt hybrid strategies.

Table 3.4 Overview of prototype strategies

	Competence lock–in	**Hybrid**	**Break–out competences**
Product/ service	Not own product range, mainly servicing products development (sometimes maintenance) and support of other organizations	Some product design capabilities and/or capabilities for integration and support in unique environments	Own product design (IP very important) and brand development and/or unique products for highly complex environments
Process	Technological change and specialisation/ competence in line with the requirements of external software producer(s)	Some technological innovations and specialized competencies	Unique own technology (but also technological innovations here)
Function	No significant movement between stages, though maybe some movement within	Moving up and/or moving down the value chain – often simultaneously in two different production dimensions	Moving closer to final markets/consumers and/or to product development
Production	Services mainly testing and/or code writing under order (or very often implementation of solutions of other companies with no substantial own intellectual property)	Some project management capabilities and/or some increasingly complicated code writing	Reduced internal production capacity, and developing chain management competences Often in consulting services or design of own systems
Market	Price competition and aspects of quality are of paramount importance in serving few non-major customers	Innovation is a key element of competitive advantage, maybe alongside secondary considerations such as design, quality or even price	Niche products/ solutions for major international players, where design capabilities and quick response are of paramount importance

Source: Enterprise Survey.

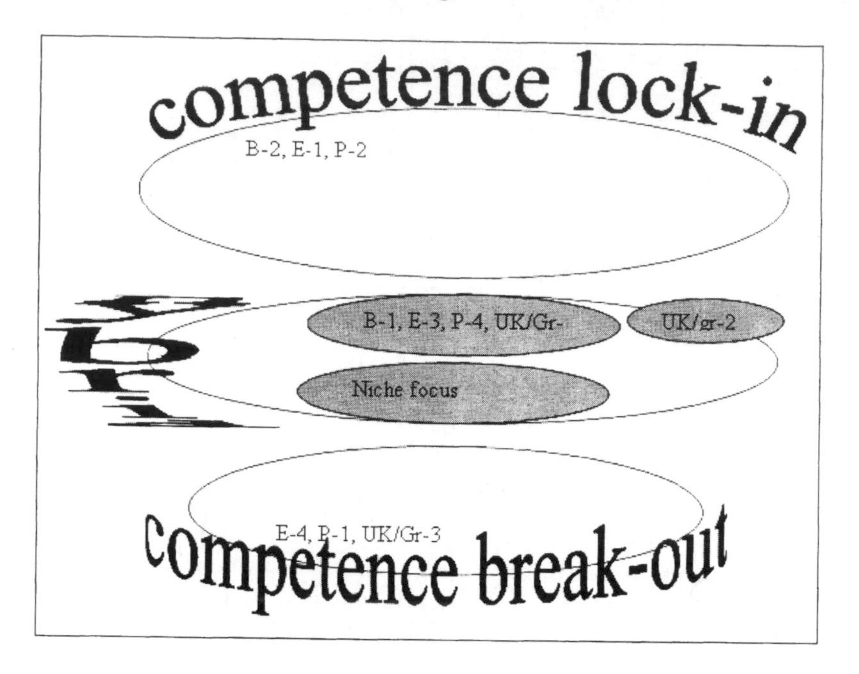

Figure 3.4 Patterns of enterprise strategies: Software
Source: Enterprise Survey.

Pattern of enterprise strategy in electronics

In analysing the electronics sector we followed the same approach as in the analysis of clothing, footwear and software. Building on the literature, we developed a matrix with three ideal-type strategies (Table 3.5).

Table 3.5 Overview of prototype strategies

	Competence lock-in	Hybrid	Break-out competences
Product/ service	Not recognisable products, integration into network product architectures	A combination of not recognisable and recognisable products. Design capabilities are present. IP does not play a substantial role	Own recognisable product (design and brand development) and/or unique products for highly complex environments where IP is very significant
Process	Narrow technological specialization/ competence that may be cutting edge but is an integral part of a wider process	Some technological innovations and competences, often developed in cooperation with major customers	Unique own technology, and/or ability to combine and apply different technologies in a new and unique way
Function	No significant movement between stages, though maybe some movement within	Moving up and/or moving down the value chain – often simultaneously in two different production dimensions	Moving closer to knowledge-intensive parts of the supply chain often simultaneously with a move towards final markets/consumers
Production	Production remains at the heart of the company	Production remains importance, but often combined with new services	Reduced internal production capacity, and developing chain management competences and/or services

Source: Enterprise Survey.

In contrast to the other three industries, the strategies in electronics much more closely reflect the three ideal-types (Figure 3.5). Companies in three of the five countries, Bulgaria, Estonia, and Greece, developed lock-in strategies, with Greek companies focusing exclusively on the flexible-response segment of the market (Greece 2), while there were Estonian companies focusing on the flexible (Estonia 2) and the price-sensitive segments (Estonia 3).

Figure 3.5 Patterns of enterprise strategies: Electronics
Source: Enterprise Survey.

There were no Bulgarian companies that adopted break-out strategies; Bulgarian companies developed either lock-in or hybrid strategies. Bulgarian companies that developed hybrid competences were located in either the price-sensitive (Bulgaria 1) or the flexible response (Bulgaria 2) segments of the market.

The strategies adopted by Greek and Estonian companies were polarized and were either lock-in (Estonia 2, Estonia 3, Greece 2) or break-out (Estonia 2, Greece 1). However, the Greek companies that developed lock-in strategies were only located in the flexible-response segment of the market (Greece 2). Interestingly it was UK and Polish (rather than Greek) companies that demonstrated similar behaviour and adopted either hybrid (Poland 1, UK 1) or break-out competences (Poland 2, Poland 3, UK 2, UK 3).

A Kaleidoscope of Strategies

The analysis of the primary data on a sector-by-sector and country-by-country basis provided us with a total of 55 groupings. Whilst there was significant diversity – especially in clothing and electronics – there were also similarities. Some of these similarities are found around our prototype strategies, but not only there. These similarities enable us to identify six main strategy patterns, presented below in linkages as well as performance. Thus, we derive the following groupings: 1) competence lock-in, 2) competence lock-in for markets that require flexibility, 3) hybrid focusing on price-sensitive segments of the market, 4) hybrid, 5) competence break-out, and 6) outliers.

The groupings are presented in a centripetal manner (with those adopting competence lock-in strategies in the periphery, and those adopting break-out strategies in the core). Each industry occupies a different colour and side of the rectangle, whilst those groupings that fall in the same pattern are linked or overlap (see Figure 3.6).

The incidence of these groupings varies significantly from country to country and from sector to sector. Thus there is a considerable concentration of enterprises that adopt a competence lock-in strategy in Bulgaria, whilst no firm in the UK falls into this grouping (Figure 3.7). In contrast, nearly half the enterprises in the latter country adopt a competence break-out strategy (none of the firms in Bulgaria fall into this grouping). Interestingly, in Estonia competence lock-in strategies are invariably linked with flexible markets. As far as sector is concerned, there is a greater incidence of enterprises that adopt competence lock-in strategies in clothing, whilst this is much less the case in software (Figure 3.8). In the latter sector, hybrid strategies are more commonly present, and, to some degree, break-out. However it is in electronics that competence break-out is most commonly apparent. Lastly, in footwear, hybrid for price-sensitive markets and competence-lock for other (than price-sensitive) markets are of considerable importance. These disparities in the national and sectoral composition of each strategy pattern may assist in interpreting variation in external linkages as well as performance.

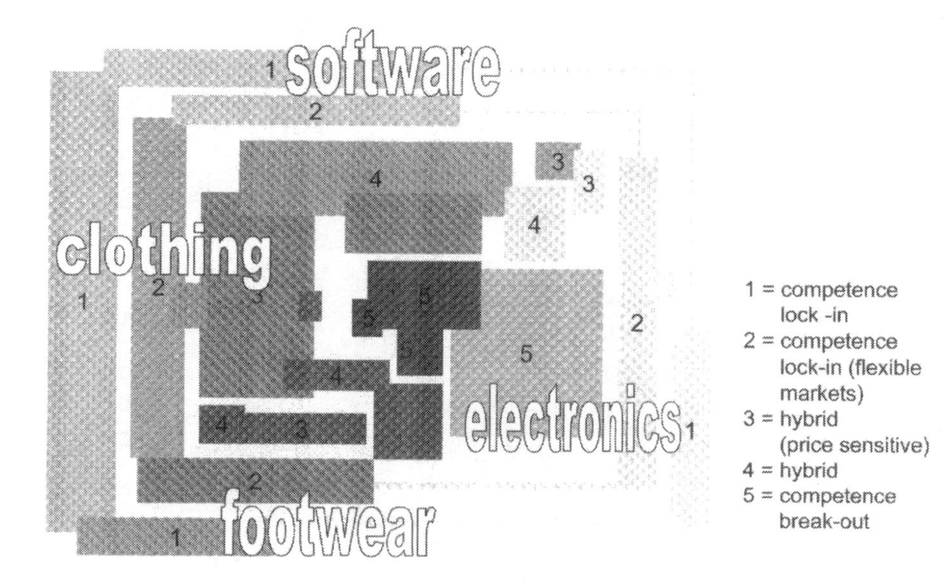

Figure 3.6 Kaleidoscope of strategies
Source: Enterprise Survey.

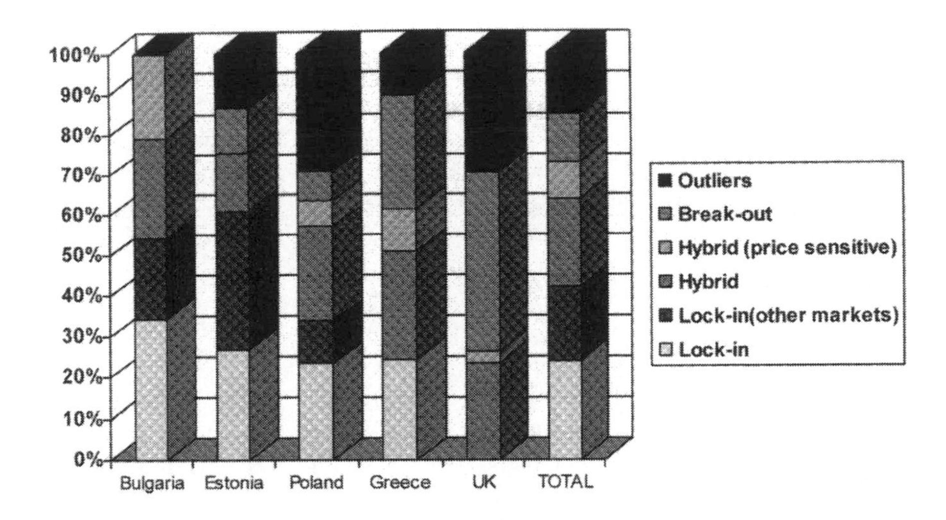

Figure 3.7 Strategy patterns by country
Source: Enterprise Survey.

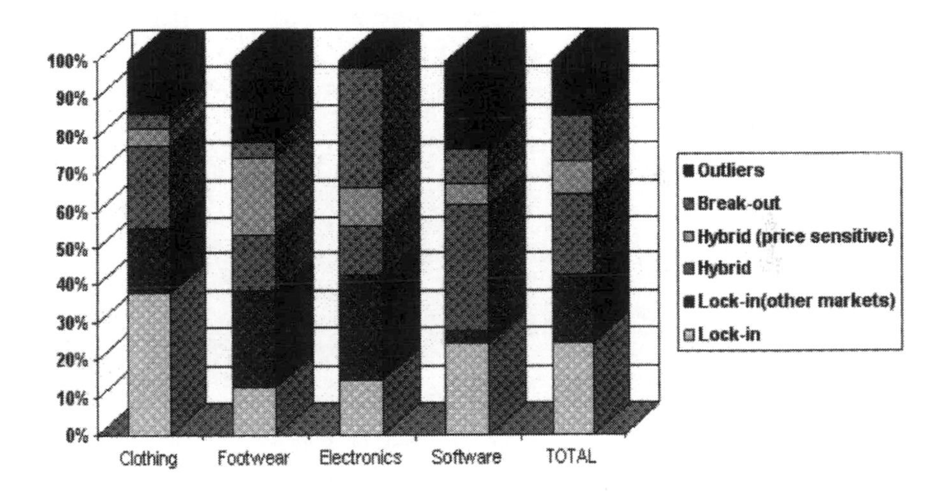

Figure 3.8 Strategy patterns by sector
Source: Enterprise Survey.

There are some disparities in the characteristics of the enterprises adopting different patterns of strategy. As can be seen from Table 3.6, enterprises that adopt competence break-out strategies tend to be larger and well established units, which do not rely heavily on international markets (less than half their sales). Outliers are also established firms with lower export orientation – interestingly these are most common in the two largest markets (UK and Poland), but they tend to be somewhat smaller units. Export orientation and relatively recent establishment are invariably linked to enterprises that adopt competence lock-in strategies, and are most commonly found in the two smaller Eastern European countries (namely Bulgaria and Estonia). These characteristics raise an interesting question for future research: that is, to what extent is there a progression (through time) from one strategy pattern to another (as the age and size evidence indicates), or to what extent do firms not radically alter their patterns of strategy? The latter case would mean that the lifespan of firms that adopt competence lock-in strategies may be shorter than those aiming to break-out.

Table 3.6 Enterprise characteristics by strategy pattern

Strategy group		Total employment-units (V 49)	Export as per cent of total sales in 2004 (V 95)	Age of firm
Outliers	Mean	103.0303	45.0440	27.5960
	N	99	91	99
	Std Deviation	181.45216	37.75268	31.71555
Competence lock-in	Mean	110.5549	60.8500	14.1220
	N	164	160	164
	Std Deviation	214.07424	37.57947	12.36896
Competence lock-in (other markets)	Mean	194.6393	73.7723	13.8033
	N	122	119	122
	Std Deviation	324.42040	35.07494	13.87943
Hybrid (price sensitive markets)	Mean	183.7333	60.2200	19.8500
	N	60	50	60
	Std Deviation	259.78954	35.63780	16.90635
Hybrid	Mean	126.1233	50.1115	16.7103
	N	146	139	145
	Std Deviation	244.73496	34.59282	15.52772
Competence break-out	Mean	948.6750	46.9531	28.3951
	N	80	64	81
	Std Deviation	4091.67853	35.79990	29.01408
Total	Mean	234.5887	57.1355	18.8465
	N	671	623	671
	Std Deviation	1448.51998	37.36754	20.74008

Source: Enterprise Survey.

Enterprise Strategies and External Linkages

In order to explore the impact of space upon enterprise strategies, we developed three new measures that capture proximity/distance – influenced heavily from the literature. The first concerns physical proximity, and is a three-category variable. So, in instances where the main international market of the enterprise is in a country that borders with the country of origin, the measure is low (1=high proximity); a market elsewhere in EU is moderate (=2), and elsewhere in the world high (3=low proximity). The findings are presented in Table 3.7. This indicates some interesting disparities by strategy pattern. Overall, European markets are of paramount importance for nearly 90 per cent of enterprises. Interestingly, it is enterprises that opted for competence break-out strategies, outliers, and those adopting hybrid strategies that appear able to work in markets outside their immediate vicinity. In the case of the last pattern, this may be linked to the importance of USA dominance in the software sector. Interestingly, immediate physical proximity appears to be more important for enterprises that used a competence lock-in strategy for markets where competition occurs primarily in terms others than price – although this is linked with the over-representation of Estonian firms (which rely heavily on the Finnish market) in this pattern. This also appears to be the case, though to a somewhat lesser degree, among enterprises which adopted competence lock-in strategies for price-sensitive segments of the market.

Table 3.7 Geographical proximity by strategy group

			Geographical proximity			Total
			1	2	3	
Strategy group	Outliers	Count % within strategy group	32 38.1	33 39.3	19 22.6	84 100
	Competence lock-in	Count % within strategy group	60 40.0	76 50.7	14 9.3	150 100
	Competence lock-in (other markets)	Count % within strategy group	57 52.3	50 45.9	2 1.8	109 100
	Hybrid (price sensitive)	Count % within strategy group	16 30.8	29 55.8	7 13.5	52 100
	Hybrid	Count % within strategy group	58 43.9	50 37.9	24 18.2	132 100
	Competence break out	Count % within strategy group	19 42.2	17 37.8	9 20.0	45 100
Total		Count % within strategy group	242 42.3	255 44.6	75 13.1	572 100

Source: Enterprise Survey.

The second measure of proximity/distance used for our purposes includes institutional proximity (Table 3.8). This captures the ease of transacting internationally on account of national and international institutions. The EU market is viewed as the one with the highest institutional proximity (=1), other advanced market economies follow (2=moderate), whilst countries in the former Soviet Union are viewed as institutionally distant (=3). Disparities on this measure are modest by strategy grouping, primarily on account of the overwhelming importance of the EU markets. Again, however, enterprises that adopt hybrid strategy patterns (including those for price-sensitive markets) opt more frequently for institutional distance.

Table 3.8 Institutional proximity by strategy group

			Institutional proximity			Total
			1	2	3	
Strategy group	Outliers	Count	60	17	7	84
		% within strategy group	71.4	20.2	8.3	100
	Competence lock-in	Count	129	14	7	150
		% within strategy group	86.0	9.3	4.7	100
	Competence lock-in (other markets)	Count	99	3	8	110
		% within strategy group	90.0	2.7	7.3	100
	Hybrid (price sensitive)	Count	43	1	8	52
		% within strategy group	82.7	1.9	15.4	100
	Hybrid	Count	100	17	14	131
		% within strategy group	76.3	13.0	10.7	100
	Competence break-out	Count	36	9	0	45
		% within strategy group	80.0	20.0	.0	100
Total		Count	467	61	44	572
		% within strategy group	81.6	10.7	7.7	100

Source: Enterprise Survey.

The third measure captures organizational proximity, which can also be viewed as a proxy for patterns of integration (Table 3.9). We view as the highest level of organizational proximity instances where there is a direct ownership link, that is, FDI or JV. A moderate level of organizational proximity is perceived in cases where there are at least five years of continuous relationship with a single buyer, and this buyer accounts for at least 40 per cent of sales. A low level of organizational proximity occurs in all other instances. Enterprises that adopted a hybrid strategy (both patterns) did not opt for a high organizational proximity approach. This is not particularly surprising, as these are often local businesses that engage in international markets as a means of strengthening their position in the domestic market. The reverse is the case regarding enterprises which adopted a competence lock-in strategy, especially for other than price-sensitive markets. This may be influenced by sector, as there is a considerable incidence of clothing firms within this strategy pattern.

Table 3.9 Organizational proximity by strategy group

			Organisational proximity			Total
			1	2	3	
Strategy group	Outliers	Count % within strategy group	24 37.5	10 15.6	30 46.9	64 100
	Competence lock-in	Count % within strategy group	36 26.5	30 22.1	70 51.5	136 100
	Competence lock-in (other markets)	Count % within strategy group	52 46.8	13 11.7	46 41.4	111 100
	Hybrid (price sensitive)	Count % within strategy group	13 26.0	6 12.0	31 62.0	50 100
	Hybrid	Count % within strategy group	20 18.9	22 39.3	64 60.4	106 100
	Competence break-out	Count % within strategy group	22 39.3	6 10.7	28 50.0	56 100
Total		Count % within strategy group	167 31.9	87 16.6	269 51.4	523 100

Source: Enterprise Survey.

In exploring the nature of the relationships between strategy patterns, we deployed five indicators (Table 3.10). The first and the second are viewed as measures of dependence: the number of foreign companies serviced and the percentage of sales going to the main international customer. The third and fourth are viewed as measures of strength of the relationship: the number of years of continuous transaction and their view of the degree of mutual confidence. The final one is a Lickert-type measure of the balance of power in the relationship (the lower the index, the more power resides with the enterprise surveyed).

Table 3.10 Nature of relationships

Strategy group		How many companies did you service on a subcontracting basis last year (V125)	Main customer as a per cent of total (V135)	Average number of years of continuous relationship (V144)	Balance of power (V148)	Mutual dependence (V149)
Outliers	Mean	4.8542	43.9219	7.0000	1.91	1.42
	N	48	32	45	93	92
	Std Deviation	4.51470	33.75149	5.88527	1.960	1.626
Competence lock-in	Mean	5.4857	55.66327	6.8154	3.40	3.19
	N	140	113	130	148	148
	Std Deviation	9.63692	33.11681	3.80897	1.354	1.347
Competence lock-in (other markets)	Mean	6.681	58.1818	6.5182	3.69	3.30
	N	109	88	110	121	120
	Std Deviation	8.59446	35.24138	3.29228	1.489	1.515
Hybrid (price sensitive markets)	Mean	55.4182	54.9048	6.5364	3.55	3.00
	N	55	42	55	60	60
	Std Deviation	5.82361	32.20499	3.88117	1.241	1.340
Hybrid	Mean	6.9135	45.3596	6.3552	3.28	2.85
	N	104	89	116	137	137
	Std Deviation	11.82979	29.66297	4.00435	1.670	1.570
Competence break-out	Mean	15.7097	59.4231	8.2326	2.54	1.95
	N	31	26	43	76	76
	Std Deviation	37.53860	35.84765	5.25917	1.963	1.664
Total	Mean	6.6407	53.0769	6.7509	3.12	2.71
	N	487	390	499	635	633
	Std Deviation	13.09542	33.26648	4.13636	1.719	1.645

Source: Enterprise Survey.

Enterprises that adopted a competence break-out strategy depend heavily upon their main buyer, who accounts for more than half their sales turnover (the highest of all). However, they also posses greater customer base, which in turn diminishes their vulnerability. This is an interesting finding that indicates that developing competences that may enable a firm to break-out do not necessarily mean that the company operates in isolation from other businesses in the sector. In fact, the relationships developed by enterprises falling into this grouping are both long-lasting (mean duration of 8.2 years – the longest of all groupings). Moreover, these relationships rely less upon mutual confidence but the fact that they are more symmetrical in terms of power is also significant.

However, the most significant dependence upon a small number of international buyers and the main buyer among them is apparent in the case of companies that adopt competence lock-in strategies (both for price-sensitive and flexible response markets). Overall, the durability of relations is considerable (more than six years), and there is recognition of the mutual confidence that this creates, as well as the asymmetry of power.

Those enterprises that adopted a hybrid strategy appear to rely to a much lesser degree upon a single international buyer: a trend reinforced by the fact that in most instances a significant percentage of sales is directed in the domestic market. However, this does not impact upon the durability of relationships but somewhat diminishes the sense of mutual confidence and power asymmetries.

Lastly, the strategy pattern that reports the lowest level of dependence on buyers involves the outliers. They rely less on the main customer, have a wider customer base, and a lower sense of mutual dependence and asymmetrical power than firms in any other strategy pattern.

These findings are influenced to a considerable degree by sector and country because of the considerable disparities in the incidence of strategy patterns in these categories.

Statistical analysis regarding the variables that capture the nature of emerging relationships provides some interesting results, presented in Table 3.11. Not unexpectedly, there is a statistically significant (at $p<.01$) relationship between the number of international buyers and the importance of the main buyer. This relationship is negative (the more number of the buyers, the lower the importance of the main buyer) and moderately strong. A positive relationship appears to exist between the significance of the main buyer and the degree of mutual confidence. More interestingly, there appears to be a strong and statistically significant ($p<.01$) correlation between the balance of power and mutual confidence.

Table 3.11 The nature of relationships

		Main customer as a per cent of total (V135)	How many companies did you service on a subcontracting basis last year (V125)	Average number of years of continuous relationship (V144)	Balance of power (V148)	Mutual dependence (V149)
Main customer as a % of total (V135)	Pearson	1	-.229**	.039	.091	.281**
	Sig. (2-tailed)		.000	.430	.057	.000
	N	437	416	412	435	433
How many companies did you service on a subcontracting basis last year (V125)	Pearson	-.229**	1	.057	-.001	-.044
	Sig. (2-tailed)	.000		.195	.982	.301
	N	416	549	510	546	544
Average number of years of continuous relationship (V144)	Pearson	.039	.057	1	.125**	.180**
	Sig. (2-tailed)	.430	.195		.003	.000
	N	412	510	563	561	559
Balance of power (V148)	Pearson	.091	-.001	.125**	1	.665**
	Sig. (2-tailed)	.057	.982	.003		.000
	N	435	546	561	714	712
Mutual dependence (V149)	Pearson	.281**	-.044	.180**	.665**	1
	Sig. (2-tailed)	.000	.301	.000	.000	
	N	433	544	559	712	712

Note: ** Correlation is significant at the 0.01 level (2-tailed) *Source*: Enterprise Survey.

Table 3.12 Year of foreign involvement by strategy group

Year of involvement (V75)		Strategy group						Total
		Outliers	Competence lock-in	Competence lock-in (other markets)	Hybrid (price sensitive markets)	Hybrid	Competence break-out	
Pre 70	Count	0	0	0	0	0	2	2
	% within strategy group	.0	.0	.0	.0	.0	10	1.2
71/88	Count	0	1	1	0	2	1	5
	% within strategy group	.0	3	1.9	.0	6.9	5	2.9
1989	Count	1	0	1	0	0	1	3
	% within strategy group	4	.0	1.9	.0	.0	5	1.7
1990	Count	0	2	0	0	4	1	7
	% within strategy group	.0	6.1	.0	.0	13.8	5	4.1
1991	Count	1	0	3	0	0	2	6
	% within strategy group	4	.0	5.7	.0	.0	10	3.5
1992	Count	1	1	3	0	2	3	10
	% within strategy group	4	3	5.7	.0	6.9	15	5.8
1993	Count	0	3	4	1	1	0	9
	% within strategy group	.0	9.1	7.5	8.3	3.4	.0	5.2
1994	Count	1	1	4	0	0	1	7
	% within strategy group	4	3	7.5	.0	.0	5	4.2
1995	Count	2	2	1	2	3	0	10
	% within strategy group	8	6.1	1.9	16.7	10.3	.0	5.8

Table 3.12 continued

Year of involvement (V75)		Strategy group						Total
		Outliers	Competence lock-in	Competence lock-in (other markets)	Hybrid (price sensitive markets)	Hybrid	Competence break-out	
1996	Count	3	2	8	2	1	1	17
	% within strategy group	12	6.1	15.1	16.7	3.4	5	9.9
1997	Count	0	3	6	2	3	4	18
	% within strategy group	.0	9.1	11.3	16.7	10.3	20	10.5
1998	Count	1	2	4	1	5	1	14
	% within strategy group	4	6.1	7.5	8.3	17.2	5	8.1
1999	Count	2	4	1	1	2	0	10
	% within strategy group	8	12.1	1.9	8.3	6.9	.0	5.8
2000	Count	6	0	2	0	3	0	11
	% within strategy group	24	.0	3.8	.0	10.3	.0	6.4
2001	Count	3	4	2	0	0	0	9
	% within strategy group	12	12.1	3.8	.0	.0	.0	5.2
2002	Count	0	1	5	1	1	1	12
	% within strategy group	.0	3	9.4	8.3	3.4	5	7
2003	Count	4	4	2	0	1	1	13
	% within strategy group	16	12.1	3.8	.0	3.4	5	7.6
2004	Count	0	3	6	2	1	1	13
	% within strategy group	.0	9.1	11.3	16.7	3.4	5	7.6
Total	Count	25	33	53	12	29	20	172
	% within strategy group	100	100	100	100	100	100	100

Source: Enterprise Survey.

Data regarding the timing of the processes at work are weak. Table 3.12 captures the timing of FDI and JV creation. This process peaked in 1997 – and appears to be a consequence of the creation of sub-contracting linkages. Following a modest decline, the pace of integration accelerated again in 2004 and 2004.

Figure 3.9 attempts to explore the combined effects of performance change (for employment, sales and profits) by strategy pattern. A number of interesting patterns emerge. The evidence suggests that enterprises that adopt competence lock-in strategies record the highest incidence of relatively poor performance (decline or profitless expansion) and weak organic growth. This is despite the fact that these enterprises are often concentrated in the lowest labour-cost countries (Bulgaria and Estonia) among the five surveyed here. It is the outliers and hybrid strategies that record robust performance. The case of enterprises adopting competence break-out strategies is interesting. Nearly two-thirds of such firms report jobless growth, itself linked with the fact that they are located in the UK and Greece.

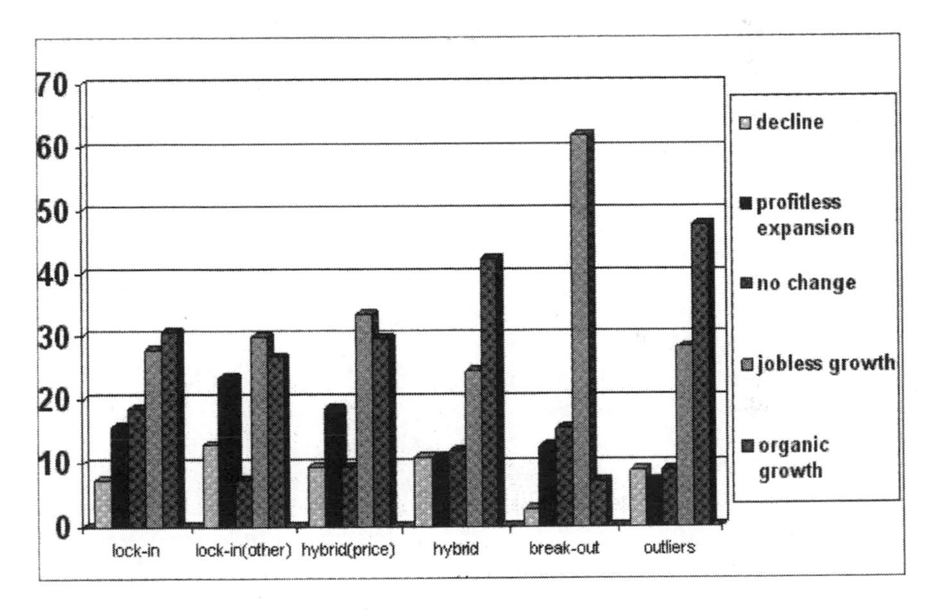

Figure 3.9 Combined performance by strategy group

Source: Enterprise Survey.

Concluding Remarks

The approach adopted in this chapter enhances our understanding of enterprise strategies in labour-intensive industries in two ways. Firstly, conceptually: enterprise strategy is viewed as multidimensional, influenced by a number of factors at work (industrial context, local context, enterprise characteristics and competences). Therefore there is greater complexity of outcomes that capture more accurately diversity in the real world than is the case in the literature, especially the GCC approach. Secondly, methodologically: through a move beyond ideal-type strategies, again increasing our ability to capture diversity. However, this move is carried out in a structured manner that allows for the identification of dimensions of similarity and diversity. By exploring the micro-dynamics of industrial change, this chapter seeks to open up the discussion around the notion of upgrading. We argue that while upgrading is very useful in describing tendencies on a high level of abstraction, it fails to capture the complexity of actual strategies and the diversity of choices and possible paths that companies face.

Drawing on the lock-in/hybrid/break-out conceptualization, we demonstrate that competence lock-in strategies are not linked exclusively with the price-sensitive segment of the market. In fact, this strategy is often important in markets that require flexible response (especially in sectors such as clothing and footwear). Thus a shift to markets for complementary orders – often viewed as an intermediate position between cut-throat mass-produced items and the design, own-label segment of the market – does not involve a fundamental change in strategy. This fact means that this is attainable, but does not fundamentally alter the competences of the firm. Further, competence break-out strategies revolve around the development of competences that are not relationship-specific, and thus, can facilitate multiple or changing relationships. However, this situation does not mean that enterprises that adopt such strategies engage in weak relationships or rely to a lesser degree than other firms upon their main buyer. Instead, what appears to be the case is that the advantages gained by developing competences that are not relationship-specific are not readily exercised by firms but are a means of diminishing vulnerability.

All of the above brings us to one of the main findings of the present chapter: dependence and asymmetrical power endowments may co-exist with mutual confidence. This phenomenon can be explained using a 'voice, exit, loyalty' argument. Indeed, Hirschman (1972) argues that voice can be deployed in circumstances when one participant in a relationship believes that the other party will seriously consider his or her voice. In the context of labour-intensive industries, this situation may exist when there are significant power inequalities, with the most powerful agent feeling confident that the less powerful one will consider the former's voice. Interestingly, this chapter captures the view of the less powerful agent in the relationship.

Our findings reintroduce the importance of the domestic market in this field of scholarly inquiry. While national and local market may also be influenced by global dynamics, they remain a distinct and more accessible market for local enterprises. As such, these markets may be used as a key element in the emergence of hybrid strategies that use global integration as a way of strengthening the competences, and, consequently, the position of individual firms. This position is particularly important in countries with large domestic markets, such as the UK and Poland. Evidence regarding performance indicates that this pattern may be effective in specific settings. Companies combined these from different positions: in order to develop production competences and/or a recognisable brand name, but also as short-term tactics in order to address fluctuation in demand for example. In making these choices, companies are also faced with a constantly evolving set of available paths that are only open for a limited period of time (and differ between contexts); these paths are not equally accessible for all firms, and commitment to any one of them often carries strong 'weight of legacies' that makes it difficult to move back onto a different path of development. Time, therefore, is a key dimension for understanding both the evolving structure of opportunities that companies face within a specific context as well as the path-dependent nature of these choices given the choices that companies have already made in the past.

Therefore, the implications of strong relationships for future performance merits further investigation. In the literature, and implicitly in this chapter, we adopt a positive view. Whether strong relationships increase the vulnerability of firms, however, especially when buyers are faced with competitive pressures or even default, remains an open question.

In terms of performance, the evidence presented here lends support to two complementary arguments. Firstly, there are disparities in performance between enterprises adopting diverse strategies. Thus, the performance of enterprises that adopt competence lock-in strategies compares unfavourably with that of firms that adopt hybrid strategies or with outliers. However, the disparities are not conclusive. Moreover, there are cases of success among competence lock-in strategies and instances of decline among enterprises adopting hybrid strategies. This brings us to the second argument: that there is no single recipe for success. Success is not only conditional on strategy, but also upon its appropriateness in the industrial and regional context, the competences of the firm, and, of course, how effectively the strategy is implemented.

Lastly, as far as policy implications are concerned, the findings of this chapter lend support to the view that state policies can only have a limited impact on the direction of change in the LII, and thus debates about the benefits of market versus strong state intervention do not satisfactorally capture the real choices that states have available. The increasing complexity of the economic, political and social environment that states face with both supra-national and sub-national players becoming ever more prominent makes a strong case for an active and enabling, though not necessarily only and always directly intervening, state.

References

Abernathy, F.H., Volpe, A. and Weil, D. (2006), 'The Future of the Apparel and Textile Industries: Prospects and Choices for Public and Private Actors', *Environment and Planning A* 38:12, 2207–32.

Anselin, L. (2003), 'Spatial Externalities', *International Regional Science Review* 26:2, 147–52.

Assmann, D. and Punter, T. (2004), 'Towards Partnership in Software Subcontracting', *Computers in Industry* 54:2, 137–50.

Audretsch, D.B. (2003), 'Innovation and Spatial Externalities', *International Regional Science Review* 26:2, 167–74.

Bair, J. (2005), 'Global Capitalism and Commodity Chains: Looking Back, Going Forward', *Competition and Change* 9, 153–80.

Bair, J. (2006), 'Regional Trade and Production Blocs in a Global Industry: Towards a Comparative Framework for Research', *Environment and Planning A* 38, 2233–52.

Bair, J. and Gereffi, G. (2003), 'Upgrading, Uneven Development and Jobs in the North American Apparel Industry', *Global Networks* 3, 143–69.

Bazan, L. and Navas-Aleman, L. (2001), 'Comparing Chain Governance and Upgrading Patterns in the Sinos Valley, Brazil', Paper presented at workshop 'Local Upgrading in Global Chains' at IDS, 14–17 February 2001.

Buckley, P. and Ghauri, P. (2004), 'Globalisation, Economic Geography and the Strategy of Multinational Enterprises', *Journal of International Business Studies* 35:2, 81–98.

Burt, R. (1992), *Structural Holes: The Social Structure of Competition* (Cambridge, MA: Harvard University Press).

Burt, R. (2004), *Brokerage and Closure an Introduction to Social Capital* (Oxford: Oxford University Press).

Carlyle, A. (2001), 'Developing Organized Information Displays for Voluminous Works: A Study of User Clustering Behavior', *Information Processing and Management* 35, 677–99.

Castells, M. and Hall, P. (1994), *The Technopoles of the World: The Making of the Twenty-first Century Industrial Complexes* (London: Routledge).

Christopher, M. (2000), 'The Agile Supply Chain Competing in Volatile Markets', *Industrial Marketing Management* 29, 37–44.

Coe, N., Hess, M., Yeung, H., Dicken, P. and Henderson, J. (2004), 'Globalizing Regional Development: A Global Production Network Perspective', *Transactions of the Institute of British Geographers* 29:4, 468–84.

Cooke, P. (1996), 'Reinventing the Region: Firms, Clusters and Networks in Economic Development', in Daniels, P.W. and Lever, W.F. (eds), *The Global Economy in Transition* (Harlow: Longman), 310–27.

Corso, M. and Pavesi, S. (2000), 'How Management Can Foster Continuous Product Innovation', *Integrated Manufacturing Systems* 11:3, 199–211.

Curran, J., Rutherford, R. and Smith, S.L. (2000), 'Is there a Local Business Community?: Explaining the Non-participation of Small Business in Local Economic Development', *Local Economy* 15:2, 128–43.

Deardorff, A. and Djankov, S. (2000), 'Knowledge Transfer under Subcontracting: Evidence from Czech firms', *World Development* 28:10, 1837–47.

DTI (2005) 'The Government Manufacturing Strategy', <http://www.dti.gov.uk/manufacturing/strategy.pdf>, accessed 20 December 2005.

Dunford, M. (2002), 'The Changing Profile and Map of the EU Textile and Clothing Industry', <http://www.geog.susx.ac.uk/research/eggd/ege/pdf/02_t.pdf>.

Dunford, M. (2006), 'Industrial Districts, Magic Circles, and the Restructuring of the Italian Textiles and Clothing Chain', *Economic Geography* 82:1, 27–59.

Fujita, M., Krugman, P.R. and Venables, A. (1999), *The Spatial Economy: Cities, Regions and International Trade* (Cambridge, MA: MIT Press).

Gereffi, G. and Memedovic, O. (2003), 'The Global Apparel Value Chain: What Prospects for Upgrading by Developing Countries', *Sectoral Studies Series* (Vienna: UNIDO).

Gereffi, G. and Mayer, F. (2004), *The Demand for Global Governance* (Durham: Duke University WP, SAN 04-02).

Gereffi, G., Humphrey, J. and Sturgeon, T. (2005), 'The Governance of Global Value Chains', *Review of International Political Economy* 12:1, 78–104.

Gibbon, P. (2001), 'At the Cutting Edge: UK Clothing Retailers and Global Sourcing', CDR Working Paper 01.4, August 2001.

Harrigan, K.R. (1985), 'An Application of Clustering for Strategic Group Analysis', *Strategic Management Journal* 6, 55–73.

Henderson, J., Dicken, P., Hess, M., Coe, N. and Yeung, H. (2002), 'Global Production Networks and the Analysis of Economic Development', *Review of International Political Economy* 9:3, 436–64.

Hirschman A.O. (1972), *Exit, Voice and Loyalty Responses to the Decline of Firms* (Cambridge: Harvard University Press).

Huang, C. (2002). 'Distributed Manufacturing Execution Systems: A Workflow Perspective', *Journal of Intelligent Management* 13:6, 485–97.

Huggins, R. (2000), 'The Success and Failure of Policy-implanted Inter-firm Network Initiatives: Motivations, Processes and Structure', *Entrepreneurship and Regional Development* 12:2, 111–35.

Humphrey, J. and Schmitz, H. (2001), 'Governance in Global Value Chains', *IDS Bulletin* 32:3.

IMF (2004), 'Global Monitoring Report 2004: Policies and Actions for Achieving the MDGs and Related Outcomes', Background Paper.

Jack, S.L. and Anderson, A.R. (2002), 'The Effects of Embeddedness on the Entrepreneurial Process', *Journal of Business Venturing* 17, 467–587.

Kalantaridis, C. (2000), 'Globalisation and Entrepreneurial Response in Post-Socialist Transformation: A Case Study from Transcarpathia, Ukraine', *European Planning Studies* 8:3, 285–99.

Kalantaridis, C., Slava, S. and Sochka, K. (2003), 'Globalization Processes in the Clothing Industry of Transcarpathia, Western Ukraine', *Regional Studies* 37:2, 173–86.

Karlsson, C. and Dahlberg, R. (2003), 'Entrepreneurship, Firm Growth and Regional Development in the New Economic Geography: Introduction', *Small Business Economics* 21, 730–76.

Kirat, T. and Lung, Y. (1999), 'Innovation and Proximity Territories as Loci of Collective Learning Processes, *European Urban and Regional Studies* 6:1, 27–38.

Laschewski, L., Phillipson, J. and Gorton, M. (2002), 'The Facilitation and Formalisation of Small Business Networks: Evidence from the North-East of England', *Environment and Planning C: Government and Policy* 20, 375–91.

Lazzeretti, L., De Propris, L. and Storai, D. (2004), 'Impannatori and Business Angels: Two Models of Informal Capital Provision', *International Journal of Urban Regional Research* 28:4, 839–54.

Lehtinen, U. (1999), 'Subcontractors in a Partnership Environment: A Study on Changing Manufacturing Strategy', *International Journal of Production Economics* 60:1, 165–70.

Levitt, T. (1995), 'The Globalisation of Markets', in P.N. Chauri, and S.B. Prasad, (eds), *International Management: A Reader* (London: Dryden Press).

MacKenzie, R. (2002), 'The Migration of Bureaucracy: Contracting and the Regulation of Labour in the Telecommunications Industry', *Work, Employment and Society* 16:4, 599–616.

Malberg, A. and Maskell, P. (2002), 'The Elusive Concept of Localization Economies: Towards a Knowledge-based Theory of Spatial Clustering', *Environment and Planning A* 34:3, 429–49.

Messner, D. (2002), 'The Concept of the "World Economic Triangle": Global Governance Patterns and Options for Regions', IDS Working Paper 173 (Brighton: IDS-Sussex).

Neidik, B. and Gereffi, G. (2006), 'Explaining Turkey's Emergence and Sustained Competitiveness as a Full-package Supplier of Apparel', *Environment and Planning A* 38, 2285–303.

OECD (2004), *A New World Map in Textiles and Clothing: Adjusting to Change* (Paris: OECD Publications).

Ottati, G.D. (1996), 'Economic Changes in the District of Prato in the 1980s: Towards a More Conscious and Organized Industrial District', *European Planning Studies* 4:1, 35–54.

Palpaceur, F., Gibbon, P. and Thomsen, L. (2005), 'New Challenges for Developing Country Suppliers in Global Clothing Chains: A Comparative European Perspective', *World Development* 33, 409–30.

Pickles, J., Smith, A., Bucek, M., Rukova, P. and Begg, R. (2006), 'Upgrading, Changing Competitive Pressures, and Diverse Practices in the East and Central European Apparel Industry', *Environment and Planning A* 38, 2305–24.

Ponte, S. (2005), 'Quality Standards, Conventions and the Governance of Global Value Chains', *Economy and Society* 34:1, 1–31.

Rabellotti, R. (2001), 'The Effect of Globalisation on Industrial Districts in Italy: The Case of Brenta', IDS Working Paper 144 (Brighton: University of Sussex).

Raikes, P., Jensen, M., and Ponte, S. (2000), 'Global Commodity Chain Analysis and the French Filière Approach: Comparison and Critique', *Economy and Society* 29:3, 390–417.

Rama, R., Ferguson, D. and Melero, A. (2003), 'Subcontracting Networks in Industrial Districts: The Electronics Industries of Madrid', *Regional Studies* 37:1, 71–88.

Ran, A. and Kuusela, J. (1996), 'Selected Issues in Architecture of Software Intensive Products' (Joint Proceedings of the 2nd International Software Architecture Workshop: San Francisco, CA (updated 20 January 2008) <http://delivery.acm.org/10.1145/250000/243630/p147-ran.pdf?key1=243630&key2=4374799111&coll=GUIDE&dl=ACM&CFID=48545207&CFTOKEN=43468764>, accessed 5 June 2008.

Sacchetti, S. and Sugden, R. (2003), 'The Governance of Networks and Economic Power: The Nature and Impact of Subcontracting Relationships', *Journal of Economic Surveys* 17:5, 669–91.

Schiavone, F. (2005), 'Industrial Districts and Delocalisation: How to Manage Cultural Diversity?', *International Journal of Globalisation and Small Business* 1:2, 152–67.

Schmitz, H. (1999), 'Global Competition and Local Cooperation: Success and Failure in the Sinos Valley, Brazil', *World Development* 27:9, 1627–50.

Schmitz, H. (2003), 'Globalized Localities: Introduction', in H. Schmitz (ed.), *Local Enterprises in the Global Economy: Issues of Governance and Upgrading* (Cheltenham: Elgar).

Schmitz, H. (2006), 'Learning and Earning in Global Garment and Footwear Chains', *European Journal of Development Research* 18, 546–71.

Scott, A. (2002), 'Competitive Dynamics of Southern California's Clothing Industry: The Widening Global Connection and Its Local Ramifications', *Urban Studies* 39:8, 1287–306.

Scott, A. (2006), 'The Changing Global Geography of Low-Technology, Labor-intensive Industry: Clothing, Footwear and Furniture', *World Development* 34:9, 1517–36.

Smallbone, D., Venesaar, U., Rumpis, L. and Budreikate, D. (1996), 'Export Activity in Manufacturing SMEs in the Baltic States: An Empirical Analysis', paper presented in the 9th Nordic Small Business Conference (Lilehamer, Norway, May).

Smith, A. (2003), 'Power Relations, Industrial Clusters and Regional Transformations: Pan-European Integration and Outward Processing in the Slovak Clothing Industry', *Economic Geography* 79:1, 17–40.

Smith, A., Rainnie, A., Dunford, M., Hardy, J., Hudson, R. and Sadler, D. (2002), 'Networks of Value, Commodities and Regions: Reworking Divisions of Labour in Macro-regional Economies', *Progress in Human Geography* 26:1, 41–63.

Steinfeld, E. (2004), 'China's Shallow Integration: Networked Production and the New Challenges for Late Industrialization', *World Development* 32:1, 1971–87.

Tokatli, N. (2007), 'Asymmetrical Power Relations and Upgrading among Suppliers of Global Clothing Brands: Hugo Boss in Turkey', *Journal of Economic Geography* 7, 67–92.

UNCTAD (2003), *World Commodity Survey, 2003–2004: Market, Trends and the World Economic Environment* (Geneva: United Nations Publication).

UNCTAD (2004), *Commodity Atlas* (Geneva: United Nations Publication).

Uzzi, B. (1997), 'Social Structure and Competition in Inter-firm Networks: The Paradox of Embeddedness', *Administrative Science Quarterly* 42, 35–67.

von der Heydt, A. (1999,) 'Efficient Consumer Response', in von Der Heydt, A. (ed.) *Handbook of Efficient Consumer Response.* Handbook ECR (Vahlen Verlag: Munich).

Watanabe, C., Zhu, B., Griffy-Brown, C. and Asgari, B. (2001), 'Global Technology Spillover and its Impact on Industry's R&D Strategies', *Technovation* 21:5, 281–91.

World Bank (2004), *The World Bank Annual Report* (Washington: World Bank Group).

Wortman, M. (2003), 'Structural Change and Globalisation of the German Retail Industry', Discussion Paper SPIII 2003–202b (Berlin: WZB).

Wrigley, N. and Lowe, M. (2002), *Reading Retail. A Geographical Perspective on Retailing and Consumption Spaces* (London: Hodder Arnold).

Yeung, H., Liu, W. and Dicken, P. (2006), 'Transnational Corporations and Network Effects of a Local Manufacturing Cluster in Mobile Telecommunications Equipment in China', *World Development* 34, 520–40.

Chapter 4

Social Consequences of Delocalization in Labour-Intensive Industries: The Experience of Old and New Members of the EU

Krzysztof Gwosdz, Bolesław Domański[1]

Introduction

The notion of delocalization commonly refers to downsizing or closure of plants and/or companies, which is related to a transfer of activities to a foreign country (outward relocation). At the same time, it implies that some countries gain new activities that were formerly located abroad, by expansion of existing establishments or greenfield investment (inward relocation).

Delocalization has become a hot topic, primarily due to its social consequences in DCs where jobs are lost through outward relocation. Whereas a huge section of the debate focuses on the number of jobs, it also raises important issues of the effects of increasing mobility of capital on wages and skills, and concerns about a 'race to the bottom' in labour standards and 'social dumping'.

The issue is not entirely new and is closely linked to the extensive debate on the relationship between globalization, deindustrialization and the labour markets. Studies of social consequences of delocalization are conducted within two broad contrasting frameworks:

- Contestation of globalization as a process detrimental especially to countries/regions where jobs are lost;
- Belief in an inevitable process of industrial restructuring, where the decreasing number of industrial jobs in advanced economies is seen as a natural stage of economic change, which enhances long-term national/regional competitiveness.

1 The authors are grateful for helpful comments on the earlier drafts of this chapter to several members of the MOVE project: Robert Guzik, Margarita Ilieva, Christos Kalantaridis, Athanasios Kalogeresis, Janusz Kornecki, Lois Labrianidis, Poli Roukova, Stoyan Totev and Ivaylo Vassilev.

The multidimensional and uneven nature of the processes under discussion in various sectors and areas, and difficulties in separating the social consequences of delocalization from the effects of other processes such as technological change, make it impossible to reduce these consequences to simple general statements. Three types of questions can be addressed here: what (types of consequences), who (social groups affected) and where (places). Popular perceptions tend to focus on the short-term direct effects of jobs lost or gained as a result of downsizing or expansion of plants, their closure or opening. There may be positive and negative social consequences of both outward relocation and inward relocation. We need insights into both direct and indirect effects. A long-term perspective is necessary to assess the real nature of processes and consequences fully. As delocalization may take different forms, that is, FDI or subcontracting, social consequences will vary accordingly. The fundamental issue is that the same process produces different effects depending on the industrial and spatial contexts within which it operates (see Figure 4.1).

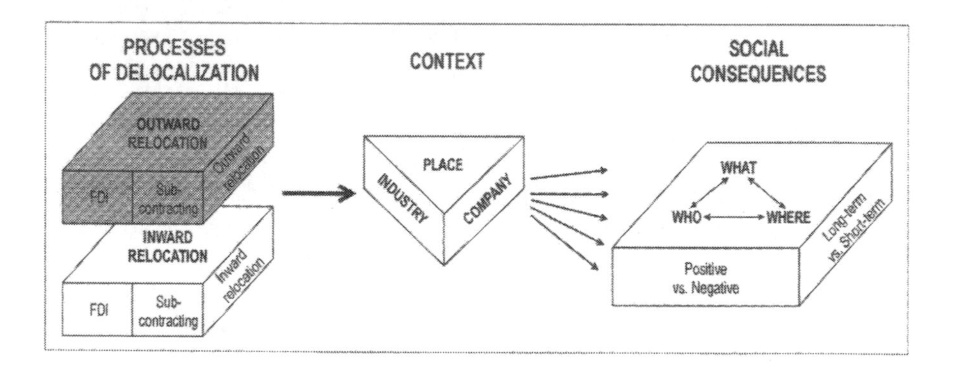

Figure 4.1 The complex nature of the social consequences of delocalization
Source: Authors' elaboration.

The debate on the social consequences of delocalization

Not surprisingly, the role of delocalization in job losses in advanced industrialized countries has attracted most attention. The serious difficulty here lies in isolating changes resulting from delocalization from the effects of other factors, including technological advancements.

Many studies argue that the effect of trade flows linked to delocalization is too small to account for significant labour market changes in DCs. Fluctuations in domestic demand and increases in labour productivity related to technological changes are considered to be much more important (Krugman and Lawrence 1994; Morcos 2003). Rowthorn and Ramaswamy (1997) claim that about two-thirds of the decline in the share of manufacturing between 1970 and 1994 in the USA can be explained by higher productivity in manufacturing than in services. This is questioned by Feenstra (1998), who argues that if we take into consideration trade in both final and intermediate inputs, the impact of globalization on employment is equivalent to changes induced by technological innovation. This debate is sometimes framed as one of 'trade' versus 'technology' explanations. Helg and Tajoli (2004, 8) maintain that this distinction is artificial, and that it is better to see delocalization as another 'distinct cause of shift in labour demand'.

Estimates based on the data of the European Foundation for the Improvement of Living and Working Conditions indicate that delocalization is responsible for 9 per cent of the annual job losses in the EU (*Restructuring and Employment* 2006). The report, *Relocation: An Element of Industrial Dynamics* (2000), based on a survey of 3000 Belgian companies, attributes 28 per cent of employment loss through collective layoffs in industry in the period 1990–1995 to relocation. The majority of transfers of activities took place towards and from other DCs.

The pressure of potential relocation may pose a threat and contribute to less favourable employment conditions. People still working in the industry may have to accept lower pay, more involuntary part-time work and temporary jobs (Waddington and Parry 2003; Hadjimichalis 2006). The role of large industrial companies which tend to pay higher wages becomes more limited vis-à-vis smaller companies (Glasmeier and Jensen 2001). Mingione (1997, 1) argues that although numerous job opportunities are created in the tertiary sector, they 'do not reflect the traditional standards of social regulation and therefore entail a weakening of the mechanisms of social integration and a growing risk of exclusion'.

Eliason and Storrie (2003) found that the workers displaced due to plant closures in Sweden from the mid-1980s suffered substantial losses in earnings. Hijzen et al. (2003) claim that outsourcing accounts for about half the increase in domestic wage inequality in the UK. Feenstra and Hanson (2001) maintain that the activities that are outsourced use a large amount of unskilled labour and consequently cause growing inequality between high-skilled and low-skilled workers in DCs. Opponents argue that the shift from unskilled to skilled workers in DCs is a natural market-driven phenomenon underpinned by the competitiveness of nations and the resources in which they are better endowed.

It is often suggested that the main threats of job losses and worsened employment conditions concern low-skilled blue-collar workers, poorly educated people, women and ethnic minorities. A study by the European Commission (EC) (2006) reports that if a woman is unemployed one year, her transition probability towards employment is half that of a man.

Negative social consequences go beyond the workplace. Decline in income may result in impoverishment, loss of identity, disrupted family and social life and worsening life prospects for children. Further indirect consequences attributed to delocalization include increasing tax competition, which may lead to reduction in government expenditures vital to the poor (International Labour Organization 2004).

On the other hand, delocalization in Western European countries also has positive employment effects. Several studies show a positive correlation between outward foreign investment and exports in delocalizing industries and other related sectors. The foreign site may allow the firm to continue a low-end activity that would no longer be profitable in the home country (Crestanello and Dalla Libera 2003). Many scholars argue that relocation is often the least bad solution (*Relocation* 2000).

So far, we have discussed the issue from the point of view of places of outward relocation. Location of labour-intensive activities in countries of the periphery brings about various positive effects, the most obvious being new jobs. According to Ghose (2003), net employment created in manufacturing industries in LDCs has been larger than the net employment loss in DCs.

In addition, we can find growth in productivity, knowledge spill-overs, training programmes and upgraded skills. Outward relocation may also produce multiplier effects in the region, including creation of new indigenous companies (Morcos 2003).

Many studies suggest that relocation of production has affected working conditions and wages in a positive direction (Gradev 2001). This trend has been shown for Bulgaria by Begg et al. (2000). According to Lorentowicz et al. (2005), in Poland outsourcing contributes roughly 35 per cent to the change in the relative wage for skilled manufacturing workers. The winners may include women in LDCs, who gain access to wage employment and improved social status.

On the other hand, there is concern about the inferior quality and possibly temporary character of new jobs created in low-cost countries as a result of delocalization. The low-paid, low-skill jobs may represent a sort of 'social dumping'.

Main aims and research questions

Despite the growing body of literature, understanding of various aspects of the social consequences of delocalization is far from satisfactory. There is an especially limited number of studies approaching the issue from the broader perspective of both the countries of outward relocation and those of inward relocation.

This chapter attempts to make a contribution to the current debate on the social consequences of delocalization on the basis of an empirical study of four labour-intensive industries in five countries – both old and new members of the EU: the UK, Greece, Estonia, Poland and Bulgaria. These include three sectors that are very sensitive to labour costs – clothing, footwear and electronics – and one that is particularly sensitive to labour skills – software. The UK represents the traditional core of European manufacturing. Greece could provide a relatively cheap location for labour-intensive activities in the EU, which might be threatened by its recent enlargement. The three post-communist countries differ in size and in level of economic development. The social consequences are explored in the context of the ongoing process of European integration.

The focus is on three main groups of issues:

- Impact on the number of jobs and unemployment levels
- Effects in the quality of jobs
- Long-term consequences for local, regional and national social wellbeing.

These issues are explored both for areas which are experiencing decline and for countries and regions which report growth in labour-intensive activities from the point of view of three major types of determinants: industry-specific, company-specific and place-specific.

Social consequences may be studied using two approaches – tracing the careers of people who lost their jobs due to delocalization (which requires long-term data on individuals) or analysing quantitative and qualitative changes in the labour markets and social wellbeing on different geographical scales. The latter approach is applied in this chapter.

The vital problem concerning the quantitative effects on the labour market is not simply the number of jobs lost or gained, but the impact of delocalization on employment and unemployment levels. This reflects the capability of the labour market to absorb job-losers and new entrants. There is also the question of the role of labour shortages in delocalization processes.

The analysis of the quality of jobs has to take into account issues of feasible alternatives, of winners and losers and the possible segmentation of the labour market.

Finally, the salient question is the long-term impact of delocalization on the social wellbeing on various geographical scales. Does the decline in labour-intensive industries contribute to the erosion of the economic strength and social wellbeing of countries and regions? Is the growth of labour-intensive activities a rising path to sustainable, competitive economy and thus enhanced quality of life? The concepts of path dependence, embeddedness and localized capabilities created in territories, on the basis of which a company's competences are built, may also be useful here.

The Quantitative Impact on the Labour Market

Impact of delocalization on the number of jobs and unemployment trends

The study provides some insight into the creation/destruction of jobs under the influence of delocalization in four industries and five countries. The trends revealed reflect the varied position of countries in delocalization.

In Bulgaria, which may be a benchmark for inward relocation, 62 per cent of the companies investigated increased employment compared to the situation when they were first involved in delocalization within the last 15 years. In Poland and Estonia, the increase occurred in 57 per cent of enterprises; however, 23 per cent of Polish companies showed diminishing employment. A similar share of Greek companies experienced decline, with 32 per cent of enterprises growing and a large number unchanged. In the UK the proportion of companies reducing and expanding their workforce was 53 to 25 per cent respectively (see Table 4.1).

Table 4.1 Changes in the level of employment of the surveyed companies involved in delocalization

Country	Number of surveyed companies	Decrease in employment	No change	Increase in employment
		per cent of companies		
UK	75	53.4	21.9	24.7
Greece	80	23.8	43.8	32.5
Estonia	200	9.8	33.5	56.7
Poland	200	23.4	19.8	56.9
Bulgaria	200	20.0	18.0	62.0

Source: Company Survey.

There is no direct link between delocalization and rise in regional unemployment. This can be illustrated by the lack of correlation between changes in industrial employment and changes in unemployment rate in 198 NUTS-2 regions of the EU–15, 1999–2004 (Figure 4.2). There are regions where a decline in industrial employment leads to higher unemployment, but a vast number of regions have experienced falling unemployment despite the job losses. This fact shows that industrial jobs are replaced by workplaces in other sectors.

The relationship between a decreasing number of industrial jobs and regional employment/unemployment rates can be analysed in greater detail for the UK. Between 1995 and 2006, the country experienced another wave of deindustrialization and 1.24 million manufacturing jobs were lost (25.7 per cent). At the same time, the unemployment rate fell from 9 per cent to 5 per cent, and the total number of jobs increased by 2.5 million. The employment rate also increased from 71.3 to 74.3 per cent.

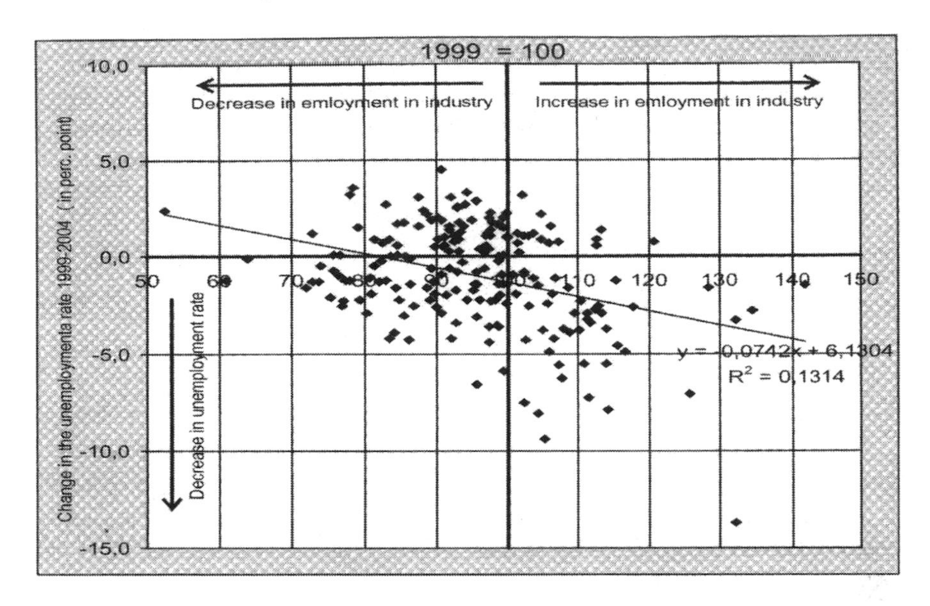

Figure 4.2 The relationship between changes in industrial employment and unemployment rate in NUTS – 2 regions of the 15 countries of the EU, 1999–2004

Source: Authors' calculations based on Eurostat data.

The analysis of the 330 UK districts shows an interesting picture. In the areas where manufacturing employment remained relatively unchanged, the rate of male employment[2] grew significantly more than in the areas that lost 25 per cent of manufacturing jobs or more (see Table 4.2). The female employment rate grew faster than that of males and was less related to the decrease of jobs in manufacturing. This is even more striking if we take into account the fact that more female jobs disappeared in manufacturing after 1990 (women accounted for 40 per cent of the workforce in 1990 and 34 per cent in 2004).[3] Women proved to be more successful in finding new jobs in services.

The impact of delocalization varies among sectors. The employment shifts are most evident in the clothing trade. Between 1995 and 2004, these shifts were most dramatic in Bulgaria (77.7 per cent increase), on the one hand, and in the UK and Greece (70.3 and 43.5 per cent decrease respectively), on the other. The changes in Estonia and Poland were relatively moderate.

2 The employment rate is a good indicator of the strength of the labour market, reflecting the inflow to and withdrawal from this market (Danson 2001).

3 This difference can be partly explained by outsourcing of non-production activities.

The growth in the Bulgarian clothing industry is a conspicuous example of delocalization. The sector has revived significantly since the crisis of the early 1990s. The principal growth in clothing took place in peripheral districts of the country, especially those adjacent to Greece: Blagoevgrad, Kurdzhali, and Smolyan.

Table 4.2 Deindustrialization and the change in the employment rate in the UK by districts 1995–2006

			Relative change in employment rate (in percentage points)		
Change in the manufacturing employment	**Number of districts**	**Unemployment rate in 2006 UK = 100**	**All workers**	**Male**	**Female**
Decrease by 50 per cent and more	24	106.7	2.3	1.0	3.9
Decrease by 25–50 per cent	177	103.4	2.4	1.2	3.7
Decrease by 5–25 per cent	88	101.0	2.9	1.9	4.0
No major change (+ – 5 per cent)	22	95.5	4.4	3.4	5.5
Increase by 5 per cent and more	19	95.8	2.3	1.6	3.3

Note: 78 districts were not included in the analysis due to lack of data.
Source: Authors' calculations based on Nomis official labour market statistics.

On the other hand, clothing was at the forefront of the manufacturing decline in Greece. In the heyday of the late 1980s, the industry employed 81,000 people; the number dropped to 60,000 by 1995 and to 15,000 in 2002.[4] On the whole, Greece lost 41,000 manufacturing jobs between 1993 and 2003 (7 per cent). Clothing was heavily concentrated in central Macedonia, where total employment increased by 6.3 per cent 1999–2005. In terms of its unemployment rate, central Macedonia performed similarly to the national economy. Thus, delocalization of clothing did not affect the regional labour market (1 per cent of all jobs now). Still, some small towns in central and eastern Macedonia could have been affected.

4 Data for 1988 and 1995 cover all companies, while data for 2002 refers only to entities with more than 10 people (smaller companies represented half of employment in clothing in 1988).

In the UK, the current decline in clothing and footwear represents only the tail-end of a process that began several decades ago. According to the key informants interviewed, if companies had not relocated their production to cheaper sites, they would have gone bankrupt and their workplaces been lost anyway. The labour market effects are fairly limited because the process occurred over a long period of time and was accompanied by growth in other sectors.

The impact of a universal decline in employment in footwear in the countries studied is limited due to the small size of this industry and its geography. In Greece it represents 0.5 per cent of all employment in manufacturing and is concentrated in and around the two largest Greek cities of Athens and Thessaloniki, meaning that the impact on the labour market is negligible. In the UK, the footwear industry employed 6,400 people in 2003, that is, 0.2 per cent of total manufacturing. In Poland roughly 10–15 per cent of 40,000 footwear jobs are linked to subcontracting and foreign subsidiaries. About half the Polish companies involved in internationalization and 70 per cent of their Bulgarian counterparts show an increase in employment.

In electronics the main carriers of internationalization are TNCs, and plants are often large. In the UK the sector declined rapidly after a period of growth induced by public subsidies. This mainly affected the northeast of England, Scotland and south Wales. Almost 55 per cent of the UK companies surveyed had reduced their employment in recent years, with one-fifth showing an increase. Plant closures in electronics, however, are closely connected with rapid technological change and do not always represent outward relocation.

Despite the lack of growth in employment in electronics in post-communist countries in general, more than half the companies increased the number of jobs. The spectacular growth of some parts of the electronics industry in Poland is fuelled by the expansion of Asian and American companies, which enter the European market in this way. The location of large new plants and expansion of older ones have had a significant impact on some medium-sized towns, for example, Mława and Kwidzyn. There is also some shift from Nordic countries to Estonia. Bulgaria and Greece are not important players in this industry.

The picture in software is radically different. The dominant process is expansion rather than relocation. Employment is growing significantly in all the countries. The share of companies that reduced their workforce in recent years is below 10 per cent, except for the UK, where it is 23 per cent (but half the British companies increased employment). According to key informants, the impact of delocalization on British software is visible in simple code writing, which has made some people redundant. The spatial concentration of software companies in national capitals and other major cities in Poland, Bulgaria and Estonia reinforces their economic position in the country.

Table 4.3 The extent of delocalization and effects on the labour market in clothing sector

Country	Change in the number of jobs (1995–2004) 1995 = 100	The role of delocalization in job creation/destruction	Impact on regional unemployment
UK	29.7	Significant	Insignificant
Greece	56.5	Significant	Modest
Estonia	100.8	Significant	Moderate
Poland	83.5	Significant	Moderate
Bulgaria	177.7	Significant	Significant

Note: Scale = insignificant, modest, moderate, significant.
Source: Authors estimations based on the data of national statistics offices, company survey and key-informant interviews.

Table 4.4 The extent of delocalization and effects on the labour market in footwear sector

Country	Change in the number of jobs (1995–2004) 1995 = 100	The role of delocalization in job creation/destruction	Impact on regional unemployment
UK	28.2	Significant	Insignificant
Greece	49.2	Moderate	Insignificant
Estonia	86.0	Modest	Insignificant
Poland	54.4	Modest	Insignificant
Bulgaria	74.4	Insignificant	Insignificant

Source: Authors estimations based on the data of national statistics offices, company survey and key-informant interviews.

Table 4.5 The extent of delocalization and effects on the labour market in electronics sector (NACE 30–33)

Country	Change in the number of jobs (1995–2004) 1995 = 100	The role of delocalization in job creation/destruction	Impact on regional unemployment
UK	68.3	Significant	Moderate
Greece	122.5*	Insignificant	Insignificant
Estonia	n.d.	Significant	Moderate
Poland	97.5	Significant	Moderate
Bulgaria	57.8	Insignificant	Insignificant

Note: *1995–2002.
Source: Authors estimations based on the data of national statistics offices, company survey and key-informant interviews.

Table 4.6 The extent of delocalization and effects on the labour market in software sector

Country	Change in the number of jobs (1995–2004) 1995 = 100	The role of delocalization in job creation/destruction	Impact on regional unemployment
UK	252.8	Modest	Insignificant
Greece	311.7*	Insignificant	Insignificant
Estonia	153.6**	Modest	Insignificant
Poland	258.8	Modest	Insignificant
Bulgaria	182.1	Modest	Insignificant

Note: *1995–2001, **2001–2004.
Source: Authors estimations based on the data of national statistics offices, company survey and key-informant interviews.

Labour scarcity and its effects

One of the challenges for labour-intensive industries is the difficulty faced by companies in labour recruitment. This difficulty may stem from limited labour supply in general, scarcity of people with specific skills or alternative employment opportunities.

Many of the managers interviewed and key informants in the UK raised the issue of the poor image of factory work in general, especially among young people. 'Kids see factories as dark, noisy and smelly', says a footwear industry manager. Young people are less willing to take mundane factory jobs and follow work routines than older workers. This is related to competition from less demanding job opportunities elsewhere and/or rising aspirations. Uncertainty caused by delocalization discourages the entry of young workers and entrepreneurs even further.

A negative feedback mechanism is at work, which accelerates the shrinking of the traditional labour-intensive industries. The companies are sensitive to cost, pay low wages and offer a limited number of new jobs. Accordingly, new entrants to the industry are few, and the pool of 'within the industry' candidates is contracting due to factory closures in the past and an ageing workforce. Labour skills in clothing and footwear are increasingly lacking from the local market, and existing companies face mounting problems in recruitment. The important point is that once skills are lost, they are hard to replace, and technical competences of towns and regions wither away. Therefore, the delocalization of activities requiring specific industrial skills may be an irreversible phenomenon.

In 'transition economies' the supply of skilled labour for clothing and footwear is also becoming insufficient in certain places, as more alternative job opportunities emerge and aspirations grow. Low pay, a stricter working regime than in many services and seasonal fluctuations make factories relatively less attractive. This is accompanied by a diminishing supply of vocational school-leavers. Labour migration to Western Europe makes matters worse in Poland, Estonia and Bulgaria.

Companies use different strategies to tackle labour recruitment problems. A quite obvious action would be to raise wages and offer extra-wage benefits and better employment conditions. However, for companies engaged in delocalization in labour-intensive activities, often competing on price, this action may be difficult and has to be part of a broader restructuring towards higher value-added products. Another option is to train people without the required qualifications, which is costly, takes time and may adversely affect product quality. The recruitment of foreign migrants as factory employees is yet to take place in CEE on a large scale, in contrast to the UK.

Last but not least, producers may choose location in areas with more ample labour supply and/or lower wages. This fact probably underlies the increased share of peripheral regions of Bulgaria in the apparel industry. In Estonia the clothing industry is growing in Ida-Varruna, where salaries are 25 per cent lower than in Tallinn. The ultimate solution may be to subcontract production abroad or invest in a foreign country endowed with a cheaper labour force – that is, to delocalize. Several successful Polish clothing companies have already moved their production to the Ukraine or China.

All these strategies except for the last may bring about positive effects for workers, providing jobs to more peripheral areas as well as training and/or higher wages.

What Kinds of Jobs are Lost and Gained?

Quality of jobs

Western and Southern Europe There is little doubt that a vast number of jobs lost in outward relocation countries were poor jobs in terms of wages. In British clothing and footwear, earnings range from 47 per cent in clothing to 73 per cent in footwear in relation to the manufacturing average, with the respective figures for Greece being 64 and 69 per cent.

Wages and salaries in British clothing, footwear and electronics increased 1996–2004 in parallel with the decreasing number of employees. The typical pattern is that 'better' jobs stay at home longer. For example, many large electronics companies have relocated their manufacturing abroad, but training and R&D facilities have remained in the UK.

The survey reveals that 37 per cent of the UK companies involved in delocalization and more than 43 per cent of the Greek enterprises offered wages higher than the average in the industry, with 47–49 per cent paying about the average, and only 14 and 9 per cent respectively below the average. This may confirm the generally positive impact of delocalization on the efficiency of companies[5] and/or the fact that higher paid jobs are maintained at home.

This finds further support in the fact that half the British companies have increased the proportion of white-collar employees, compared to the situation when they were first involved in delocalization within the last 15 years; no company reports a decrease in this respect. White-collar workers now represent nearly 50 per cent of the total workforce in the electronics companies surveyed and roughly 40 per cent in clothing and footwear. The percentage of white-collar workers and of those with tertiary education has risen, especially in traditional industries.

There is a striking contrast between Greece and the UK here. The vast majority of Greek companies (85 per cent) report no change in their employment structure; only 12.5 per cent have increased the share of people with tertiary education and of white-collar staff. The latter comprise just 14–17 per cent of the workforce in electronics and clothing. This may reflect a stronger market position, larger size (in electronics) and higher competitiveness of British companies in comparison to their Greek counterparts. The latter may use delocalization more as a survival rather than an expansion strategy. In all four sectors, between 80 and 90 per cent of British enterprises moved to a competitive advantage based on design and product development, while 85 per cent of Greek clothing companies and all footwear companies declare that their competitive advantage rests on labour-intensive products.

There is little support for concerns about deterioration of employment standards of people who remain in labour-intensive activities in the developed economies.

5 We must bear in mind that the survey was conducted among existing enterprises, so those that went out of business were not represented.

An increase in temporary employment is found in 15 per cent of surveyed British companies, mostly in software and electronics. In addition, only 9 per cent of enterprises have extended part-time employment; the figure is twice as high among clothing companies. The share of Greek companies that have experienced a rise in temporary and part-time employment is 5–7 per cent. About 10 per cent of British electronics enterprises increased their use of rented employment.

However, there are two possible negative consequences. First, the jobs that were lost provided employment to people who were generally in a weak position in the labour market. Second, the quality of jobs undertaken by the former employees of labour-intensive industries is not necessarily good, though this has not been studied here.

Central and Eastern Europe Jobs that appear in labour-intensive industries in post-communist countries are generally inferior to those in other sectors, but their evaluation is more positive if we take the perspective of the industries themselves or the local labour market situation.

About one third of the companies surveyed in Estonia and Poland involved in delocalization show higher wages than the sector average; the share is 40 per cent in Bulgaria. Relatively high wages are particularly characteristic of Estonian and Polish software business and of Estonian electronics. Clothing companies in Estonia more frequently offer wages below the average than above it. In Poland a similar situation is characteristic of electronics, while the share of clothing companies with higher and lower wages is the same. By contrast, clothing and footwear in Bulgaria show the highest share of enterprises with above-average earnings. In all countries, large clothing and footwear companies (250 employees and more) tend to pay more than SMEs, whereas big Polish and Estonian electronics companies offer wages above the average less often.

The differences revealed can probably be attributed to the fact that the economic performance of Polish and Estonian clothing companies dependent on their foreign partners for inputs and markets compares unfavourably with that of successful companies developing their own brands and targeting the domestic market. The widespread positive impact of subcontracting on wages in Bulgaria may reflect its generally greater role in local clothing and the smaller share of successful domestic companies. The assembly-type operations of foreign affiliates and location of factories in peripheral areas may lie behind lower wages in electronics in Poland, whereas electronics producers in Estonia represent more advanced establishments vis-à-vis other local businesses.

The proportion of the workforce with tertiary education and of white-collar employees has increased in 25 per cent of clothing companies in Poland and Estonia, and slightly less in Bulgaria. Such a rise has occurred in half of all electronics enterprises in Poland. White-collar workers constitute 31 per cent of total employment in electronics in Poland, 25 per cent in Estonia and 13 per cent in Bulgaria (similar results to Greece). The figure is considerably lower in clothing: 11–12 per cent in Poland and just 5 per cent in Bulgaria.

There is a widespread increase in the role of skill-intensive products in the three countries, which is accompanied by the manufacturing of higher value-added goods. In addition, a greater number of enterprises in Poland and Estonia declare a competitive advantage based on design and product development. Estonian companies reveal a clear move towards services such as design and marketing.

Growing permanent employment is accompanied by an increase in temporary employment in 30 per cent of the Estonian and Polish companies. In both countries this trend is most common in electronics and software, in Estonia also in clothing, while in Poland clothing companies more often reduce temporary employment. It plays a marginal role in Bulgaria (4 per cent of companies). The growth in part-time employment is taking place primarily in Estonia (27 per cent of companies), rarely in Poland (12 per cent) and Bulgaria (2 per cent). In addition, rented employment is used by 38 per cent of Estonian clothing enterprises and 15 per cent of Polish electronics companies; in other sectors and countries it is almost non-existent. All in all, worse types of jobs are more typical of labour-intensive activities in more developed, post-communist economies, which may reflect companies' attempts to maintain lower costs and flexibility.

FDI usually creates better employment conditions than subcontracting undertaken by domestic enterprises. Foreign-owned companies offer higher wages and salaries than their domestic competitors in the same industry and on the local labour market. Both the survey and the key informant interviews show that foreign companies provide more training and hence contribute to improved skills. Former employees of foreign-owned clothing companies in Bulgaria or of Italian footwear subsidiaries in Poland enjoy a good reputation and are sought after in the labour market.

On the whole this indicates some progress in the standard of employment and skills at the low end of industrial jobs and hardly supports the 'social dumping' argument. The new CEE members of the EU are not the flexible labour markets characterized by inferior labour standards compared to those in Western Europe. This is related to their implementation of EU legislation prior to accession.

The winners and the losers: The segmentation of labour markets

The losers of delocalization in the DCs have to be primarily sought among older redundant employees with particular industrial skills. They are often middle-aged or older women, who move to low-skill jobs in services or retire and hence have limited impact on unemployment rates. A high proportion of migrants and minority groups among the employees of the UK clothing industry means that women from these groups may be among the people most affected. Also, in Greece the outward relocation of this industry has mostly harmed women. Many of them were close to retirement age, as since the mid–1980s there was little young blood injected into this occupation. Paradoxically, male workers losing their jobs in electronics in towns in the north of England and Scotland may be the least successful in finding alternative employment.

In CEECs, many companies prefer to recruit young people without professional experience, because they are more flexible and do not have 'bad' habits acquired in a previous career. The better-educated youth benefit from increased demand for white-collar staff. At the same time, the limited number of young people willing to undertake manual work in the footwear and clothing industries may become a barrier to their development in the future. Women are preferred in footwear, clothing and parts of the electronics industry. They are believed to be more accurate in manual work and more likely to accept low wages. It is also women above middle age who profit from internationalization in clothing and footwear sectors. In Bulgaria, women who previously worked in agriculture and food processing find new full-time manufacturing jobs and training.

The expansion of traditional labour-intensive industries largely takes place outside the developed areas of CEECs. There are also big electronics factories located in medium-sized peripheral towns in Estonia and Poland. Jobs are created or maintained in communities that live in peripheral areas and are often plagued by high unemployment and weak alternative employment opportunities. They include areas inhabited by ethnic minorities, especially Turks in Bulgaria (Pickles 2001). Companies hire and train relatively less-skilled people, contributing to progress in local skills and capabilities. This sometimes leads public authorities to invest in education, for example, on the Estonian island of Saaremaa, where three international electronics companies are situated.

Delocalization may contribute to the segmentation of the labour markets. This segmentation may take different forms:

- The sectoral segmentation of the entire labour market with cleavages between industries providing high-quality and low-quality jobs
- Divisions within the particular industry based on the different position of companies in the value chain and related labour conditions
- Internal segmentation within the company.

Clothing and footwear are usually regarded as inferior segments of the labour market; the same is true of assembly operations or simple production in electronics. From this perspective, the growth of these sectors may not be a favourable change. At the same time, the manufacturing of advanced electronics products may belong to a superior submarket, which is also typically true of software activities.

The processes of upgrading towards higher value-added products and non-production services (R&D, design and marketing) described earlier in the case of British electronics and clothing indicate an increasing share of the superior segment of the labour market in these industries. The tendency to move up the value chain with parallel changes in the employment structure and improvement of labour conditions is to some extent observed in all labour-intensive activities in CEE, meaning that the lower segments of the labour market improve.

The quantitative analysis of employment structures and trends in delocalizing companies in labour-intensive industries in the UK and Greece does not indicate a significant increase in the level of segmentation of the job market in terms of temporary, part-time and agency workers.

The explicit internal segmentation of the company labour market can be identified in large foreign-owned consumer electronics factories in Poland. There are three distinct segments that differ in employment stability: 'core' staff employed on permanent contracts, part-time workers employed on more flexible conditions and rented employees. These submarkets are associated with performance of certain functions and production of particular final or intermediate goods, where 'good' employment is related to activities regarded as core to the company. Access to the privileged submarket is constrained.

The Long-term Impact on National, Regional and Local Social Wellbeing

The fundamental question concerns the overall long-term effects of delocalization on social wellbeing. In the long run, social wellbeing is determined by economic development on various geographical scales; thus it is necessary to consider how delocalization affects competitiveness of local, regional and national economies. This approach means that the diversification of the economy, the role of viable economic activities, local embeddedness of large companies and the development of localized capabilities may be important intermediating factors here. The significance of structural features for the sustainable development of the economy is obvious. The embeddedness of large companies, especially TNCs, in economies where they carry out their activities is a popular concept in studying the relationship between the global and the local (Ettlinger 1999; Phelps 2000). The concept implies that the company is planted in local networks, a fact which affects its impact on the host economy. The notion of localized capabilities was introduced in the evolutionary, competence theory of the company. Maskell (2001) argues that companies' competences are built on 'created localized capabilities'. They are a dynamic product of interaction between the company and the territory (Domański 2005) and may be helpful in interpreting spatial differences in the effects of delocalization. These concepts can be used to discuss how the long-term development trajectory of localities, regions and countries changes under the influence of delocalization of the labour-intensive industries being studied and hence affects broader social wellbeing.

The comparison of the old industrial districts of Durham and Northampton is very telling. The former, where a policy of attracting inward investment by subsidies was employed to tackle the declining economic base, has experienced a serious decline in electronics since the mid-1990s. The current unemployment level is 7.3 per cent, 2.3 percentage points above the national average, and the male employment rate has fallen by 3.9 percentage points (pp) as compared to 1995. Northampton, a traditional centre of the footwear industry in the UK, lost a quarter of its manufacturing workplaces, but the effects were very modest. The unemployment rate is equal to the national average and the female employment rate has increased by 9.3 pp (as compared to 3.9 pp in the whole country). Durham's sensitivity to industrial decline is even better reflected in Gross Domestic Product (GDP) indicators. In 1995, GDP per capita in the area represented 75 per cent of the UK average; ten years later it had decreased to 62 per cent. By contrast, the GDP of Northamptonshire grew from 96 to 107 per cent of the UK average 1995–2005. Thus the standard of living in the Durham area has been negatively affected by delocalization. The difference can be attributed to lower diversification of the local economy, poorer development of various alternative activities, especially services, and the disembedded nature of large electronics factories attracted by public subsidies (Hudson 1989, 2005).

In Greece, the negative impact of the contraction of the clothing industry was rather modest on the regional level. However, in some peripheral prefectures of central Macedonia, for example, Imathia, Pella, Serres, and even in Thessaloniki, GDP per capita dropped by more than 10 pp as compared to the national average 1995–2005. The growth of clothing in Greece represented a relatively short-term phenomenon based on the advantages of a cheap location for subcontracting within the EU. These advantages have been eroded with the advent of new low-cost competitors in post-communist countries, so the long-term development trajectory of places with inherent structural weakness has not been altered.

A simple typology of regions/localities can be based on two criteria: internal features of a place and the strength and type of delocalizing activities. In the case of outward relocation and strong regions/localities, delocalization is a form of 'creative destruction', where old industries are replaced by new viable activities. Thus social consequences for the community are negligible even in the short term, and in the long run the overall effects are clearly positive. However, in weak regions (Northeast England, Northern Greece) delocalization brings about negative consequences, if alternative dynamic sectors do not emerge or represent temporary solutions only, triggering a need for cyclical restructuring without reinforcing the position of a region vis-à-vis strong communities. Thus low-road strategies may contribute to the deepening of structural weaknesses, if they do not facilitate the development of generic local capabilities. Still, the main reason is not delocalization, but the economic weaknesses of the region/community.

For the new EU member states, which are gaining or maintaining new jobs in labour-intensive industries, the vital problem is the long-term viability of these activities and their impact on broader competitiveness and sustainable economic development.

The overall effects of delocalization may be different in areas where a given industry is already concentrated and in peripheral places where it creates a new economic base. It seems that the established areas of labour-intensive industry have better opportunities for success, as their enhanced industry-specific and localized capabilities may stimulate upgrading towards higher value-added and/or niche products, and hence they escape from a lock-in in low-value-added activities. There are stronger local linkages (embeddedness) and non-production competences. The overall economic development of other sectors matters too. A good example is

Słupsk in northern Poland, where an Italian-Polish JV company established in the early 1990s triggered positive changes in many locally-owned companies, strengthening the position of the area in footwear production in Poland.

In peripheral areas, which are at least temporary winners of delocalization of labour-intensive industries, for example, southern Bulgaria and some medium-sized towns in Poland and Estonia, they often contribute to the diversification of the local/regional economy, but may also lead to excessive dependence on new activities. The long-term success of such economies depends on several factors. Low value-added character of activities developed here, inferior position in the value chain, lack of strategic-decision and other non-production capabilities as well as weak local linkages may undermine upgrading of the industry and lead to a lock-in situation. Nevertheless, some regions and communities may succeed in enhancing their position, if the current growth of labour-intensive industries creates generic localized capabilities conducive to further development of other economic activities.

All things considered, different scenarios will emerge in the future evolution of regions and localities dependent on labour-intensive industries, probably reflecting the 'strong' and the 'weak' nature of their economies.

Conclusion

The public debate on the social consequences of delocalization of labour-intensive industries is clouded by common misinterpretations.

First, the social effects of delocalization are generally more limited than is sometimes maintained. There seems to be a geographical fallacy whereby phenomena that are significant locally become generalized on the national level. Analysis shows that the social consequences of delocalization of labour-intensive industries are mainly observed on a local scale, to a lesser extent on the regional level, and are almost negligible in entire national economies, with the notable exception of Bulgarian clothing. Moreover, widespread emphasis on job losses ignores the fact that this decline usually has no direct impact on unemployment levels. There are intermediating factors, such as social and economic features of the region/locality and national labour regulations, which determine whether the impact is strong or weak. Neither clothing nor footwear contributed to higher unemployment in the UK; adverse effects of the closure of electronics plants can be found in some towns. The collapse of the Greek clothing industry dating back to the late 1980s is an evident case of relocation of production to neighbouring low-cost countries, but its negative effects also manifest themselves on the local scale alone.

There is little doubt that the social consequences of delocalization are not only connected to industry and enterprise characteristics, but are largely place-dependent. The balance of negative versus positive effects is to a large measure determined by the role of the sector/employer on the labour market and the structure and overall performance of the regional/local economy. The main problem is not delocalization itself, but how to overcome the 'weaknesses' of certain regions and localities.

A frequently neglected element in the mechanisms of delocalization is labour shortages. Something of a vicious circle exists in shrinking labour-intensive industries, such as clothing and footwear. The downsizing of production entails a continuous contraction of the pool of skilled labour, workforce ageing and limited number of new entrants. This is underpinned by low attractiveness of jobs and negative perceptions of them, particularly by young people. As labour skills become scarce, the existing companies find themselves under further pressure to move out. The erosion of local capabilities and the decline of production capacities are inseparably linked with one another, and may lead to the demise of a particular economic activity in certain places.

The net employment effects of delocalization within the EU are rather positive, at least in the mid-term. First, thanks to the Europeanization of labour-intensive industries more jobs remain within Europe, instead of moving to other parts of the world. Second, delocalization facilitates lower unemployment in the new member states to a far greater extent than it contributes to higher joblessness in the developed areas, where more alternative employment opportunities exist. Finally, a substantial part of manufacturing jobs and related improvements in skills and capabilities stemming from relocation from Western to CEE go to peripheral regions of the latter and to underprivileged social groups, especially women.

On the whole, the social characteristics of delocalization processes can hardly be interpreted as the 'race to the bottom' in terms of wages and employment conditions in the labour-intensive activities in the EU which are the subject of this research. There is also little evidence for the 'social dumping' hypothesis concerning CEE countries. Earnings in the industry are generally on the rise, temporary and part-time employment has little significance, although slightly more in electronics than elsewhere. This trend may be primarily interpreted as an effect of the regulated environment of the EU, which prevents a 'race to the bottom'. EU and national regulations create a stable environment, which entails additional costs for companies but provides them with the favourable conditions that allow them to avoid costs of uncertainty and instability. In addition, this situation may be supported by the high level of economic development of the EU–15 and overall improvement in the new EU member states since the 1990s, which leads to greater social expectations and pressure on companies and public authorities. Finally, the enterprises may also tend to 'behave' better in Europe/the EU ('at home') than in LDCs in other parts of the world. The low-road approach of suppressing wages and employment standards would not stop delocalization in the situation of low attractiveness of work in labour-intensive activities and increasing competition of cheaper producers from outside the EU.

References

Begg, R., Pickles, J. and Roukova, P. (2000), 'A New Participant in the Global Apparel Industry: The Case of Southern Bulgaria', *Problemi na Geografiata* 3/4, 121–52.

Crestanello, P. and Dalla Libera, P. (2003), 'International Delocalization of Production: The Case of the Fashion Industry of Vicenza. Modena', paper presented at the Conference on Clusters, Industrial Districts and Firms: The Challenge of Globalisation. Conference in honour of Professor Sebastiano Brusco Modena, Italy, 12–13 September 2003.

Danson, M.W. (2001), 'The True Scale of the Regional Problem in the UK', *Regional Studies* 35:3, 241–46.

Domański, B. (2005), 'Transnational Corporations and the Postsocialist Economy: Learning the Ropes and Forging New Relationships in Contemporary Poland', in C. Alvstam and E. Schamp (eds), *Linking Industries across the World: Processes of Global Networking* (Aldershot: Ashgate), 147–72.

Eliason, M. and Storrie, D. (2003), 'The Echo of Job Displacement', William Davidson Institute Working Paper No. 618, <http://ssrn.com/abstract=455420>, accessed 5 June 2008.

Ettlinger, N. (1999), 'Local Trajectories in the Global Economy', *Progress in Human Geography* 23, 335–57.

European Commission (2006), *Employment in Europe 2006* (Luxembourg: The European Commission).

Feenstra, R.C. (1998), 'Integration of Trade and Disintegration of Production in the Global Economy', *Journal of Economic Perspectives* 12:4, 31–50.

Feenstra, R.C. and Hanson, G. (2001), 'Global Production and Rising Inequality: A Survey of Trade and Wages', National Bureau of Economic Research, Working Paper No. 8372.

Ghose, A. (2003), *Jobs and Incomes in a Globalizing World* (Geneva: ILO).

Glasmeier, A.K. and Jensen, J.B. (2001), 'Big Firms and Economic Development: Revisiting Works by Bennett Harrison', *Antipode* 23, 49–71.

Gradev, G. (ed.) (2001), *CEE Countries in the EU Companies' Strategies of Industrial Restructuring and Relocation* (Brussels: European Trade Union Institute).

Hadjimichalis, C. (2006), 'The End of Third Italy as we Knew it?', *Antipode* 32:1, 82–106.

Helg, R. and Tajoli, L. (2004), *Patterns of International Fragmentation of Production and Implications for the Labour Markets* (Michigan: Gerald R. Ford School of Public Policy, The University of Michigan).

Hijzen, A., Görg, H. and Hine, R. (2003), *International Fragmentation and Relative Wages in the UK*, Discussion Papers 717 (Bonn: Institute for the Study of Labor).

Hudson, R. (1989), *Wrecking a Region: State Policies, Party Politics and Regional Change in North East England* (London: Pion).

Hudson, R. (2005), 'Rethinking Change in Old Industrial Regions: Reflecting on the Experiences of North East England', *Environment and Planning A* 37, 581–96.

International Labour Organization (2004), *A Fair Globalization: Creating Opportunities for All* (Geneva: World Commission on the Social Dimension of Globalisation).

Krugman, P. and Lawrence, R. (1994), 'Trade, Jobs and Wages', *Scientific American* April, 22–27.

Lorentowicz, A., Marin, D. and Raubold, A. (2005), 'Is Human Capital Losing from Outsourcing?: Evidence for Austria and Poland', University of Munich: Discussion Paper 2005, 22.

Maskell, P. (2001), 'The Firm in Economic Geography', *Economic Geography* 77, 329–44.

Mingione, E. (1997), 'The Current Crisis of Intensive Work Regimes and the Question of Social Exclusion in Industrialized Countries', Discussion Paper Wissenschaftszentrum Berlin Für Sozialforschung Social Science Research Center Berlin, FS I 97-1.

Morcos, J. (2003), *International Subcontracting Versus Delocalization?: A Survey of The Literature and Case Studies from the SPX Network* (Vienna: United Nations Industrial Development Organization).

Phelps, N.A. (2000), 'The Locally Embedded Multinational and Institutional Culture', *Area* 32, 169–78.

Pickles, J. (2001), 'There are No Turks in Bulgaria: Violence, Ethnicity, and Economic Practice in the Border Regions and Muslim Communities of Post-Socialist Bulgaria' (Halle/Saale: Max Planck Institute for Social Anthropology, Working Paper No. 25).

Relocation, an Element of Industrial Dynamics: A Study about Relocation, Innovation and Employment (2000), Universities of KUL and UCL and the Federal Planning Bureau, <http://www.plan.be/admin/uploaded/200605091448005.OPDE20001en.pdf>.

Restructuring and Employment in the EU: Concepts, Measurement and Evidence (2006), European Foundation for the Improvement of Living and Working Conditions, <http://www.emcc.eurofound.eu.int/publications/2006/EF0638EN.pdf>, accessed 5 June 2008.

Rowthorn, R. and Ramaswamy, R. (1997), *Deindustrialization – Its Causes and Implications* (Washington: International Monetary Fund).

Waddington, D. and Parry, D. (2003), 'Managing Industrial Decline: The Lessons of a Decade of Research on Industrial Contraction and Regeneration in UK and Other EU Coal-producing Countries', *Mining Technology: IMM Transactions Section A* 112:1, 47–56.

Chapter 5

Between Policy Regimes and Value Chains in the Restructuring of Labour-Intensive Industries

Ivaylo Vassilev, Grzegorz Micek, Artemios Kourtesis

Introduction

The notion of governance has come into prominence in the context of global economic, social and political restructuring where one of the key changes is that coordination is no longer the exclusive domain of states. Indeed broad social processes are becoming increasingly embedded in much more complex institutional arrangements that are organised around diverse spatial scales (subnational, national, supranational). From the perspective of industrial production, this phenomenon raises challenges of coordination across spatially and institutionally distant sites, while from the perspective of the state, the challenge is to establish, within its territory, relatively stable couplings of the increasingly globally-mobile capital flows and the largely immobile labour.

Thus the notion of governance has been discussed in relation to spatial embeddedness of social relations, the challenges of integration in the context of global production and distribution, value creation and inequalities in its appropriation and distribution, upgrading and development, and even more broadly in relation to the restructuring of global capitalism. The notion of governance could be broadly defined as a set of mechanisms and practices that shape collective decision-making roles, procedures and relationships, within the framework of multiple and interrelated domains. These relationships could be highly formalized, as in the case of national jurisdictions, but could also be informal, as in the case of interorganizational agreements, codes of practice, and so on. In the context of this study, we define governance as 'any form of coordination of interdependent social relations that could include three general forms: anarchy of exchange (markets), hierarchy of command (for example, state) and heterarchy of self-organization (for example, horizontal networks)' (Jessop 2002).

The main aim of this chapter is to address questions related to the governance of the processes of delocalization of labour-intensive industries, with a specific emphasis on the mechanisms of public governance and especially the changing role of states, the EU and other non-state actors. Further, we view governance as a multi-level process, where boundaries are often shifting and thus actors as well as objects of governance are constantly being created and reshaped. The EU, for example, is a prime example of a multi-level governance structure where authority and policy-making are decentralized and operate across multiple social domains, subnational, national and supranational (Marks et al. 1996).

We argue that while delocalization constitutes a key economic conundrum as well as a political and social concern, delocalization, as such, is not an appropriate object of governance. While the powers of the states to influence processes within their own territories are reduced, at the same time states are also acquiring new powers of coordinating, or steering, that influence other levels of governance (both subnational and supranational) (what Jessop (2002) calls meta-governance). Thus we argue that delocalization-related issues need to be addressed within a broader social and economic agenda where the role of an active state remains crucial.

In our analysis we draw on a broad set of literatures (economic geography, development studies, GVCs, economic sociology and political economy) and study the interrelationships of state-centred (territorialized) and industry-centred (networked/de-territorialized) perspectives. In addition, while we focus on the same set of key players (states, TNCs, EU and global governing bodies), we argue that the two perspectives offer us different insights into the significance of these players for the coordination of the relations in the four industries that we study. There are a couple of analytical distinctions that we use in the analysis: public/private and facilitative/regulator/compensatory governance. We apply these distinctions through the interpretation of a combination of secondary and primary data.

In the first part of this chapter we address some of the key issues related to questions of governance and delocalization, and in the second part we offer a discussion based on original empirical research. The study is based on the analysis of secondary data, in-depth interviews with key representatives of business, politicians and academics and a survey among managers of enterprises in four labour-intensive industries – clothing, footwear, software and electronics. The main issues addressed in the chapter relate to employment conditions, taxation and tariff barriers and state aid schemes. We ask about the ways in which different players respond to existing governance structures, which actors are responsible for setting and enforcing rules, how the rules are related to trade, production, consumption, labour relations and the environment, and at what levels they operate. Policy implications, particularly on the national and EU levels, are central to our analysis but we discuss these questions in relation to the broader process of economic, political and social restructuring.

Literature Review

Restructuring of the state

The dominance of the nation state over the past two centuries has equipped it with an impressive array of both responsibilities and tools of governance. National institutions are actively involved in the wage-setting process; they regulate working conditions, working hours and the cost of overtime, vacations, holidays and sick leave; they restrict child and forced labour, and ensure non-discrimination; they provide unemployment, disability and retirement income insurance, and in many countries health insurance; they set the conditions for hiring and firing, unionization and collective bargaining (OECD 2003; World Bank 2004; European Commission 2005a). After WWII and until the mid-1970s, the state had a central role in economic restructuring and 'modernisation' (Stone 1965).[1] After the oil crises of the 1970s and especially after the collapse of the Soviet block in the late 1980s, the state-centred model came under attack, particularly through the growing influence of international organizations such as the IMF, World Bank and OECD, which actively promoted neo-liberal approaches to addressing both economic and social issues. The immense power of international organizations to influence national policies through the levers of financial assistance were evident in the adoption of overwhelmingly neo-liberal reforms by post-socialist governments in the first part of the 1990s. While the neo-liberal agenda continued to be dominant in the later periods, it was moderated through an active involvement on the part of the EU, which favoured a stronger institutional approach in which markets needed to be built rather than assumed to emerge naturally, thus addressing questions of embeddedness and path-dependence.[2] Therefore, as internationalization and deregulation advance at a global level, the centrality of the nation-state is destabilized and governance issues are 'spilling over' its boundaries.

Jessop (2000) conceptualises these changes at the level of the state as a move from a Keynesian Welfare National State (KWNS) towards a Schumpeterian Workfare Post-National Regime (SWPR). Thus he argues that the restructuring of the state can be discussed in relation to: 1) the objectives of state regulation (from a state focused on intervention (Keynesian) to a state focused on creating conditions for competitiveness (Schumpeterian); 2) the move from being the main provider of welfare towards shifting responsibility to individuals; for example, linking benefits to work and/or actively looking for work (workfare); 3) the move from centrality of regulation within national (state) boundaries towards the growing significance of different levels of governance that can be subnational (such as

1 The state led development was not only imitative, but also built on a response to local circumstances. The case was made for restructuring the economy towards 'inward directed' industrial development on the basis of import substitution (Todaro 1994).

2 On issues related to the role of the EU, see for example Kohsaka (2004) and Rodrik (2004).

regions or cities for example,) and supranational (such as the EU, IMF, World Trade Oganisation (WTO) and others).

Within this context the state is becoming one among many other centres of governance (although still probably the most important one). Jessop suggests conceptualising this change as a move from 'national' (space) to 'post-national' (space), where the centrality (sovereignty and other issues) of the state is destabilised and therefore states are increasingly moving towards what he calls 'regimes' of governance, where the powers of the states to influence processes within their own territories are reduced (thus a move from government towards governance). However, the latter process is paralleled by another in which states gain significance and acquire new powers of coordinating, or steering the new levels of governance (both subnational and supranational), and he calls this meta-governance. Within this context, governance concerns the establishment and shaping of local and global mechanisms in addition to national ones, as well as the interrelation between local, national and global institutions. Furthermore, the increasingly complex dependencies between different scales (Jessop 1998) as well as the wide variety of actors operating at different levels create new challenges and require the adoption of new strategies on the part of the key players.

State strategies of adaptation

Depending on how conflicts are negotiated, Jessop offers a typology of four types of state restructuring: neo-liberal, neo-statist, neo-communitarian and neo-corporatist (Jessop 2002; see Table 5.1).

These strategies are incorporated within, for example, active regional policies on the part of the states (as well as on the part of the EU). Thus in many countries there is an explicit national policy for the development of their less-developed regions, aiming to improve infrastructure and create the conditions that could encourage investment. These policies usually include infrastructure works, state-aid schemes and investment incentives, and the provision of tax and social insurance benefits to potential investors. Further, the role of regions has substantially increased in almost all EU member states over the last two decades. Regional (but in certain cases also local) authorities have become more active in shaping and implementing their own policies, particularly with a view to attracting investment and promoting innovation in their regions. As competition between regions (or local areas) gradually increases, local actors adopt more ambitious approaches to improve their competitiveness (Wallis 1996; OECD 2001b) particularly by stressing the role of knowledge and innovation (Cooke 1996; Lundvall 1997), R&D (regional) technological infrastructure, support for higher education, and so on.

Table 5.1 Governance types in SWPR

Neoliberal	Neocorporatist	Neostatist	Neo-communitarian
Promote free competition	New balance between competition and cooperation	Moving away from state control and towards regulated competition	Limit the role of free competition
Reduce role of law and state	Decentralized 'regulated self-regulation'	Guiding the national strategy rather than planning at the top	Enhance the role of the third sector
Sell off public sector	Widen range of private, public and other 'stakeholders'	Auditing performance of public and private sectors	Expand the social economy
Market proxies in residual public sector	Expand role of public-private partnerships	Public-private partnerships but under state guidance	Emphasis on social use-value and social cohesion
Free inward and outward flows	Protect some core-economic sectors but in an open economy	Neo-mercantilist protection of core economy	Fair trade rather than free trade; think Global, act Local
Lower direct taxes and increase consumer choice	High taxation to finance social investment	Expanding the role for new collective resources	Redirect taxes: citizens' wage, carers allowances

Source: Adapted from Jessop 2002.

Restructuring of EU institutions

The EU is also characterised by multi-level governance (Marks 1993; Hooghe 1996), where policy-making and sovereignty are shared across different levels (subnational, national and the EU). The rather complex nature of EU governance mechanisms is reflected in the diversity of solutions adopted in different policy areas. Thus while in areas such as competition, trade or agriculture, for example, the European Commission (EC) is the responsible body for the implementation of the respective policies, in other areas (for example, education) member states still retain their independence in formulating and implementing national policies.[3] Given the strong reluctance of certain member states to yield more powers to supranational institutions, the Commission adopted a new form of governance, the open coordination mechanism. According to this, the EC provides an analysis of the

3 These are obviously not the only possible cases. For example, monetary policy for the Euro-zone is left exclusively in the hands of the European Central Bank, while in a number of areas there is shared responsibility between the member states and the EU institutions (environment, employment, structural policies, and so on).

situation in the areas concerned and sets specific, quantified targets for the member states to meet. It is then left to the member states to formulate and implement the policy measures that they deem appropriate in order to achieve these targets, while the Commission ensures proper monitoring and facilitates benchmarking between the various national policies.

In the case of delocalization of firms, a number of policy areas (for instance, competition, structural policies, social protection and labour-market issues) are important. In the case of the industrial policy the role of the Commission is rather limited, while policy formulation and implementation lie primarily with the member states, given that industrial policies are still under their competence. The open method of coordination now offers the framework in which national policy performance can be discussed, developed and improved. In this context, and with a view to support the process of structural change in the EU, the Commission has outlined a set of specific measures which cover regulatory framework, community policy and combine sector level policies.[4]

Similarly, employment policy remains primarily in the hands of national governments and the EU intervenes under the open coordination mechanism. The Commission has proposed a European employment strategy around three priorities: 1) boosting labour-market participation; 2) improving the adaptability of workers and companies; and 3) investing more in human capital (European Commission 2005b). In the context of delocalization, the critical issue is the creation of new activities and jobs and the shifting of resources from declining sectors to sectors where the EU can sustain a comparative advantage.

Given that differences in tax systems can influence the decision of firms to delocalize, taxation policy is an area that merits some further discussion. Thus for example, while there is no explicit EU taxation policy, there is an on-going process of co-ordination and harmonization of national tax policies among countries within the EU. To date, major steps toward harmonization have been achieved in the field of indirect taxation, notably the abolition of customs duties, the introduction of the Community Customs Code and common VAT system and the harmonization of the most important excise duties. In the area of direct taxation, the 12 new EU members have an average corporate tax rate of about two-thirds of the old 15 EU members, which could (only) partly explain the delocalization of industries from Western to CEE.

4 Regulatory framework arguments support an idea that burdens on industry must be reduced to the bare minimum of what is strictly necessary to achieve objectives of regulation, and a balance must be struck between industrial competitiveness and the need for regulation. Community-policy arguments stress that synergies must be better exploited to improve the policies' impact on industrial competitiveness (developing a knowledge-based economy and strengthening cohesion). The combination of these policies means that the EU must continue to develop the sectoral dimension of industrial policy while ensuring that its sector policies strengthen industrial competitiveness (European Commission 2004).

Global governance deficit and private governance

Globalization of trade and production on the one hand and their integration across the globe on the other, raise the question of coordination of activities across space (Feenstra 1998; Gereffi et al. 2005). Within this context the role of TNCs becomes increasingly important. For example, transfers of technology within large TNCs is not a mechanical process but requires adaptation to different legal frameworks and cultural practices, which in turn are important factors for the success of their internationalization (Johansson and Vahlne 1977, 1990; La Porta et al. 1998; Djankov et al. 2002; Glaeser and Shleifer 2002). The dominance of the nation-state has led to a situation in which governance at both the subnational and supranational level is less regularized, which in turn means that private actors (mostly, but not solely, enterprises) are dominant. Gereffi et al. (2005) analyse different modes of relationships along the value chain (market, modular, relations and hierarchy), where the focus is on TNCs and their suppliers. However, the governance of the value chain is not restricted to the role of companies and involves a broader set of actors (Humphrey and Schmitz 2000, 2001, 2002; Nadvi and Waltring 2002) including, *inter alia*, NGOs, governments, standard organizations, among others. This ranges from a forum of discussion (for example, the OECD (see OECD 2001a, 2005)), voluntary agreements (such as in the UN 'Global Contract') and consultations (for example, International Labour Organization [ILO] conventions [see ILO 2004, 2005]) to binding agreements (for example, the WTO and IMF). Thus the WTO, for example, plays a central role in governing the global institutional trade order, and certain WTO agreements penetrate deeply into the realm of jurisdiction and influence domestic rules and regulations directly. This situation often creates tension between the WTO system and national sovereignty (Matsushita 2004). The IMF and the World Bank are more narrowly focused, offering stabilization and structural adjustments programs, although these are always conditional on adopting certain policies.

Focusing on governance mechanisms along global commodity chains (GCCs), Gereffi and Mayer (2004) suggest that governance can be usefully discussed in relation to the main domain within which they operate,[5] private or public, and the main functions that they serve within the organization of GCCs, facilitative, regulatory and compensatory (see Table 5.2). Here the facilitative function is associated with, among others, the establishment of property rights, enforcing contracts, establishing rules of fair competition and providing information. The regulatory function is concerned with controlling the negative externalities of markets such as environmental pollution, exploitation of workers and so on. Finally, compensation refers to mitigation of the tendency of markets to produce highly unequal distribution of outcomes and the term includes social insurance,

5 Humphrey and Schmitz (2001) address the question of governance by asking how parameters are set and then enforced: are they firms within the chain (for example, the lead firm) or external entities that are enforcing them?

health care, education and retraining, progressive tax systems and other welfare policies. Nevertheless, we would like to point out that the boundaries between these functions are not always clear. For example, competition policy, classified under the 'facilitative' function, could also be part of the 'regulatory' function, as it provides the overall legislative framework for both private actions (for example, mergers and acquisitions) as well as public policies (for example, state-aid schemes, investment incentives law and so on).

Table 5.2 Modes and realms of governance

Governance modes	Governance domains	
	Public domain	**Private domain**
Facilitation	Property rights Banking and commercial law Competition policy	Market ideology Professional codes and norms
Regulation	Labour law Environmental and health and safety regulations	Voluntary codes of conduct Corporate social responsibility commitments and practices Pressure groups and consumer boycotts
Compensation	Social insurance Education and retraining programmes Public health policies	Collective bargaining arrangements Philanthropy

Source: Adapted from Gereffi and Mayer 2004.

More specifically, Gereffi and Mayer (2004) argue that economic globalization is associated with a growing governance deficit. The 'thinness' of institutions becomes evident by observing the strengthening of facilitative institutions both internationally (for example, the GATT and IMF) and at the national level, compared to the weakness of regulatory and compensatory institutions globally and their gradual erosion in DCs. Furthermore, facilitative institutions are mostly associated with the regulation of capital flows and are not applicable to labour, where migration flows are generally governed at the level of the state. They argue that the tendency, however, is for an extension of the functions of global governance mechanisms, or what they call thick (as opposed to mainly facilitative functions in thin) governance.

In the following section, drawing on Gereffi and Mayer (2004), we offer some further in-depth discussion of the ways in which these tendencies vary across industries and countries. Moreover, building on Jessop's distinction between KWNS and SWPR we discuss different aspects of state strategies in relation to labour markets and industrial structure of the five countries studied.

State Policies and Governance

Labour markets and state policies

Wages and labour cost Table 5.3 shows the average labour cost and the minimum wage per month in each of the countries covered by the study, for selected years in the period 1996–2006. As expected, the old member states have a significantly higher labour cost than the new ones, with the UK average labour cost being five to six times higher than that of Poland and Estonia and up to 18 times higher than Bulgaria. Greece has an average labour cost of slightly less than €2000 per month, which is roughly three times higher than Poland and Estonia and nine times higher than Bulgaria. The large discrepancies observed in the average labour cost between old and new member states have undoubtedly played a significant role in the decision of firms to delocalize. In traditional industries in the UK and Greece, high labour costs were the most important factor in delocalizing to lower labour-cost countries (mentioned by 67.8 per cent of companies in clothing and 66.7 per cent in the footwear industry). Many Polish clothing companies were forced to become involved in outward processing trade (OPT) due to the high costs of labour and of mandatory social security in particular. Polish and Greek footwear managers argue that 'a lesser burden from social security could stimulate companies to employ new people, such as designers, sales representatives which [sic] would enable them to escape from the vicious circle of OPT dependency'. Labour-cost influence was less imminent in software and electronics. However, in electronics, the sensitivity to labour costs depends very much on where the company is situated in the value chain. Producers from the lower part of the value chain (foreign consumer electronics companies) behave similarly to companies in traditional sectors.

Table 5.3 Monthly labour cost and minimum wage, 1996–2004 and 2003–2006

Country	Labour costs, EUR						Minimum wage, EUR	
	1996	2000	2001	2002	2003	2004	2003	2006
EU–25	2254.7	2732.3	2768.7	2864.2	2892.6	2979.1	–	–
EU–15	–	3154.4	3149.8	3252.7	3330.2	–	–	–
Bulgaria	–	179.0	189.8	193.6	202.2	213.5	56	82
Estonia	275.9	429.1	496.3	562.4	608.4	650.3	138	192
Greece	1446.3	1658.1	1739.9	1849.4	1984.3	–	605	668
Poland	447.1	672.4	791.8	783.1	698.2	699.2	201	234
UK	2168.8	3676.9	3793.4	3891.3	3642.4	3848.6	1106	1269

Source: Eurostat 2006a.

Flexibility Table 5.4 presents the main trends towards flexibility in the labour markets of the five countries under consideration.

Table 5.4 Trends in labour market policies

State	1980s	1990s and 2000s
United Kingdom	Deregulated flexibility	Partially deregulated flexibility
Greece	Deregulated flexibility	Partially deregulated flexibility
Estonia	Strongly regulated anti-flexibility	Partially regulated flexibility
Poland	Strongly regulated anti-flexibility	Partially regulated flexibility
Bulgaria	Strongly regulated anti-flexibility	Mainly unregulated flexibility

Source: Adapted from Wallace 2004.

In the UK, the flexibility regime is deregulated or partially regulated, and working time and employment conditions are negotiated in the workplace (Wallace 2003). There are very few restrictions to employing people on part-time and on short hours (Boje et al. 2007). Many new regulations introduced by the New Labour in the UK are minimal and originate from EU directives. New member states moved from a high degree of state control, where the policy was one of deliberate rigidity, towards different degrees of regulated flexibility. The heretofore achieved, regulated flexibility is low in the cases of both Poland and Estonia, while in Bulgaria there is a lack of coherent policies, and flexibilization is largely unregulated (Wallace 2004).

In the Bulgarian case, the positive role of employment conditions has been most important in clothing (for 30.0 per cent of enterprises) and the footwear industry (29.5 per cent positively).

Estonia and Poland belong to a different group of countries: traditional industries have been more sensitive to employment conditions but their impact is divided into positive and negative influences. This means that labour costs became a reason also for outward delocalization from these countries, more in Estonia and less in Poland. Labour conditions were less important in software and the electronics industry:

> I think, I feared it much more than the reality but in that the force of having a minimum wage imposed, being forced to accept flexible working, I would have seen as being limiting our ability to do this up here, before they arrived but in fact they are quite consistent with the changes that globalisation has demanded on the business anyway, and because of this and the way the business has changed we did abort the minimum wage and we are inclined to allow the flexibility because that's good for the business, the way that it is today. It would have been bad for business the way that it was five years ago but by the time it had become reality. So I think that is probably less of a negative than we would have perceived it was going to be. But I think the Government could do a lot less with this. (UK, electronics)

A main dividing line between northwestern EU countries and new member states (including Greece in this case) is the high degree of informal flexibility that applies in the latter group. This tendency is especially strong in countries such as Bulgaria, where labour markets are less regulated and additional contracts, casual work, and other types of 'atypical' employment are not unusual (Wallace 2004). In particular, the situation in certain peripheral areas can be quite different than what appears in the formal regulatory framework of the country.

Three of the industries that we study (clothing, footwear and to a lesser degree electronics) are located mainly in peripheral areas. The lack of alternative work opportunities in peripheral locations pushes the wage level down but also weakens the implementation of existing regulation. In such cases, the role of external buyers in the improvement of working conditions can be very important (private governance mechanisms).

Within a context where formal regulation is lacking or is poorly enforced, informal arrangements on the company level become essential. Thus, for example, many Asian companies invest in Poland in order to enter the EU market, to get access to cheap labour (although in reality finding appropriate people is difficult) and to learn. However, many of these companies (especially the large Korean, Indian and Chinese consumer electronics companies) dictate labour conditions and do not follow certain labour regulations.[6] More specifically there are cases where:

6 We have managed to establish a balanced and objective point of view through interviewing trade union representatives, company managers and local labour inspectors.

- When workers try to protect their rights, they are moved to low-paying Asian subcontractors of TNC which follow their main customers.
- In practice a working day lasts on average 2–4 hours more than a maximally permitted workday.
- During unplanned production breaks of several hours, workers must stand up, can not sit or talk.
- A temporary job is characterized by the absence of regulations in relation to time requirements, while selection is not related to qualifications.
- Common blackmail by medium-level Polish management exists: trade unionists are promised promotion if they resign from the union.

The above discussion clearly points to the strong private governance in this sector, where neither labour inspectors nor trade union representatives can influence company practices. TNCs shape the delocalization processes when entering CEECs, coming with their Asian subcontractors, who usually offer worse working conditions.

The degree to which private arrangements are significant also depends on the size of the informal economy, which varies between sectors (it is more important in clothing and footwear) and countries (it is more significant for new member states). In the new member states, for example, there is a clearly identifiable two-tier structure between companies operating in the formal and the informal economy, with significant differences in terms of opportunities, degree and process of enforcement of regulation. Thus while the legal framework regulating working conditions in the new member states is comparable to those of other EU countries, the presence of a large informal sector means that these regulations are often not implemented.

In this context it is often standards imposed by buyers that play a major regulating role, but poor working conditions are more often associated with international buyers that are only competing on price and operating on very slim profit margins. In this sense, poorer standards (as much as they guarantee lower labour costs) are one of the reasons that push the relocation of such buyers to new member states.

> An increasing number of companies is making efforts to introduce improvements, since these conditions form part of the order assignment contract. (Key informant, clothing, Bulgaria)

> The contractors were seriously interested in the employment conditions and working hours. (Key informant, footwear, Bulgaria)

> The Greek owners of clothing factories in Bulgaria do not adhere to the requirements of labour conditions as stipulated by law. They have come to Bulgaria to avoid the EU requirements related to labour condition. (Key informant, clothing, Bulgaria)

> The most important thing is the low cost of labour as well the existence of obedient labour (they do not complain for overtime, etc.). (Key informant, clothing, Greece)

Availability of labour Increased economic integration and free movement of labour have a strong impact on domestic labour markets, especially in small countries such as Estonia or Bulgaria, but recently also in Poland. In the case of Estonia, highly-skilled and reasonably-priced labour has been one of the cornerstones of rapid economic growth.

Companies, however, face increasing difficulties in finding appropriate employees, which may lead to either reduction or delocalization of production. All four of the analysed industries experience significant labour shortages. Thus 83.1 per cent of all companies interviewed report such shortages. The respective percentages were 97.5 per cent in the case of Bulgarian companies and only 57.3 per cent in the case of British firms. The most remarkable shortages were the the lack of seamstresses (in Bulgaria and other CEECs) and the scarcity of workers for large Asian TNCs investing in Poland. TNCs applied to regional and national authorities to obtain a permit to employ Korean workers. The lack of appropriate software developers in Poland and the need for some companies to import labour from Ukraine is a similar emerging phenomenon.

The question of labour scarcity is important in all four sectors. Sometimes it can be attributed to the level of wages, especially when there are alternative opportunities (for example, emigration); sometimes it is an issue of fast-rising aspirations and expectations. Additionally, there seems to be a decreasing number of people who would like to work in traditional industries. Against these trends, there seem to be four solutions for companies:

- Increase wages (a limited if not impossible option in industries that compete on price).
- Relocate within the country (many traditional industries locate in more peripheral areas where labour has lower expectations).
- Import labour (apply to regional authorities to import Asian labour, the cases of LG Philips in Wroclaw, Samsung in Czech Republic).
- Delocalize outside the country/EU.

Industrial restructuring and state policies

As a result of liberalization, globalization and integration of the markets, the international spill-over effects of national tax policies have increased, since differences in tax levels can have an important impact on investment flows.

Taxation and tariff barriers The tax burden is an important determinant in cost-related strategies of firms. All countries in our sample have lower than the EU–25 average tax burden. There was a tendency towards a decrease of a share of tax revenues in GDP in new EU members and an increase in Greece and the

UK. Estonia is a very clear example of a country with emphasis on taxation of consumption. The tax burden of labour is high due to high social tax. Taxation in the UK is biased towards taxation of capital, remarkably more so than in the rest of the EU countries (Table 5.5), while taxation of labour is relatively low in comparison with the other EU countries. The level of corporate income tax and of labour taxes has been considered an important factor supporting FDI and delocalization to Bulgaria, Estonia and Poland. The position of these countries' companies in the value chain also depends on the education level and R&D expenditure, a large part of which is financed through public expenditure (hence through tax revenues). Such a rationale leads to the conclusion that a low tax policy aimed at improving a country's competitiveness may create short-term advantages but may lead to potential negative consequences in the long run.

Table 5.5 Tax revenue and implicit tax rates by type of economic activity

Country	Tax revenue, per cent of GDP		Implicit tax rate on:					
			Consumption		Labour		Capital	
	1995	2004	1995	2004	1995	2004	1995	2004
EU–25*	39.7	39.3	21.1	21.9	35.7	35.9	23.1	25.8**
Euro area*	39.9	39.7	20.3	21.5	35.7	36.6	23.6	29.2**
Bulgaria	–	–	–	–	–	–	–	–
Estonia	37.9	32.6	20.3	20.8	39.2	37.6	17.9	10.3**
Greece	32.6	35.1	17.3	17.5	34.1	37.9	12.1	17.0
Poland	38.5	32.9	21.8	19.3	37.9	34.6**	–	19.4
UK	35.4	36.0	19.6	18.7	25.7	24.8	33.3	34.9

Note: * EU–25 and Euro area overall tax rates are computed on the basis of a GDP-weighted average, ** Figures for 2003.
Source: Eurostat 2006a.

Duties are also important in creating a competitive environment. Although many Polish and Bulgarian companies cry out for higher duties on Chinese products, others claim that exports and imports should become cheaper, mainly through decreased social charges. Managers state that such interventions (for example, introduction of higher duties on Chinese goods), in the absence of additional, more necessary measures related especially to the social burden on labour, will not resolve the problem of the closure of Polish footwear companies.

The Trade Barriers Regulation helps to develop trade with third-country markets. However, this tool is not very well-known: some entrepreneurs from electronics companies complaining about excessively low taxes for Chinese and

Thai goods and problems in exporting Polish products to South America were not aware of the existence of such a tool.

Regulation at the EU level is largely necessary in the case of textile products. Even relatively expensive countries such as Hong Kong (China), the Republic of Korea and Taiwan play a major role in apparel exports, because they still have access to large apparel quotas primarily issued by the USA and Western Europe (Gerreffi 2005). The role of the EC is to initiate safeguard investigations when imports rise above quotas. Additionally, punitive tariff duties should be imposed on goods after concluding that export countries had been paying hidden subsidies to their industries thereby allowing them to send goods to Europe at markedly lower prices than those in Asia. Such a procedure was successfully implemented in the case of imports of leather shoes from China and Vietnam.

Investment incentives to attract FDI and subcontracting activities Incentives affect investment decisions. The potential options include national, regional or local grants, tax credits, employment incentives, site or infrastructure improvements and others.

The results of our study show that governments play a modest role in undertaking activities to attract FDI or subcontracting. Only 18.7 per cent of respondents reported taking advantage of public initiatives. The highest positive answer rate was reported for Estonia with 37.4 per cent followed by Greece with 33.4 per cent, the UK 14.8 per cent, Poland 12.2 per cent and Bulgaria 2.0 per cent. In general, it must be stressed that it is mainly large foreign companies that exploit incentives.

By industries, the share of respondents receiving support for FDI or subcontracting was 24.3 per cent in the clothing industry, 22.4 per cent in electronics, 14.5 per cent in software and 8.8 per cent in footwear. There is no significant difference between new and old industries. The highest share of enterprises receiving government support was in Estonia's clothing (66.1 per cent) and footwear industry (45.5 per cent), the Greek electronics industry (42.9 per cent), the Polish electronics (23.8 per cent) and software industry (19.6 per cent) and the UK electronics industry (25.0 per cent). This support was not significant in terms of the amount of received funds.

Companies that take advantage of incentive schemes base their operations on the production of value-added goods significantly more often than other companies. These enterprises compete more often on quality, design and flexibility. They are more stabilized: they usually have contracts with subcontractors and less often plan to relocate activities abroad. Companies that benefited from governmental initiatives are more active as shown by (the statistically significant) higher number of service firms on subcontracting basis in 2007. On average, enterprises that take advantage of active initiatives undertaken by national/local government are more often foreign companies, which also give subcontracting to a company abroad. Statistical analysis carried out shows that foreign companies have significantly better relations with local and central authorities than indigenous firms. This fact leads to the statement commonly expressed in Poland and Bulgaria that 'it is only foreign companies which benefit from local and national incentives'. This is a *leitmotif* behind low involvement in some local and national incentives. Among foreign companies, TNCs particularly benefit from national or local programmes. These large enterprises choose among localities that offer suitable conditions and the largest exemptions. In addition, their lobbying efforts in favour of particular bills, legislation and other forms of regulation should not be neglected. This phenomenon supports the thesis about growing private governance.

Global changes and the role of the state

Governance from the GCC perspective There are similar tendencies in all four industries regarding the growth of what Gereffi and Mayer (2004) call a 'governance deficit', that is, the disparity between the degree of internationalization and global market development on the one hand, and the regulation and compensation mechanisms on the other. The findings provide indications of some relevant tendencies including, for example, the degree to which ISO standards are significant for companies, although the functions that they fulfil are difficult to assess. Thus a significant number of interviewees argued that they were disappointed by the negative effects that ISO requirements had on their businesses; their main function appeared to be meeting administrative requirements for government quotas and/or funding rather than facilitating their companies' position in the market.

The findings also suggest that the strongest private arrangements are being made on the one hand between global buyers who dominate the rules of production and trade, and on the other hand subcontractors and subsidiaries aiming to meet quality-standard requirements.

Both companies and key informants felt that governments were not making sufficient efforts to support industrial restructuring, especially in terms of funding, education and retraining, upgrading and access to external markets:

> Biggest trouble is with workers. Technical University is also not preparing enough graduates. Education is main issue and our headache. Situation is almost crazy with hiring. (Estonia, electronics)

> The government is now talking a lot about innovation and using all the right words, but they are not helping businesses to work out how they can do that; because if you think about manufacturing in the last 50 years it's gone through different phases and essentially what it has been doing over the last 20 years is all about lean and low cost and that's great now that isn't going to be enough, but that's what businesses have done and they have probably cut cost to the bone. They've cut out expensive people and capabilities, and the capabilities they have kept in are for keeping the cost down, not necessarily for innovation and developing the business and that is a real issue. (UK, electronics)

With the exception of software, the four sectors discussed here are largely considered to be of low priority for governments. In this respect, there is almost no difference between new and old EU members.

In footwear, private arrangements appear to be the most essential ones. Most interviewees strongly agree that there is a lack of specific government policies supporting the footwear industry in terms of funding, education (reskilling), upgrading and access to external markets. Like clothing, the footwear sector is not considered to be a priority in the countries studied, a situation which results in little or no government support. Footwear companies present the lowest percentage in receiving outside assistance (in electronics 58.8 per cent, in clothing 55.6 per cent, in software 45.5 per cent and in footwear 31.6 per cent). These numbers could be higher, as over one-third of managers argue they were trying to obtain EU funds, but were surprised by the 'length of the complicated process' and consequently gave up or didn't manage to get them.

In electronics there are many environmental standards that regulate the position of firms in the value chain. Some companies complain they are stuck in a low position within the value chain due to the high cost of introducing these standards. Likewise, in clothing, a significant number of interviewees argued that they were disappointed by the negative effects that ISO requirements had on their businesses. In software, it is the largest software companies (for example, Microsoft, Oracle, SAP) that dictate rules and standards with states having little power in governing the software sector. In clothing, there is a tendency for global and supranational institutions as well as TNCs to dominate the facilitative and the regulatory domains (Gereffi and Mayer 2004):

> Significant regulation is mainly on the national level, regional regulation is quite insignificant. While there may be some variations between England and Northern Ireland there won't be such differences, but they could be for rules for company registration for example. But even there now with the possibility to register as a 'European company' they can be registered in any EU country. (Key informant, UK)

While the role of governments is most significant at the level of regulation and compensation, all four industries that we studied call for different forms of state support. Thus in terms of market support for example, clothing and footwear companies mostly demand facilitation of links with external partners and support in entering foreign markets. In contrast, in the case of electronics and software, it is measures related to the domestic market that appear to be more significant. For example, procurement and stimulating cooperation between TNCs and local companies in large projects could engender capacity building and organizational learning.

Public arrangements appear to dominate the facilitative domain, while private and public arrangements both have influence on the regulatory domain across all countries studied. There is, however, a substantial apparent difference between new and old member states in terms of compensatory arrangements. Such arrangements are dominated in the case of the new member states by government regulation, while in old member states, and especially the UK, they are determined on an equal level by government regulation and private arrangements.

> Yes, I mean when they brought in the minimum wage thing, that was because it was all going to go to the minimum wage but I couldn't just pay the girls the bottom line of the minimum wage, I had to increase it, even into the admin staff. It costs us a lost more than £1 an hour in the loss of earnings; it is a lot of money. And to be fair, most, none of our girls were earning the minimum wage anyway, I had to proportion a wage increase to compensate for it. (UK, small producer)

Quality standards in all four sectors are mostly arranged privately and are dominated by TNCs, whilst the terms of trade are determined both by governments and (increasingly) by global frameworks and regional agreements. For example, government regulations concerning labour conditions, health and safety are to a large extent enforced through the demands made by TNCs. International certification also acts as a mechanism for attracting foreign buyers. The impact of certificates varies between industries with private certification from global players being a very important form in software, while in electronics there is a combination of international certification and approved partners lists. However, the latter is also significant for software.

The changing role of the state: From a KWNS to a SWPR?

While the above discussion sketches the overall tendencies in the four industries, there are differences in the specific role that individual states play and the concrete strategies that they adopt. These differences depend on a number of factors, such as the degree to which governments see their role as one of controlling outcomes (delocalization, for example) as opposed to intervention on the supply side, the mechanisms in place of negotiation (for example, presence of forums for negotiation between government, business and labour, and the balance of power), the definition of core and peripheral industries and regions, the degree to which policies of social cohesion are considered important and the degree and forms of inequalities that are deemed acceptable.

The availability of incentives is not sufficient to have an effect on its own. Other important factors are active interest in such incentives and the ability to utilize them. For example, footwear attracts a certain type of FDI in Bulgaria, where the labour force is very small in a mature industry. Furthermore, programmes to create incentives to train people for clothing and footwear need to take into consideration the lack of interest of young people in working in those industries, and finally the use of funds available to companies depends also on the managerial and entrepreneurial abilities of the managers. Thus government policies are embedded in structural factors and depend on existing attitudes within the country and the perceptions of the country from outside. These elements vary significantly between the countries and industries. This situation further emphasizes the significance of specific, highly targeted policies that are built on a well-developed, ongoing communication between involved parties as well as the existence of and ability to implement highly localized ad hoc solutions.

This sort of analysis runs counter to suggestions that the market should be allowed to rule while states should limit their involvement to a position of providing overall stability of the business conditions. While there is a significant shift away from the state that intervenes directly, this move does not necessarily lead to a passive state. While such a position could be taken (what Jessop calls neo-liberal SWPR), it is neither the only one available nor the most efficient one.

In addition, we argue that macroeconomic parameters alone cannot determine competitiveness. While states can choose to prioritize certain macroeconomic criteria, there is nothing intrinsically and universally 'good' about them. Instead, such decisions can be part of a strategy, but they only lead to competitiveness if they lead to differentiation from the conditions that operate in other countries. This distinction is crucial, because it shifts the question from 'what is best' to 'if we do that what follows next', more specifically, what sort of businesses is an open economy likely to attract, how long are they likely to stay, what are the consequences for the areas where they operate, among others?

All five countries in our study gravitate around liberal and neo-liberal state strategies. Poland and Bulgaria have mostly moved between different KWNS models: from statist towards a liberal strategy, Greece is a combination of KWNS-liberal and SWPR-neo-liberal strategy. It is the UK and Estonia that seem to have the strongest orientation towards supply-side intervention, with the UK demonstrating some forms of state guidance for certain market segments (see Figure 5.1).

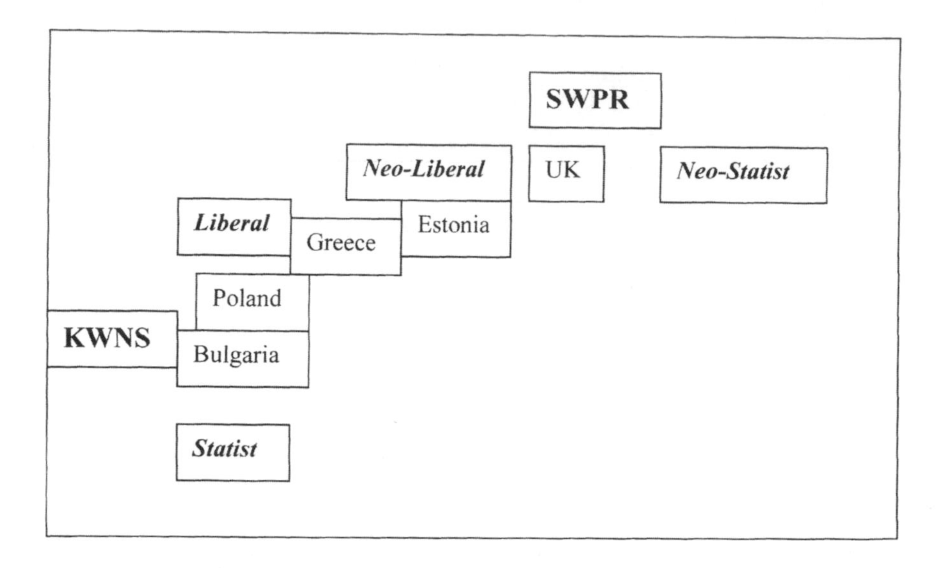

Figure 5.1 State strategies
Note: This Figure draws on Bob Jessop's distinction between KWNS and SWPR that was discussed earlier in the chapter.
Source: Compiled by the authors.

Conclusions and Policy Recommendations

The notion of governance has come into prominence in the context of global economic, social and political restructuring, where one of the key changes is that coordination is no longer the exclusive domain of states. We have argued that when applied to the process of delocalization these changes can be usefully discussed: first, from the perspective of industrial reorganization in terms of the new challenges of coordination of production across spatially and institutionally distant sites; and second, from the perspective of the state, where the challenge is to establish relatively stable couplings of the increasingly globally mobile capital flows and the largely immobile labour within its territory. We hold that governance needs to be discussed as a dynamic and multi-level process where actors and their strategies are interdependent and constantly reshape each other. In this sense, issues related to delocalization and its consequences need to be addressed within a broader social and economic agenda where the role of an active, though not necessarily only and always directly intervening, state is crucial.

More specifically, delocalization trends could be influenced, and to some extent shaped, through active industrial, social, labour and tax policies. In addition, while addressing the negative consequences of delocalization remains an important policy objective, increasingly such interventionist measures need to be combined with policies that are focused on the supply side and are more about enabling adaptation rather than about short-term responses to crises. Active policies in terms of education and retraining are a prime example of such policies.

Employment conditions continue to be mainly regulated on the national level. Traditional industries such as clothing and footwear are more likely to experience a strong negative impact of worsening employment conditions, and thus employees in these sectors tend to work in a much more uncertain environment compared to employees in electronics, and especially software, companies.

It is possible to argue that major market economies (the UK and Greece) rely less on EU funding and institutional support but get the predominant part of their support from domestic sources. Regarding industries, electronics was more strongly supported in terms of value, but the difference with other industries was not large in terms of the number of companies. The support was more country- than industry-specific.

As expected, our data demonstrates that, depending on whether a country is mainly involved in outward or inward delocalization, its priorities and the expectations of its business community differ. Thus support for product development and marketing is most important in the case of outward delocalization, while inward delocalization is more closely related to concerns over the creation of new workplaces, infrastructure development and training. The impact of trade unions and business associations on company decision to delocalize is practically absent in Poland, Bulgaria and the UK, and very limited in Estonia and Greece. In addition, non-tariff barriers have only a modest influence on company decisions. Our analysis clearly indicates that there is a growing significance of private governance mechanisms, particularly in terms of the great impact that TNCs have in shaping the labour markets and regional competitiveness. This tendency is strongest in the consumer electronics sector. This conclusion is consistent with what Gereffi and Mayer (2004) call 'a governance deficit' at the global level, where mostly contingent, private arrangements are very significant. There is, however, an ongoing process of thickening of governance mechanisms, which is apparent at all levels of governance. Interdependencies are especially visible in, for example, the case of TNC investment in CEECs where regulation is simultaneously coordinated at the global, regional and national levels, as well as by the EU. EU institutions also have an important role to play in extending the scope of supranational governance mechanisms to include regulation, as well as extension, of the links between different institutional levels within the EU. In all these measures, the EU has an important role to play in offering a socially engaged alternative to the dominant neo-liberal form of globalization and state restructuring.

Stability and predictability of the business environment are two of the most often quoted reasons for choosing a location within the EU. It must be argued that stable labour and duty regulations reduce uncertainty for foreign clothing, electronics and software companies coming from outside the EU.

It has been demonstrated that mainly TNCs benefit from government incentives and outside assistance. Our evidence suggests that subsidising foreign companies in labour-intensive industries may only have short-term positive effects. Some of the strategies that appear to be successful include offering support to companies that have the potential to upgrade, implementing and strengthening after-care policies and enhancing the local links. Such solutions, however, should not be taken at face value and mechanically implemented, as they raise both questions of economic rationality and social justice. This point raises the broader question regarding the appropriate and socially acceptable mechanisms of relatively stable reconciliation that states might establish within their territories, between the increasingly globally mobile capital flows and the largely immobile labour force.

The potential of subcontracting and outsourcing has already reached its limits in footwear in Greece and Poland as well as in clothing in Greece, Estonia and Poland. In other cases it is hampered, for example, in Bulgarian footwear, Polish and Bulgarian software Estonian and Polish electronics. Thus, from the perspective of strengthening longer-term competitiveness, the key question for CEECs is about attracting higher value-added activities. While capacity-building as opposed to different forms of protectionism is usually considered to be the superior option, especially over the longer-term, our earlier discussion suggests that state policies can only be effective if they are highly context-sensitive. This need for sensitivity may mean that different types of strategies may be appropriate in different environments.

References

Boje, T.P., Ejrnæs, A., Torres, A., Trifiletti, R. and Salmieri, L. (2007), 'Social, Family, and Labour Market Policies: Types of Work-Care Arrangements', in I. Vassilev, and C. Wallace (eds), *WORKCARE: Social Quality and Changing Relationships between Work, Care and Welfare in Europe: Literature Review* (updated May 2007) <http://www.abdn.ac.uk/socsci/research/nec/workcare/reports.php>, 88–116, accessed 5 June 2008.

Cooke, P. (1996), 'Regional Innovation Systems: An Evolutionary Approach', in H. Baraczyk, P. Cooke and R. Heidereich (eds), *Regional Innovation Systems* (London: UCL Press).

Djankov, S., La Porta, R., Lopez-de-Silanes, F. and Vishny, R. (2002), 'The Regulation of Entry', *Quarterly Journal of Economics* 117:1, 1–37.

European Commission (2004), *Fostering Structural Change: An Industrial Policy for an Enlarged Europe* (Brussels: European Commission).

European Commission (2005a), *European Innovation Policy*, <http://eur-lex.europa.eu/LexUriServ/site/en/com/2005/com2005_0488en01.pdf>.

European Commission (2005b), *Cohesion Policy in Support of Growth and Jobs: Community Strategic Guidelines, 2007–2013* (updated 22 December 2007) <http://ec.europa.eu/regional_policy/sources/docoffic/2007/osc/050706osc_en.pdf>.

Eurostat (2006), *EU Integration Seen Through Statistics* (Brussels: Eurostat).

Feenstra, R.C. (1998), 'Integration of Trade and Disintegration of Production in the Global Economy', *Journal of Economic Perspectives* 12:4, 31–50.
Gerreffi, G. (2005), *The New Offshoring of Jobs and Global Development* (Kingston: ILO Social Policy Lectures).
Gereffi, G. and Mayer, F. (2004), *The Demand for Global Governance* (Durham: Duke University WP, SAN 04-02).
Gereffi, G., Humphrey, J. and Sturgeon, T. (2005), 'The Governance of Global Value Chains', *Review of International Political Economy* 12:1, 78–104.
Glaeser, E. and Shleifer, A. (2002), 'Legal Origins', *Quarterly Journal of Economics* 117:4, 1193–1230.
Hooghe, L. (1996), 'Building a Europe with the Regions. The Changing Role of the European Commission', in L. Hooghe (ed.), *Cohesion Policy and European Integration: Building Multi-level Governance* (Oxford: Oxford University Press).
Humphrey, J. and Schmitz, H. (2000), 'Governance and Upgrading: Linking Industrial Cluster and Global Value Chain Research', IDS Working Paper No. 120 (Brighton: IDS).
Humphrey, J. and Schmitz, H. (2001), 'Governance in Global Value Chains', *IDS Bulletin* 32:3.
Humphrey, J. and Schmitz, H. (2002), 'How Does Insertion in Global Value Chains Affect Upgrading in Industrial Clusters?', *Regional Studies* 36:9, 1017–27.
ILO (2004), *A Fair Globalization. The Role of ILO*. Report of the Director-General on the World Commission on the Social Dimension of Globalization. International Labour Conference, 92nd Session (Geneva: International Labour Office).
ILO (2005), *About the Employment Intensive Investment Programme (EIIP)* (updated 26 May 2005) <http://www.ilo.org/public/english/employment/recon/eiip/about/index.htm>, accessed 5 June 2006.
Jessop, B. (1998), 'The Rise of Governance and the Risks of Failure: The Case of Economic Development', *International Social Science Journal* 155:1, 29–45.
Jessop, B. (2000), 'The Crisis of the National Spatio-Temporal Fix and the Ecological Dominance of Globalizing Capitalism', *International Journal of Urban and Regional Studies* 24:2, 323–60.
Jessop, B. (2002), *The Future of the Capitalist State* (Cambridge: Polity).
Johansson, J. and Vahlne, J.E. (1977), 'The Internationalisation Process of the Firm – A Model of Knowledge Development and Increasing Market Commitments', *Journal of International Business Studies* 8:1, 23–32.
Johansson, J. and Vahlne, J.E. (1990), 'The Mechanism of Internationalization', *International Marketing Review* 7:4, 11–24.
Kohsaka, A. (2004), 'National Economies under Globalisation: A Quest for New Development Strategy', in A. Kohsaka (ed.), *New Development Strategies. Beyond the Washington Consensus* (Hampshire and New York: Palgrave), 63–84.

La Porta, R., Lopez-de-Silanes, F., Schleifer, A. and Vishny, R. (1998), 'Law and Finance', *Journal of Political Economy* 106, 1113–55.

Lundvall, B.-A. (1997), *The Globalising Learning Economy: Implications for Innovation Policy*. TSER Report, EUR 18307 (Brussels: DG Research, European Commission).

Marks, G. (1993), 'Structural Policy and Multi-level Governance in the European Community', in A.W. Cafruny and G. Rosenthal (eds), *The State of the European Community* (Boulder: Lynne Rienner), 391–410.

Marks, G., Hooghe, L. and Blank, K. (1996), 'European Integration from the 1980s: State-Centric vs Multi-Level Governance', *Journal of Common Market Studies* 34:3, 341–78.

Matsushita, M. (2004), 'Governance of International Trade Under World Trade Organization Agreements – Relationships Between World Trade Organization Agreements and Other Trade Agreements', *Journal of World Trade* 38:2, 183–210.

Nadvi, K. and Waltring, F. (2002), 'Making Sense of Global Standards', INEF Report 58 (Duisburg: Institut fur Entwicklung und Frieden, Gerhard-Mercator University).

OECD (2001a), *Reviews of National Policies for Education* (Paris: OECD).

OECD (2001b), *Local Partnerships and Better Governance* (Paris: OECD).

OECD (2003), *Labour Markets and Social Policies in the Baltic Countries* (Paris: OECD).

OECD (2005), *Information Technology Outlook 2004* (Paris: OECD).

Rodrik, D. (2004), 'Development Strategies for the Twenty-first Century', in A. Kohsaka (ed.), *New Development Strategies. Beyond the Washington Consensus* (Hampshire and New York: Palgrave), 13–38.

Stone, D.C. (1965), 'Government Machinery Necessary for Development', in M. Kriegsberg (ed.), *Public Administration in Developing Countries* (Washington DC: The Brookings Institution), 49–67.

Todaro, M. (1994), *Economic Development* (London: Longman).

Wallace, C. (ed.) (2003), 'HWF Country Contextual Reports: Demographic Trends, Labour Market and Social Policies' (updated 22 December 2007) <http://www.hwf.at/project_report02.html >, accessed 23 December 2007.

Wallace, C. (ed.) (2004), 'HWF Comparative Contextual Report: Demographic Trends, Labour Market and Social Policies (updated 22 December 2007) <http://www.hwf.at/project_report05.html> accessed 23 December 2007.

Wallis, A.D. (1996), 'Regions in Action: Crafting Regional Governance under the Challenge of Global Competitiveness', *National Civic Review* 85:2, 15–24.

World Bank (2004), *The World Bank Annual Report* (Washington: World Bank Group).

PART 2
Industrial Analysis of the European Experience

Chapter 6
The Impact of Internationalization on the Clothing Industry

Christos Kalantaridis, Ivaylo Vassilev, Grahame Fallon

Introduction

Clothing is the paramount global commodity, having some of the highest levels of import penetration as well as volumes of trade and supply chain internationalization. Trade in clothing is among the longest established in the world, yet there has recently been a dramatic increase of global interdependence in production and consumption, with producers from different countries taking turns at occupying centre stage over the past forty years. The location of offshore production has shifted constantly, including Japan (in the 1960s); Southern Europe, Hong Kong, South Korea, Singapore and Taiwan (1970s); mainland China, Sri Lanka and Southeast Asia (late 1980s); and South Asia and North Africa (1990s) (Gibbon 2002; Gereffi and Memedovic 2003). The significance of Latin America, Eastern Europe, Turkey and the Middle East have also increased (Begg and Pickles 2000; Gereffi and Memedovic 2003).

Accelerating global integration in the clothing industry can be explained by three main factors. Firstly, it has been stimulated by advances in technology – especially telecommunications, transportation and IT (Giddens 1990; Castells 1996) – combined with the predominance of neo-liberal ideology and free markets (Jessop 2002). Secondly, global integration has been facilitated by the sector's low entry barriers (in capital and skills terms) and high labour content – amounting to 60 per cent of total costs (OECD 2004) – making relocalization to countries with cheap and flexible labour relatively easy (Hanzl-Weiss 2004). Third, integration has benefited from the rise (and increasing convergence) of global buyers such as retailers, branders and marketers (Gereffi 1999).

This chapter sets out to examine the impact of processes of global integration upon inter-organizational relationships and enterprise strategy in the clothing industry, drawing upon the results of extensive fieldwork investigation in five European countries (the UK, Greece, Poland, Estonia and Bulgaria). The next section of the chapter comprises a review of the literature, forming the basis of a number of testable hypotheses. The main body of the chapter discusses the findings of the empirical research, while the final two sections provide an overview of the findings, followed by concluding remarks.

The Literature

The industry

The clothing industry has undergone substantial recent changes, with exports growing rapidly between the 1970s and late 1990s, and a significant decline in sectoral employment in DCs (Dunford 2002; Bair and Gereffi 2003). However, the impact on the overall employment in the supply chain has been less dramatic as employment numbers have risen in the pre-assembly and retail parts of the sector (Pye 2004).

Despite high levels of offshoring and decreasing employment and output, clothing continues to be significant for DCs, where the number of companies has not changed dramatically and their diversity has actually increased (Pye 2004). The EU clothing industry is still heavily concentrated in the DCs (specializing in high value-added products). Eastern European countries experienced a rapid decline in clothing employment in the early 1990s followed, however, by a revival during the mid- to late 1990s (when their clothing production rose in volume and value terms).

Overview of trade governance

Governance comprises the actions of governmental and non-governmental institutions that both encourage and constrain the behaviour of market actors (Gereffi and Mayer 2004). Governance systems have three main effects on markets: facilitative, regulatory and compensatory.

Trade quotas, and especially the MFA adopted in 1974, have been the major mechanisms regulating world trade in clothing and textiles over the recent past. An agreement became effective in 1995, providing for a ten-year transition period (ending 2004), during which WTO member countries gradually abolished quotas (Nathan Associates 2002). A second tier of regulatory mechanisms, governing OPT, preferential trading agreements (PTA) and free trade agreements (FTA) still remain, however, protecting producers in the USA, Japanese and EU trading blocks (Kwan and Qiu 2003).

Regulations also govern the production and consumption of clothing, covering product quality, environmental, health and working conditions, ethics and social responsibility (Humphrey and Schmitz 2002). While standards mostly operate at the national level, international standards and codes of practice have also gradually been introduced since the 1950s (Nadvi and Waltring 2002), including ISO9000 (quality management), ISO14000 (environment protection), SA8000 (social accountability), and WRAP (socially responsible global standards for clothes manufacturing) (Yeung and Mok 2004).

Buyers in developed countries (DCs)

Clothing industry restructuring has been associated with the rise of large buyers including retailers, marketers and branders (Gereffi 1999; Gibbon 2002). A 'retail revolution' occurred from 1965–1980, marked by the rise of giants such as Wal-Mart, K-Mart and Target, and the growth of specialized marketers and assemblers such as Nike (Appelbaum 2004). Buyers are also becoming increasingly similar, as retailers develop their own brands, while branders increasingly abandon production altogether (Bair and Gereffi 2003).

Concentration and increasing buyer control have also taken place in a number of national markets. In the USA, Wal-Mart and K-mart control 25 per cent of national clothing sales (by unit volume), while the top five account for 68 per cent of sales (Gereffi and Memedovic 2003). The UK clothing sector has also consolidated since the 1980s, being currently dominated by a small number of large, specialized retailers (Dunford 2002), none of which have significant manufacturing activities (Gibbon 2002). Similar processes of consolidation and concentration have occurred in Germany, France, Italy and Japan (Gereffi and Memedovic 2003).

Relationships

Research Question 1: What is the impact of emerging (chronological and sequential) patterns of global integration on inter-organizational relationships in the clothing industry?

Links between buyers and manufacturers vary in terms of 1) *what part of the process is outsourced*, and 2) *what types of relationships are involved in bringing the product to the buyer*. Suppliers can undertake:

1. The assembly of imported inputs (OPT),
2. Full package production – where they supply finished products to buyers' specification, or
3. Triangular manufacturing – where the lead supplier only co-ordinates different aspects of the production process (assigned to different subcontractors (Gereffi 1999)).

These different buyer-supplier relationships necessitate differing levels of upgrading on the part of suppliers, including the development of managerial know-how, design capabilities, fabric procurement, property rights protection, export financing and expertize in trade formalities handling, together with the possible development of own brands and retail outlets (OECD 2004).

Pre-assembly is the highest value-added stage of production, which is commonly carried out in-house by major clothing companies (Abernathy et al. 1999). Marketing, branding and retailing are also highly capital- and knowledge-intensive, making them difficult steps for the upgrading producer to climb. Developing own brands for the national market may not guarantee long-term survival, unless the product is internationally competitive (Karagozoglu and Lindell 1998). The choice of suppliers and their location also depends on the segment within which buyers operate. Gereffi (1999) and Gibbon (2002) distinguish between 'upper market segments' (supplied mainly by flexible enterprises, offering high quality services and located mostly in DCs) and 'basic' clothing (frequently relying on developing country sources).

Hypothesis $(H)_1$: There is a relationship between country and the market segment focus of clothing manufacturers – with those nearer the EU core focusing more on design and flexible response segments.

Three main forms of outsourcing relationship occur in the clothing industry: subcontracting, JVs and FDI. FDI is relatively insignificant in the clothing sector (Hanzl-Weiss 2004) since outsourcing is mostly done through subcontracting and (to a lesser extent) by JVs.

H_2: FDI and JVs are positively linked to working outside the price-sensitive segment of the market (with either a flexible response or design focus).

H_3: There is a positive relationship between country and the incidence of FDI and JVs.

Buyers can either choose to deal directly with manufacturers or via intermediaries. The latter can serve as importers and transmitters of production and organizational expertise (Schmitz and Knorringa 2000) and can add value through their knowledge of globally dispersed production capacities and demand and by their capacity to deal efficiently with logistics, diverse national contexts and subcontractors, and internal design issues (Enright et al. 1997).

Buyer/supplier relationships may evolve over time and are often fluid and complex (Tokatli and Eldener 2004). Suppliers may, for example, upgrade and develop a second-tier subcontracting capacity while retaining some production themselves (Appelbaum et al. 2000a). They may alternatively decide to play the role of intermediaries (Labrianidis and Kalantaridis 2004), or to deal directly with producers instead (Kalantaridis et al. 2001). Most buyers are likely to develop hybrid solutions, using both intermediaries and direct contacts; buyers/supplier links may remain very strong, especially if buyers choose to use existing suppliers to expand into new markets (Enright et al. 1997).

H_4: Flexible response and design focus of the markets are characterized by the incidence of strong relationships.

Strategies

Research Question 2: What is the degree of diversity (or commonality) of emerging strategies in the clothing industry, and to what extent are they linked with performance?

Many enterprises seek to reduce production costs by *informalization* (Anielo 2001; Appelbaum et al. 2005). The pursuit of cheaper labour by outsourcing assembly (often by subcontracting) to low cost areas is typical of the clothing industry; DC manufacturers which depend entirely on cost-reduction are most vulnerable to clothing sector restructuring but see Taplin et al. (2003). Other forms of product-related adjustment may be less risky, including increasing productivity, product range diversification, faster new product development, movement up the value chain and introducing flexible production (Zeitlin and Totterdill 1989).

H_5: Operating in price-sensitive segments of the market is negatively linked to successful adjustment.

Labour flexibility is crucial for adjustment (especially in DCs); thus it is proving increasingly necessary to treat employees as a resource rather than a cost (Taplin et al. 2003). Improved product quality and better product differentiation require high quality work and highly skilled workers, necessitating investment in training and long-term contracts. High labour turnover can impede such a strategy (Appelbaum et al. 2000a), however, and low levels of education and training on the part of clothing industry workers may be particularly problematic (Winterton and Winterton 2002), leading to a need for government support in upgrading worker skills (Husband and Jerrard 2001). Increasing flexibility and cost reductions may also require the adoption of systems for minimizing inventory levels centring on 'lean retailing' (Abernathy et al. 1999; Nordas 2004).

Clothing company strategies must also be embedded in firms' broader institutional and physical environments. Higher priority is, for example, given to the maximization of 'shareholder value' in the UK than in mainland Europe, leading to significant differences in the day-to-day operational management and strategic choices of individual companies (Palpacuer et al. 2005), while state policies and path dependencies also exert a substantial influence on the strategies chosen (Tully and Berkley 2004).

Successful restructuring strategies are likely to involve a move away from manufacturing towards branding and consolidation at the retail end of the chain. It is increasingly typical for leading clothing companies to design and market but not to make their products, earning profits from research, design, sales, marketing and the sale of financial services, rather than from scale, volume and technological innovation (Bair and Gereffi 2003). There is a danger, however, in focusing excessively on global buyers and their power, leading to a disregard of local networks, markets and retailing structures in LDCs (Hassler 2004).

H_6: Reduced manufacturing capacity in the EU is positively linked to successful adjustment.

H_7: Increased manufacturing capacity in Eastern Europe is positively linked to successful adjustment.

Findings

Overview

The criteria used for the selection of the enterprises surveyed indicate an apparent international orientation among clothing firms, although significant inter-country disparities emerge from the findings. In the case of Bulgaria, more than 90 per cent of sales were directed to international markets, compared with 15 per cent for the UK. Subcontracting was the primary means of export activity (accounting for over 80 per cent of exports in Bulgaria, Poland and Estonia), although marginal or non-existent in Greece and the UK. OPT follows a similar pattern (see Figure 6.1).

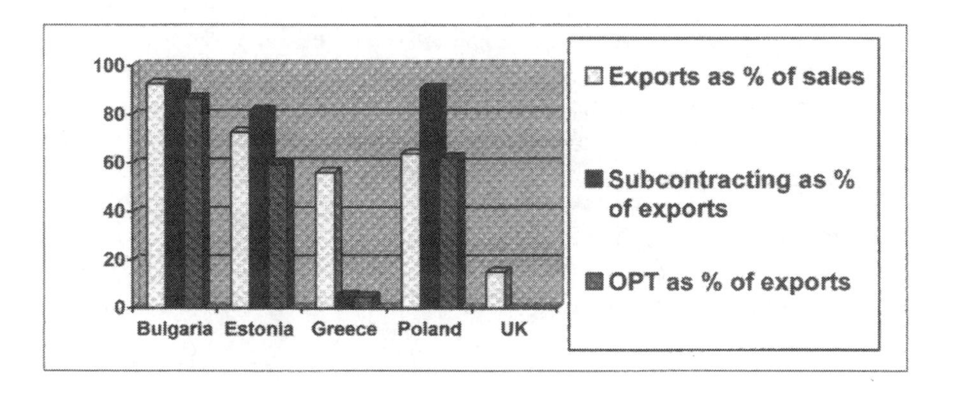

Figure 6.1 Exports, subcontracting, intermediate products and OPT by country
Source: Enterprise Survey.

There is a (predictable) relationship between ownership structures and international development (see Figure 6.2). Enterprises with direct ownership links, FDI and JVs demonstrate the greater incidence of reliance on exports, subcontracting and OPT, whereas domestically owned firms placed the least reliance upon them.

There are also substantial linkages between firm size and ownership structures. All micro enterprises in the sample and over 90 per cent of small firms are domestically owned, in comparison to just over 60 per cent of large businesses. Larger firms also exhibit more JVs (17 per cent) and FDI (21 per cent) than their smaller counterparts.

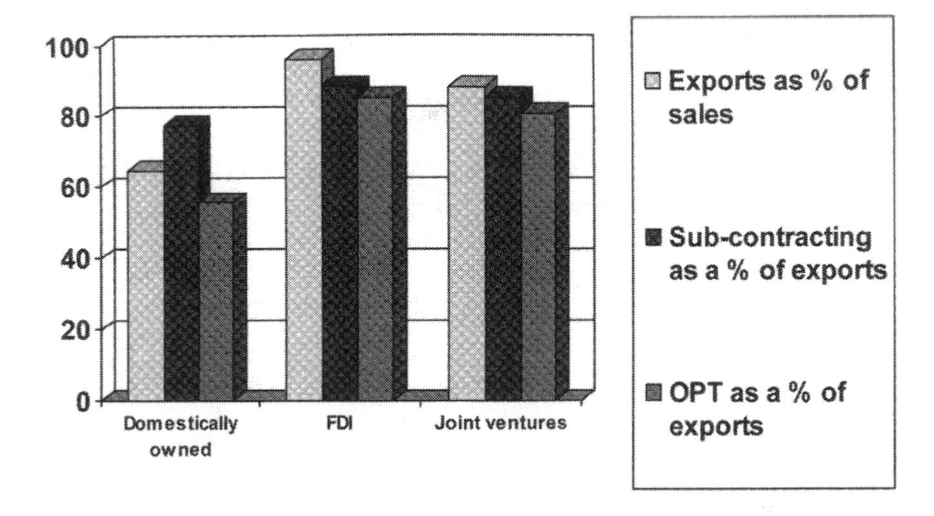

Figure 6.2 Ownership structure and international development
Source: Enterprise Survey.

On segmentation, country and ownership

H_1: There is a relationship between country and the market segment focus of clothing manufacturers – with those nearer the EU core focusing more on design and flexible response segments.

Market segmentation constitutes an apparently significant influence upon enterprise adjustment patterns. There is a statistically significant relationship between country and market segmentation, with countries nearer the EU core (such as the UK) focusing more on design and flexible response, and peripheral countries (such as Bulgaria and Poland) on price-sensitive segments. Estonian clothing firms seem to have moved further than expected towards a design focus, while Greek firms still rely surprisingly heavily on price competition (see Table 6.1).

Table 6.1 Market segments by country

			Market segment			Total
			Price-sensitive	Flexible response focus	Design focus	
Country	Bulgaria	Count (%)	49 (83.1)	9 (15.3)	1 (1.7)	59 (100)
	Estonia	Count (%)	15 (25.9)	14 (24.1)	29 (50.0)	58 (100)
	Greece	Count (%)	21 (72.4)	3 (10.3)	5 (17.2)	29 (100)
	Poland	Count (%)	54 (61.5)	28 (31.8)	6 (6.8)	88 (100)
	UK	Count (%)	2 (20.0)	3 (30.0)	5 (50.0)	10 (100)
Total		Count (%)	141 (57.8)	57 (23.5)	46 (18.9)	244 (100)

Source: Enterprise Survey.

H_2: FDI and JVs are linked to working outside the price-sensitive segment of the market (with an either flexible response or design focus).

There is no apparent relationship between these variables (see Table 6.2). The lowest incidence of foreign ownership lies interestingly in the flexible response segment of the market.

Table 6.2 Market segments by ownership

		Ownership			Total
		Domestically owned	FDI	Joint venture	
Price-sensitive	Count (%)	112 (79.4)	24 (17.0)	5 (3.5)	141 (100)
Flexible response focus	Count (%)	48 (84.2)	5 (8.8)	4 (7.0)	57 (100)
Design focus	Count (%)	36 (78.3)	7 (15.2)	3 (6.5)	46 (100)
Total	Count (%)	196 (80.3)	36 (14.8)	12 (4.9)	244 (100)

Source: Enterprise Survey.

The findings also suggest the existence of a relationship between ownership structures, market segmentation and the national context (see Table 6.3). In the case of Estonia, for example, there appears to be a substantial incidence of FDI in the price-sensitive market segment (contrary to prior expectations).

Table 6.3 Ownership structure and market segmentation by country

Country				Market segmentation			Total
				Price-sensitive	Flexible response focus	Design focus	
Bulgaria	V76	Domestic owned	Count (%)	28 (84.8)	4 (12.1)	1 (3.0)	33 (100)
		FDI	Count (%)	17 (89.5)	2 (10.5)	0 (.0)	19 (100)
		Joint venture	Count (%)	4 (57.1)	3 (42.9)	0 (.0)	7 (100)
	Total		Count (%)	49 (83.1)	9 (15.3)	1 (1.7)	59 (100)
Estonia	V76	Domestic owned	Count (%)	8 (20.5)	12 (30.8)	19 (48.7)	39 (100)
		FDI	Count (%)	6 (40.0)	2 (13.3)	7 46.7)	15 (100)
		Joint venture	Count (%)	1 (25.0)	0 (.0)	3 (75.0)	4 (100)
	Total		Count (%)	15 (25.9)	14 (24.1)	29 (50.0)	58 (100)
Greece	V76	Domestic owned	Count (%)	21 (72.4)	3 (10.3)	5 (17.2)	29 (100)
	Total		Count (%)	21 (72.4)	3 (10.3)	5 (17.2)	29 (100)
Poland	V76	Domestic owned	Count (%)	53 (62.4)	26 (30.6)	6 (7.1)	85 (100)
		FDI	Count (%)	1 (50.0)	1 (50.0)	0 (.0)	2 (100)
		Joint venture	Count (%)	0 (.0)	1 (100.0)	0 (.0)	1 (100)
	Total		Count (%)	54 (61.4)	28 (31.8)	6 (6.8)	88 (100)
UK	V76	Domestic owned	Count (%)	2 (20.0)	3 (30.0)	5 (50.0)	10 (100)
	Total		Count (%)	2 (20.0)	3 (30.0)	5 (50.0)	10 (100)

Source: Enterprise Survey.

H_3: There is a relationship between country and region (viewed here as proxies for macroeconomic stability and development of market institutions) and the incidence of FDI and JVs.

There does appear to be such a relationship (see Table 6.4), although it is somewhat different from that originally envisaged. Countries with more stable macroeconomic environments and developed market economies (the UK, Greece, Poland in descending order) do not demonstrate the greatest incidence of FDI and JVs, but rather those (riskier) countries, such as Estonia and Bulgaria, that offer greater potential opportunities for inward investors do.

Table 6.4 The incidence of FDI and joint ventures by country

			Ownership structure			Total
			Domestically owned	**FDI**	**Joint venture**	
Country	Bulgaria	Count (%)	33 (54.1)	20 (32.8)	8 (13.1)	61 (100.0)
	Estonia	Count (%)	40 (66.7)	15 (25.0)	5 (8.3)	60 (100.0)
	Greece	Count (%)	31 (100.0)	0 (.0)	0 (.0)	31 (100.0)
	Poland	Count (%)	88 (95.7)	3 (3.3)	1 (1.1)	92 (100.0)
	UK	Count (%)	12 (100.0)	0 (.0)	0 (.0)	12 (100.0)
Total		Count (%)	204 (79.7)	38 (14.8)	14 (5.5)	256 (100.0)

Source: Enterprise Survey.

On relationships

H_4: Flexible response and design focus of the markets are characterized by the incidence of strong relationships.

Firms operating in the design and, to a lesser extent, the flexible response segments appear to exhibit relatively high levels of mutual dependence and personalized relations, and a more asymmetrical balance of power (see Table 6.5). Price-sensitive firms appear to enjoy longer, continuous relationships with partner businesses, however.

Table 6.5 Market segmentation and strength of relationships

Average Linkage (between groups)		Balance of power (v148)	Mutual dependence (v149)	Personalized relations (v150)	Average number of years of continuous relationship (v144)
Price-sensitive	Mean	3.9	3.3	2.2	7.6
	N	117	117	115	115
Flexible response focus	Mean	3.9	3.5	2.4	7.1
	N	50	48	49	50
Design focus	Mean	4.2	3.7	2.8	6.3
	N	38	38	38	38
Total	Mean	4.0	3.5	2.4	7.2
	N	205	203	202	203

Source: Enterprise Survey.

The balance of power seems to rest predominantly in the hands of the buyers (the mean being over 4.00 in all cases except Poland [see Table 6.6]). Buyer power may exist alongside mutually dependent relationships, however, especially in Greek, Bulgarian and Estonian enterprises. Most manufacturer/buyer relationships are relatively impersonal, especially in Greece and (to a lesser degree) in Bulgaria and the UK. Many non-UK clothing firms service foreign companies by subcontracting.

Relationships are also influenced by firm size; the smaller the firm, the smaller the number of its customers and thus the greater its degree of dependence. Foreign ownership appears (predictably) to reduce the number of buyer options, increasing the imbalance of power between enterprises, together with mutual dependence. Market focus also appears to be a significant influence on relationships. Enterprises operating in price-sensitive segments exhibit the smallest number of buyers and the greatest buyer power, whilst those with a design focus record the least buyer power and highly developed personalized relationships.

Table 6.6 Nature of forward relationships by country

Country		Balance of power (v148)	Mutual dependence (v149)	Personalized relations (v150)	Number of foreign companies serviced by
Bulgaria	Mean	4.1	3.9	2.0	5.5
	N	61	60	60	61
Estonia	Mean	4.1	3.8	3.5	9.4
	N	58	58	58	58
Greece	Mean	4.6	4.3	2.4	10.1
	N	16	16	16	10
Poland	Mean	3.6	2.8	1.9	5.7
	N	78	77	76	81
UK	Mean	4.5	2.0	2.0	5.5
	N	2	2	2	2

On strategies

Firms were asked a sequence of qualitative questions designed to provide an overview of their strategies. The data collected were analysed using the *N-vivo* package, enabling strategic patterns to be identified for each sample country, using hierarchical cluster analysis and the Ward method. Table 6.7 and Figure 6.3 provide a summary of the prototype strategies identified.

Most emergent clusters are country-specific, suggesting that companies operating in the same market segment but different national contexts may use different strategies (for example, Poland 2, Estonia 3, and Bulgaria 3). Companies operating in the same country and market segment may adopt significantly different strategies (for example, Poland 1 and 4, Poland 2 and 3, Estonia 1 and 2, Bulgaria 2, 3, and 4). Those operating in different segments and different countries may actually adopt the same strategy (for example, Poland 2 and Estonia 1, Greece 1 and Poland 1). Some more consistent patterns emerge, however; for example, companies operating both in Greece and the UK fall into clusters 2 and 3, while clusters Poland 2 and Bulgaria 4 are identical (as are Poland 1 and Bulgaria 1).

Table 6.7 Overview of strategies

	Competence lock-in	Hybrid	Break-out competences
Product/ service	Not own product range so limited scope for action.	New product or product design for some of the product range.	New product design and brand development.
Process	Technological change (invariably in production) in line with needs of parent Enterprise.	Technological change linked to new manufacturing competences (often knowledge transfer from one dimension (OPT) to the other).	All encompassing technological change including manufacturing and/or lean retailing.
Function	Moving up or down the production chain but remaining within manufacturing.	Moving up and/or down the production chain – often simultaneously in two different production dimensions.	Moving up the production chain – often away from manufacturing towards distribution. Proximity to the consumer a key source of competitive edge.
Production	Production competences remain at the heart of enterprise strategy.		The importance of production competences and volume production decline.
Market	Serving in the main price-sensitive and to a lesser degree flexibility focused market segments.	Serving flexible response focus plus one of the other two (flexibility focus or design sensitive) market segments.	Serving in the main design focus market segments.

Source: Enterprise Survey.

Competence lock-in strategies are most common, accounting for 129 firms (over 50 per cent of the total sample). They are (predictably) apparent in the price-sensitive segment of the market (for example, Poland 2, Bulgaria 4 and Greece 1). They also appear in the flexible-response segment, however, represented by the Estonia 1 and Bulgaria 1 clusters (both with a strong export orientation and substantial FDI, but little evidence of functional upgrading). Estonia 3 also appears to adopt a very similar strategy (but for the development of design competences over some of the product range). A strong export orientation and significant foreign involvement are also typical of lock-in strategies (apart from Poland 2).

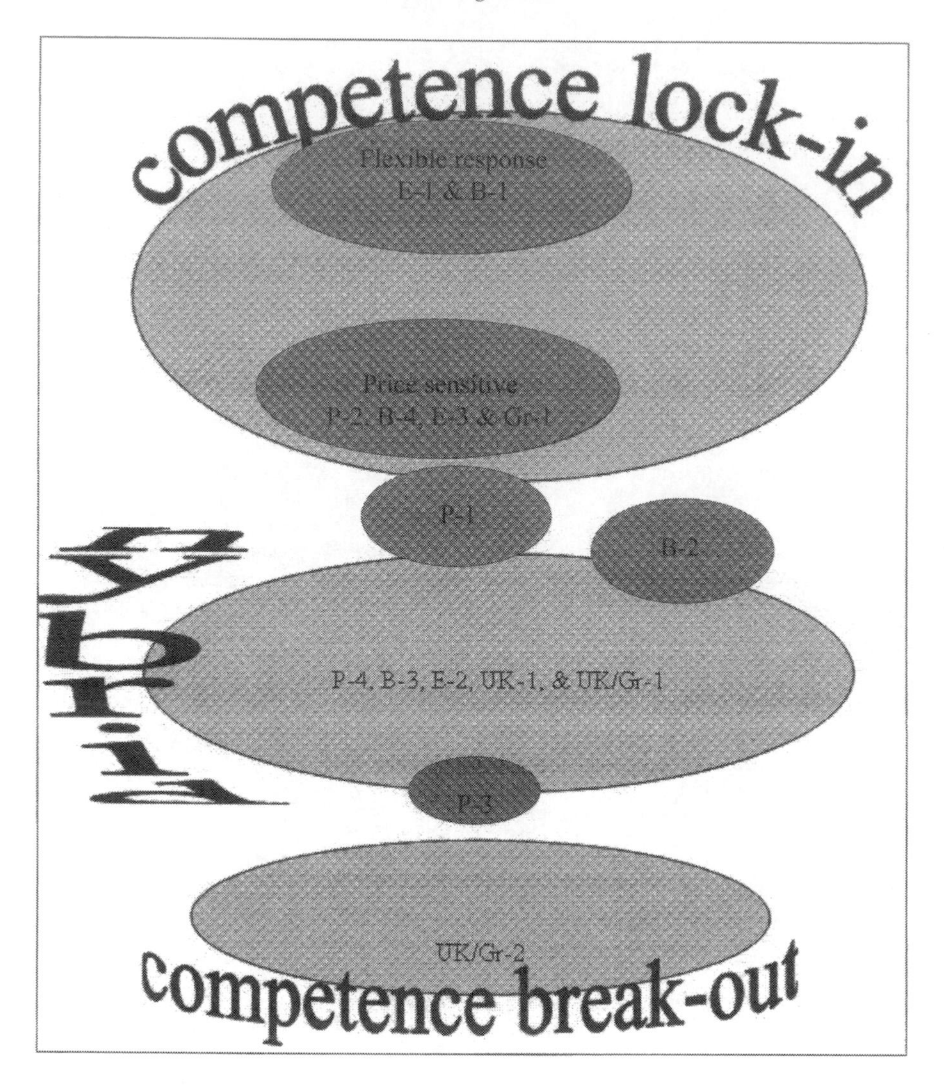

Figure 6.3 Patterns of enterprise strategies
Source: Enterprise survey.

Hybrid strategies form an interesting prototype, including 59 (just over 25 per cent) of respondents. Such firms seek to use competences developed by engagement in global production and distribution networks in order to enhance their domestic market positions. Poland (4) and Estonia (2) appear to have developed greater design competences and own-brand products than originally envisaged, while UK 1 and UK/Greece 1 are somewhat different, due to the declining importance of production. Movement upwards in the production chain appears to be common in the former grouping but less so in the latter.

One cluster (UK/Greece 2) lies within the *competence break-out category*, comprising a mere nine companies, all with a strong domestic focus. Poland 3, finally, falls between the hybrid and break-out strategies, comprising 13 brand-owning companies, showing apparent movement away from production, and further up the chain. These firms have still not completed the implementation of advanced technologies, and their focus remains on the price-sensitive segment of the market.

On performance

The degree of success (or failure) in adjusting to changes in the industry was explored, focusing on changes occurring before and after companies' integration with global markets. Four measures of performance were used: employment, turnover, profits and exports. Entrepreneurs were asked to evaluate performance on a five point, LIKERT-type scale, ranging from 1 (considerable perceived decline) to 5 (considerable perceived growth).

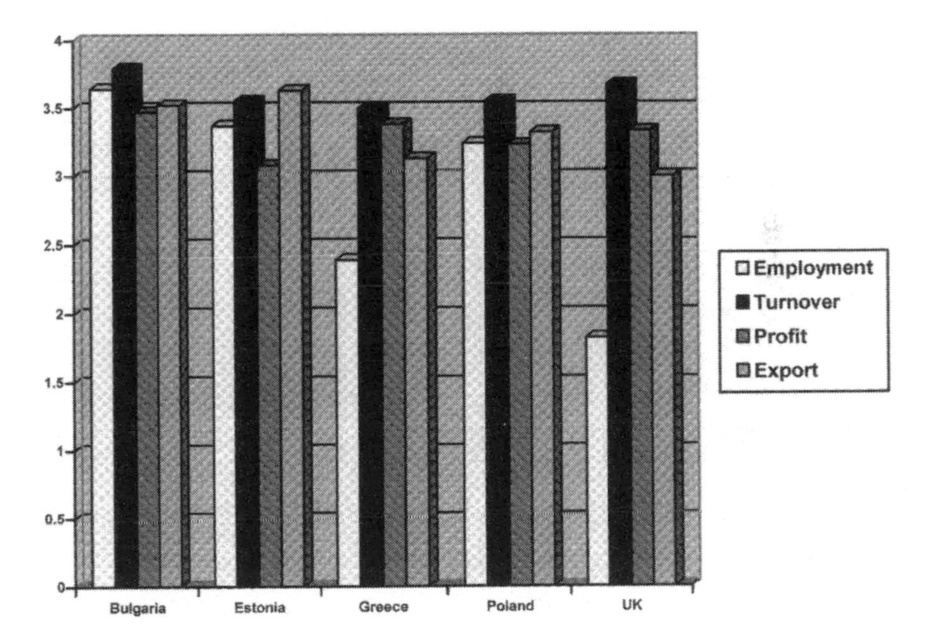

Figure 6.4 Successful adjustment by country
Source: Enterprise Survey.

Figure 6.4 captures reported adjustment success in country terms, revealing a clear distinction between the performance of Eastern European companies (strong in respect of all measures) and those based in the UK and Greece. The latter indicate a fall in employment – probably due to the relocation of production abroad – although with no apparent adverse effects on turnover and exports. Interesting findings also emerge for profitability, where Bulgarian firms perform best (with Greek enterprises in second place), while UK companies perform marginally better than Polish enterprises. These findings suggest that the clothing industry adjustment to global integration may differ between developed and LDCs.

There also appears to be a link between firm size and adjustment performance, larger firms tending to perform best in employment, turnover and profit (though not exporting) terms. This finding may be linked to some extent to country influence – since larger enterprises are found in Bulgaria, Estonia and Poland than in the UK and Greece (see Figure 6.5).

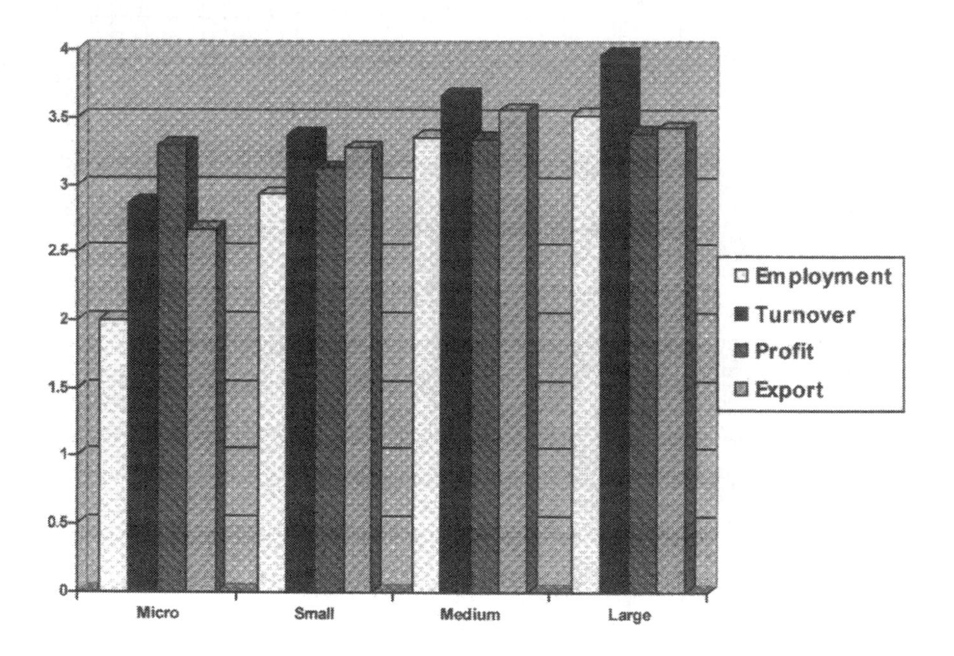

Figure 6.5 Successful adjustment by size band
Source: Enterprise Survey.

There is an apparent relationship between ownership structure and performance (see Figure 6.6), with FDI and JV companies appearing to out-perform domestically-owned enterprises on a consistent basis. This finding may be linked in part to varying levels of access to resources, such as finance (in scarce supply in Eastern European countries, where the bulk of FDI is accumulated). Country specificity may also be important, in that 48 out of 52 clothing companies with foreign ownership are located in Bulgaria and Estonia – the two best performing countries in adjustment terms.

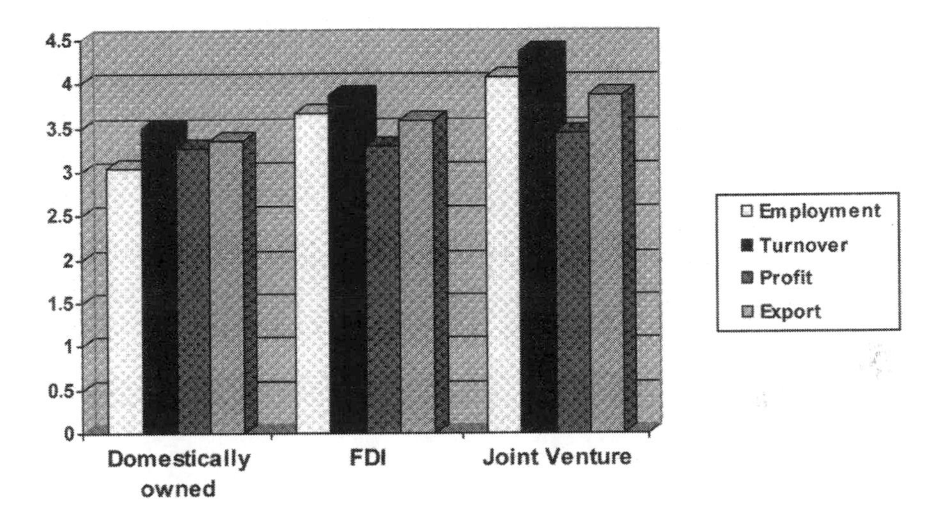

Figure 6.6 Successful adjustment by ownership structure
Source: Enterprise Survey.

H_5: Operating in price-sensitive segments of the market is negatively linked to successful adjustment.

No clear linkage between market segmentation strategies and successful adjustment emerges from the findings (see Table 6.8). Enterprises in price-sensitive segments appear to perform as well as or somewhat better than those with a flexible response and design focus in terms of turnover, profits and exporting (although rather less well in employment terms).

H_6: Reduced manufacturing capacity in the EU is positively linked to successful adjustment.

H_7: Increased manufacturing capacity in Eastern Europe is positively linked to successful adjustment.

Production-focused strategies (coded 1 in Table 6.9) appear to result in weaker performance for UK and Greek companies, although the findings are less conclusive for Eastern European firms. There is evidence in the case of Estonia (although not for Poland or Bulgaria) that a production focus may improve performance (especially turnover and profits).

Table 6.8 Market segmentation and successful adjustment

		Market segment	Annual average, total employment (v53)	Turnover progress after delocalization (v63)	Profits progress after delocalization (v64)	Export progress after delocalization (v96)
Market segment	Pearson Correlation	1	–.057	–.030	–.118(*)	–.002
	N	244	242	234	228	244
Annual average, total employment (V 53)	Pearson Correlation	–.057	1	.538(**)	.350(**)	.155(**)
	N	242	254	246	239	254
Turnover progress after delocalization (V63)	Pearson Correlation	–.030	.538(**)	1	.643(**)	.245(**)
	N	234	246	246	238	246
Profits progress after delocalization (V64)	Pearson Correlation	–.118(*)	.350(**)	.643(**)	1	.199(**)
	N	228	239	238	239	239
Export progress after delocalization (V 96)	Pearson Correlation	–.002	.155(**)	.245(**)	.199(**)	1
	N	244	254	246	239	256

Note: * Correlation is significant at the 0.05 level (1–tailed) ** Correlation is significant at the 0.01 level (1–tailed).
Source: Enterprise Survey.

Table 6.9 Production focus and successful adjustment

Country	Production		Annual average employment (v53)	Turnover progress after delocalization (v63)	Profits progress after delocalization (v64)	Export progress after delocalization (v96)
Bulgaria	1.00	Mean	3.6	3.7	3.4	3.5
		N	60	60	58	60
	Total	Mean	3.6	3.7	3.4	3.5
		N	60	60	58	60
Estonia	1.00	Mean	3.4	3.4	3.0	3.5
		N	48	48	48	48
	3.00	Mean	3.0	4.5	3.2	4.0
		N	4	4	4	4
	Total	Mean	3.3	3.5	3.0	3.6
		N	52	52	52	52
Poland	1.00	Mean	3.2	3.5	3.2	3.3
		N	67	66	60	69
	3.00	Mean	3.9	4.0	3.5	2.6
		N	13	11	11	13
	Total	Mean	3.3	3.6	3.3	3.2
		N	80	77	71	82
Greece	1.00	Mean	2.0	3.2	3.0	2.8
		N	19	18	18	19
	2.00	Mean	3.5	3.0	2.5	3.5
		N	2	2	2	2
	3.00	Mean	2.9	4.1	4.2	3.5
		N	10	8	9	10
	Total	Mean	2.3	3.5	3.3	3.1
		N	31	28	29	31
UK	1.00	Mean	1.5	3.5	2.5	2.5
		N	2	2	2	2
	3.00	Mean	2.1	3.8	3.7	3.1
		N	10	8	8	10
	Total	Mean	2.0	3.	3.5	3.0
		N	12	10	10	12

Source: Enterprise Survey.

On governance

The findings are inconclusive, but they do support Gereffi and Mayer (2004) regarding the growth of a 'governance deficit' (where internationalization leads to global market development but not necessarily to commensurate increases in regulation and compensation mechanisms). Many respondents expressed disappointment at the negative effects involved in meeting ISO requirements, whose main function appeared to be meeting requirements for government quotas and/or funding rather than facilitating their own position in the market.

> There is a completely different way of doing business in North America ... it is also the import duties there ... export licenses are creating delays and uncertainty; in the USA they do not need those, only for cases of trade embargos otherwise they can trade anywhere in the world; with other EU countries the regulation should be the same but it is not, it is not enforced; UK regulation authorities take great pleasure in enforcing regulations elimination of trading obstacles and continued positive attitude towards business. (UK based subsidiary of TNC)

> Although there is a clear need for policy towards the clothing industry ... the government has not developed any up till now. (Poland, key informants)

Strong private arrangements appear to be made between global buyers (dominating production and trade) and subcontractors and subsidiaries (seeking to meet quality standard requirements). There is concern over differences in the implementation of EU and national regulations in different countries. Different cultural predispositions emerge in relation to the varying role of government, although respondents from all countries are concerned over the perceived lack of government support for the clothing industry in terms of funding, education, reskilling, upgrading and market access, and the perceived lack of priority which governments give to the sector.

Following Gereffi and Mayer (2004), it also appears that there is a tendency for global and supranational institutions and TNCs (particular with locally embedded subsidiaries) to dominate the facilitative, regulatory and compensatory domains in the clothing sector (Table 6.10).

> Significant regulation is mainly on the national level, regional regulation is quite insignificant. While there may be some variations between England and Northern Ireland there won't be such differences but they could be for rules for company registration for example. But even there now with the possibility to register as a "European company" they can be registered in any EU country. (UK, key informants)

> Yes, I mean when they brought in the minimum wage thing, that was, because it was all going to go to the minimum wage but I couldn't just pay the girls the bottom line of the minimum wage, I had to increase it, even into the admin staff. It costs us a lot more than £1 an hour in the loss of earnings; it is a lot of money. (UK, small producer)

Table 6.10 Governance arrangements: Function

	Realms of governance	
Modes of governance	**Public**	**Private**
Facilitative	Property rights Banking and commercial law Competition policy	Market ideology Professional codes and norms
Regulatory	Labour law Environmental regulation Health and Safety regulations	Voluntary codes of conduct Corporate social responsibility Pressure and consumer boycotts
Compensatory	Social insurance Education/retraining programmes Public health policies	Collective bargaining Philanthropy
	National Governments / Global institutions	TNCs / SMEs and/or local subsidiaries

Source: Adapted from Gereffi and Mayer (2004).

Public governmental arrangements appear to dominate the facilitative domain, while private and public arrangements influence the regulatory domain in all sample countries. Compensatory arrangements are dominated by government regulation alone in new EU member states but in equal measure by government regulation and private arrangements in the UK and Greece.

> We follow all requirements, but none of these has influenced delocalization. (Bulgarian company)

Sectoral quality standards are mostly arranged privately and are dominated by TNCs, whilst the terms of trade are determined by governments but increasingly dominated by global frameworks and regional agreements. International certification also acts as a mechanism for attracting foreign buyers. Trade unions are not perceived to play any major role in governance in any of the sample countries.

> The fact that we have a quality certificate helps us. There are customers who are principally interested in that … (Bulgarian company)

On social consequences

Most UK and (to a lesser extent) Greek clothing firms have experienced a fall in employment over the last 20 years. Job numbers initially increased in firms located in the new member states, although they too have since experienced (less marked) downward pressures on employment levels. In all cases, many of the lost jobs have been low skilled.

> The groups that are mostly hit are of two types: first, semi-skilled workers, both white and from ethnic minorities and, second, low skill mainly factory workers, and mostly women; these are white working class poor and also black: mainly of Pakistani and Bangladeshi origin. (UK, key informants)

> It is difficult to make people do something, mainly because the textile sector is perceived as a declining sector, which in fact is not true … investment should be made in youth to improve their skills in order to change the image of the sector as employing rather elder people with very simple skills. (Poland, key informants)

Growing wage demands, negative worker perceptions of employment in the clothing sector and the availability of alternative forms of employment appear to be making recruitment and retention very difficult in all sample countries. Competition from the service sector and grey economy are also making it difficult to recruit new employees to low-end jobs (Aniello 2001). These are often filled by people who are less mobile and who are specifically interested in very flexible working arrangements, as well as by migrant workers (Husband and Jerrard 2001).

> Being the simplest intellectual job, the dominance of the textile sector led to vocational and social marginalisation of women … The losers include the economically weakest and least mobile groups, unable to reskill due to manual, mental and intellectual barriers, that is, mainly elderly people and women … People who lost their work in different sectors go abroad in large quantities in search for a job … However, this refers to the clothing sector only to a little extent. Most workers are women and they are much less mobile. (Poland, key informants)

> A lot of the new jobs in the UK are characterized by low pay, flexible hours, and this also means low skill requirements. There are many migrants who are filling this sort of vacancies; call centres are a good example. East Europeans are filling many of those vacancies, but my impression is that the big majority of them are not here to stay but only come for 4–5 years in order to gain some experience or save some money, the wages are still significantly higher here, and then they tend to go back to their countries of origin in order to establish themselves there. This, however, creates strong local tensions particularly in the post-industrial areas and opens space that is, exploited by the BNP … (UK, key informants)

> You need continuity though, especially with our game . I have got a girl called J.R. She is just a walking talking dictionary of clothing if you like ... she enjoys talking to people we deal with, she has worked for the company from day one. She came from another company as a specialist trainee designer there and is indispensable basically, you know With her skills she is very specialized, there aren't many jobs out there for her. On the other hand, there aren't many jobs that are out there for us ... You can't get somebody from an agency to come and do that job, so we do it so that each member of staff can cover for each other whilst on holiday. So there is not an accumulation of work for when they get back. (UK company)

Difficulties in recruiting workers to high-end jobs such as design, sales and marketing are also apparent, leading to a limiting effect on companies' abilities to upgrade and necessitating a strong emphasis on internal training initiatives.

The clothing sector has, historically, been concentrated in specific geographical regions (such as the English Northwest, Northern Greece, the Lodz area of Poland and Southwest Bulgaria), all of which have been heavily affected by the internationalization process. The process of restructuring has exercised a significant and often localized effect on these regions (compounded by the fall of communism and EU expansion in the case of Eastern European).

Overview of Findings

The methods deployed here make it difficult to capture the dynamics of the internationalization process. However, it is probable that FDI and JVs – though developing early in the process of integration (in the early- to mid-1990s) – were preceded by an initial period of lower commitment strategies, coinciding with early reforms in Eastern Europe. The timing of integration, allied to early engagement in the integration process, would therefore seem to be of particular importance in enabling all European clothing firms to exploit global opportunities.

Market segmentation appears to follow the pattern anticipated in the literature in the case of three sample countries. UK clothing firms concentrate primarily on design and flexible-response strategies, the Bulgarian firms' focus is on price-sensitive markets and Polish firms are heavily reliant on price competitiveness (whilst moving gradually up-market towards flexible-response segments). Estonian firms appear to have moved further in terms of market-segment development than Bulgaria or Poland, whereas the Greeks would seem to have been particularly slow in moving on from the price-sensitive segment. There is also a clear recognition among Greek, Bulgarian and Estonian enterprises, smaller ventures and foreign-owned firms of the significant inequality in power endowments between buyers and manufacturers.

Competence lock-in strategies occur not only in price-sensitive market segments, but also where success is conditional upon flexible response. *Hybrid strategies* are also employed, leveraging competences developed through global networks of production and distribution for domestic market advantage. There is little evidence of *competence break-out strategies* (and none for Eastern Europe). However, the findings do offer insights into the process of transition from one type of strategy (such as *competence lock-in*) towards another (*hybrid*).

Enterprise strategies are linked to successful adjustment, but distinctive strategies do not appear to be appropriate for particular market segments or countries. Successful adjustment appears rather to depend on the fit between strategy, context (market segment and country) and enterprise characteristics. Successful adjustment is reported both by enterprises located in countries that enjoy lower labour costs (such as Bulgaria), and also by firms in the UK and Greece that do not. A greater incidence of strongly performing enterprises can be identified for post-socialist countries, although there are also some strongly performing enterprises in the UK and Greece – further underlining the distinction between the 'fortunes' of the firm with those of the clothing industry as a whole.

In terms of governance arrangements, a gradual shift in emphasis appears to be taking place from the public to the private realm. This trend may be reinforced by the increasing liberalization of prevailing trade regimes, especially as (national and regional) governments appear to allocate low priority to assisting with the survival of the industry.

Public views regarding the future (or lack of one) of the sector reflect the social consequences of a long period of clothing sector decline in DCs. This, combined with a record of relatively low-paid and low-skilled jobs offering limited career progression opportunities, means that the sector is confronted with human capital constraints, as it often fails to attract young and dynamic individuals. Rather perversely, it is the history of decline that may reinforce decline in the future.

Conclusion

The findings provide useful insights into the process of integration of enterprises and regions in the global network of production and distribution. Opportunities can best be exploited by a short period of early engagement, involving low-commitment strategies, followed by an era of high-commitment strategies with significant foreign investment and JV creation. These give way later to a period of deepening global integration, when organic integration and expansion, and relationships based upon a combination of power and mutual confidence, are crucial.

A gradual shift appears to be occurring from publicly- to privately-driven forms of governance, reinforcing the importance of such relationships. Powerful agents may maintain strong relationships, providing grounds for doubting the prevailing normative views regarding enterprise strategies. Enterprise survival and growth (even in the long term) may not be attached to a relentless pursuit of up-grading (to counter a perceived relentless pursuit of cheap labour by buyers). A number of alternatives may be open to enterprises, depending on the specificities of the context and the enterprise. This more open-ended (and less deterministic) view of enterprise strategy may be deployed effectively at the micro-level.

The fortunes of the enterprise may not be linked inextricably with those of its region. Successful enterprises may be based in regions where employment in the industry has been decimated, whilst enterprises may also fail in regions experiencing rapid growth. Existing research maintains a predominantly regional focus, influenced by the greater practicality of researching different firms in one locality, rather than tracking entities across different geographical locations. Research in the latter direction, though, may provide an alternative (and complementary) approach to the study of global patterns of change in the clothing industry.

References

Abernathy, F.H., Dunlop, J.T., Hammond, J.H. and Weil, D. (1999), *A Stitch in Time: Lean Retailing and the Transformation of Manufacturing: Lessons from the Textile and Clothing Industries* (Oxford: Oxford University Press).

Aniello, V. (2001), 'The Competitive Mezzogiorno (Southern Italy): Some Evidence from the Clothing and Textile Industry in San Giuseppe Vesuviano', *International Journal of Urban and Regional Research* 25:3, 517–536.

Appelbaum, R.P. (2004) 'Commodity Chains and Economic Development: One and a Half Proposals for Spatially-Oriented Research. Global and International Studies Program', University of California, Santa Barbara, Paper 36, <http://repositories.cdlib.org/gis/36>, accessed 20 February 2007.

Appelbaum, E. Bailey, T., Berg, P. and Kalleberg, A.L. (2000a), *Manufacturing Advantage: Why High Performance Work Systems Pay Off* (Ithaca: Cornell University Press) <http://www.cornellpress.cornell.edu/>, accessed 20 February 2007.

Appelbaum, R.P., Bonacich E., Esbenshade, J. and Quan, K. (2000b), 'Fighting Sweatshops: Problems of Enforcing Global Labour Standards. Global and International Studies Program', University of California, Santa Barbara, Paper 34, <http://repositories.cdlib.org/gis/34>, accessed 20 February 2007.

Appelbaum, R.P., Bonacich, E. and Quan, K. (2005), 'The End of Apparel Quotas: A Faster Race to the Bottom?, Global and International Studies Program', University of California, Santa Barbara, Paper 35, <http://repositories.cdlib.org/gis/35>, accessed 20 February 2007.

Bair, J. and Gereffi, G. (2003), 'Upgrading, Uneven Development, and Jobs in the North American Apparel Industry', *Global Networks* 3, 143–69.

Begg, R. and Pickles, J. (2000), 'Ethnicity, State Violence, and Neo-liberal Transitions in Postcommunist Bulgaria', *Growth and Change* 31:3, 179–210.

Castells, M. (1996), *The Rise of the Network Society*, 2nd edition (Oxford: Blackwell).

Dunford, M. (2002), 'The Changing Profile and Map of the EU Textile and Clothing Industry', School of European Studies, University of Sussex, <http://www.geog.susx.ac.uk/research/eggd/ege/pdf/02_t.pdf>, accessed 20 February 2007.

Enright, M, Scott, E. and Dowell, D. (1997), *The Hong Kong Advantage* (Oxford: Oxford University Press).

Gereffi, G. (1999), 'International Trade and Industrial Upgrading in the Apparel Commodity Chain', *Journal of International Economics* 48:1, 37–70.

Gereffi, G. and Memedovic, O. (2003), *The Global Apparel Value Chain: What Prospects for Upgrading by Developing Countries* (Geneva: United National Industrial Development Organization, Sectoral Studies Series).

Gereffi, G. and Mayer, F. (2004), 'The Demand for Global Governance. Terry Stanford Institute of Public Policy', Duke University, Working Paper Series San04-02, <http://www.pubpol.duke.edu/research/papers/SAN04-02.pdf>, accessed 27 February 2007.

Gibbon, P. (2002), 'At the Cutting Edge? Financialisation and UK Clothing Retailers' Global Sourcing Patterns and Practices', *Competition and Change* 6:3, 289–308.

Giddens, A. (1990), *Consequences of Modernity* (Cambridge: Polity).

Hanzl-Weiss, D. (2004), 'Enlargement and the Textiles, Clothing and Footwear Industry', *World Economy* 27:6, 923–45.

Hassler, M. (2004), 'Governing Consumption: Buyer-Supplier Relationships in the Indonesian Retailing Business', *Tijdschrift voor Economische en Sociale Geografie* 95:2, 206–17.

Humphrey, J. and Schmitz, H. (2002), 'How Does Insertion in Global Value Chains Affect Upgrading in Industrial Clusters?', *Regional Studies* 36:9, 1017–27.

Husband, J. and Jerrard, B. (2001), 'Formal Aid and Informal Sector: Institutional Support for Ethnic Minority Enterprise in Local Clothing and Textile Industries', *Journal of Ethnic and Migration Studies* 27:1, 115–31.

Jessop, B. (2002), *The Future of the Capitalist State* (Cambridge: Polity).

Kalantaridis, K., Slava, S. and Sochka, K. (2001), 'Globalization and Industrial Restructuring in Post-Socialist Transformation', CES Working Paper No. 2.

Karagozoglu, N. and Lindell, M. (1998), 'Internationalisation of Small and Medium Sized Technology Based Firms: An Exploratory Study', *Journal of Small Business Management* 36:1, 44–59.

Kwan, Y.K. and Qiu, L.D. (2003), 'The ASEAN+3 Block', Hong Kong University of Science and Technology Business School, Working Paper Series, <http://www.bm.ust.hk/~larryqiu/ASEAN.pdf>, accessed 20 February 2007.

Labrianidis, L. and Kalantaridis, K. (2004), 'The Delocalization of Production in Labour Intensive Industries: Instances of Triangular Manufacturing between Germany, Greece and FYROM', *European Planning Studies* 12:8, 1157–74.

Nadvi, K. and Waltring, F. (2002), 'Making Sense of Global Standards', INEF Report 58, University of Duisburg, <http://inef.uni-due.de/page/>, accessed 20 February 2007.

Nathan Associates Inc. (2002), 'Changes in Global Trade Rules for Textile and Apparel: Implications for Developing Countries', Research Report prepared by Peter Manor on behalf of Nathan Associates Inc. for USAID 20 November, Washington, <http://www.nathaninc.com/>, accessed 25 February 2007.

Nordas, H.K. (2004), 'The Global Textile and Clothing Industry Post the Agreement on Textile and Clothing', Discussion Paper 5 (Geneva: WTO) <http://www.wto.int/english/res_e/booksp_e/discussion_papers5_e.pdf>, accessed 20 February 2007.

OECD (2004), 'A New World Map in Textiles and Clothing: Adjusting to Change', OECD Industry, *Services & Trade* 20, 211–54.

Palpacuer, F., Gibbon, P. and Thomsen, L. (2005), 'New Challenges for Developing Country Suppliers in Global Clothing Chains: A Comparative European Perspective', *World Development* 33:3, 409–30.

Pye, M. (2004), 'An Assessment of Skills Needs in the Clothing, Textiles, Footwear and Leather and Furniture, Furnishings and Interiors Industries', Pye Tait Limited, <http://www.dfes.gov.uk/skillsdialoguereports/docs/SD12_%20Clothing.pdf>, accessed 25 February 2007.

Schmitz, H. and Knorringa, P. (2000), 'Learning from Global Buyers', *Journal of Development Studies* 37:2, 177–205.

Taplin, I.M., Winterton, J. and Winterton, R. (2003), 'Understanding Labour Turnover in a Labour Intensive Industry: Evidence from the British Clothing Industry', *Journal of Management Studies* 40:4, 1021–46.

Tokatli, N. and Eldener, Y.B. (2004), 'Upgrading in the Global Clothing Industry: The Transformation of Boyner Holding', *Competition and Change* 8:2, 173–93.

Tully, J. and Berkley, N. (2004), 'Visualising the Operating Behaviour of SMEs in Sector and Cluster: Evidence from the West Midlands', *Local Economy* 19:1, 38–54.

Winterton, J. and Winterton, R. (2002), 'Forecasting Skills Needed in the UK Clothing Industry', *Journal of Fashion Marketing and Management* 6:4, 352–62.

Yeung, G. and Mok, V. (2004), 'Does WTO Accession Matter for the Chinese Textile and Clothing Industry?', *Cambridge Journal of Economics* 28:6, 937–54.

Zeitlin, J. and Totterdill, P. (1989), 'Markets, Technology and Local Intervention: A Case Study of Clothing', in P. Hirst and J. Zeitlin (eds), *Reversing Industrial Decline? Industrial Structure and Policy in Britain and Her Competitors* (Oxford: Berg Publishers Limited), 155–90.

Chapter 7
The Impact of Delocalization on the European Electronics Industry

Rünno Lumiste

Introduction

The electronics industry is one of most globalized activities in the world (Malecki 1997). Among the foreign-trade commodity groups, only raw materials and fuels trade value exceeds the trade of electronics products. Different from raw materials industries, the electronics industry employs a considerable number of people (Lall et al. 2003; Belderbos and Zou 2005). Labour intensity and the search for new solutions both by public and private actors have caused substantial delocalization in the electronics industry (Chang and Rosenzweig 1995).

Active delocalization in the electronics industry happens in the form of FDI, outsourcing and foreign trade (Blinder 2005). FDI has effects on both host and home economies. The electronics industry, like the car industry, creates several links between TNCs and local firms. The importance of such links in the 1960s and 1970s was already pointed out by Albert Hirschmann (1981). Further research has concentrated on technology and market spillovers (Blomström and Kokko 1998).

Recognition of the importance of such spillovers has caused the increased activity of national governments (Liagouras et al. 2006) and the EU (von Tunzelmann 2004). The interest of national governments in electronics is multidimensional. Higher employment rates, increased earnings and acceleration of learning processes are common and traditional goals for national governments. The electronics industry, however, presents greater challenges for national governments than simply increasing employment. Products and components in the electronics industry play a substantial role in national security and provide links between different knowledge-intensive industries. Military electronics is the main field of research for the development of defence systems. Electronics industry products play a major role in emerging knowledge-intensive industries such as alternatives to fossil-fuel energy generation, the medical products industry and nanotechnology.

One paradox of the electronics industry is that it calls for people with various qualifications; success requires both highly talented creative people and less qualified assembly workers. This dual-nature paradox of the workforce is the source of changes and innovations in electronics. Countries and firms with smaller knowledge bases and with large shares of manufacturing try to create their 'own' engineering capabilities, and countries and firms with strong know-how outsource manufacturing activities or try to automate labour-intensive operations.

Delocalization has been of interest to several researchers (Labrianidis and Kalantaridis 2004). There has been comprehensive research on different forms of delocalization, such as FDI, the functioning of TNCs (Caves 1996), outsourcing (Radosevic 2003), social consequences of delocalization to the regions and government policies in facing delocalization. There is also substantial research on relocations in East Asia and in the North American Free Trade Area (NAFTA) (Hennart et al. 1998; Mata and Portugal 2000; Belderbos and Zou 2005).

There has been less research on the delocalization of the electronics industry in Europe and especially in the enlarged EU. The 2004 enlargement of the EU created a totally new situation for the electronics industry. Global technological development and the reduction of trade barriers have made developments in electronics so rapid that those changes could be considered ahead of the current state-of-the-art in academic literature. One sign of such a gap between real life and research is the use of different surveys and even periodicals as a source of references, a practice which is relatively uncommon in articles dating back a decade and more (Radosevic 2003; Belderbos and Zou 2005).

Economic Position of the Electronics Industry

Supply chain(s) of the electronics industry

Two basic functions of electronic devices are the controlling and processing of data and the conversion and distribution of electric power. Generation and distribution of electrical power is a relatively stable field compared to data processing. The marketing and production process of electronics products are international by their nature. Large, integrated electronics producers offer a global nature to the industry by producing goods and marketing them globally (Hewitt-Dundas et al. 2005).

The electronics industry supply chains present constant changes and responses to technological and economic challenges. Supply chains of the industry are changing all the time. Even for the same type of product, there could be and often are supply chains with different configurations. The particular design of a supply chain depends on the product's nature, geographical location and time period.

The consumer electronics supply chain still consists of large, integrated firms that produce almost the entire product with the exception of some key components. Based on intellectual capabilities/patent portfolio, financial capabilities, business strategy and the availability of technologies, consumer electronics firms build the products from their own or sourced components.

A generalized scheme of the supply chain of consumer electronics products is presented in Figure 7.1. The principal actors in the supply chain are product-developing firms, component producers and assembly firms. The main manufacturing chain is supported by parallel chains of the service firms. Due to increasing environmental protection, a substantial part of the supply chain is the recycling of used electronics products. Knowledge-intensive parts of the supply chain could be considered the fabrication of production equipment, component production and design and testing of products.

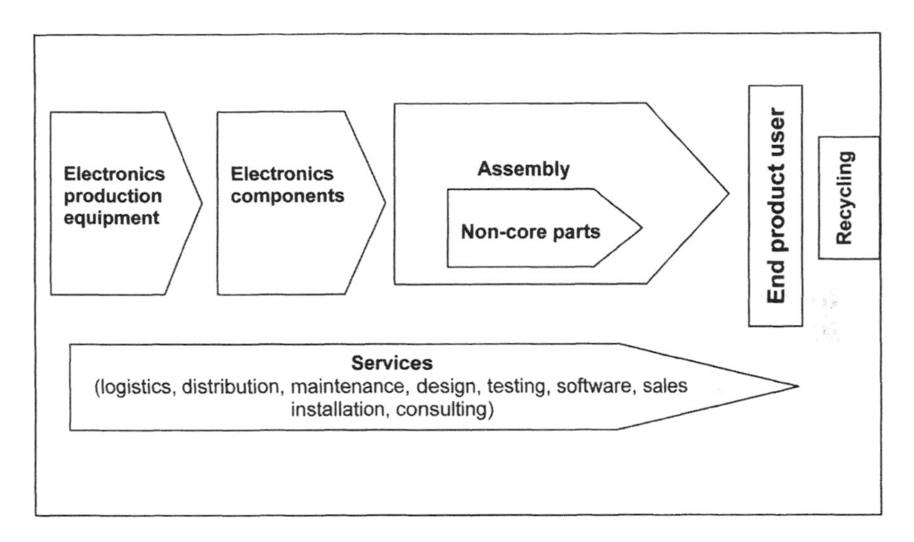

Figure 7.1 Supply chain in the electronics industry in 2004
Source: Adapted from Booz Allen Hamilton and IFC 2003.

Supply chains in telecommunications present an example that is different from those of consumer products. Supply chains in wireless communication could be considered as the most dynamic, and set the example for the entire electronics world in 2006. The significant telecommunications boom in the 1990s forced Original Equipment Manufacturer (OEM) firms to use subcontractors for extra capacity. During this period, Electronics Manufacturing Service (EMS) firms learned sourcing and assembly operations and built up production capacity. New EMS firms had very different backgrounds. Some of them were the daughters of integrated firms (Lohja-Elcoteq, IBM-Celestica), some developed from start-ups (Solectron) and some were electronics firms that specialized in offering manufacturing services. Large changes in the telecommunications supply chain occurred with the burst of the IT, dot-com bubble in 2001–2002. OEM firms were forced to reduce their personnel and to relocate their production. EMS firms that had built up substantial production capacity had to specialize, optimize the production capacity and also to relocate their production to countries with cheaper labour. The early twenty-first century has been accompanied by the extension of services offered by contracting firms. Contract manufacturers have added product design services to their portfolio (Figure 7.2). New types of firms are called Original Design Manufacturers (ODM). From the telecommunications markets, ODM firms are now expanding to medical instruments, industrial electronics markets and other sectors (Booz Allen Hamilton and IFC 2003).

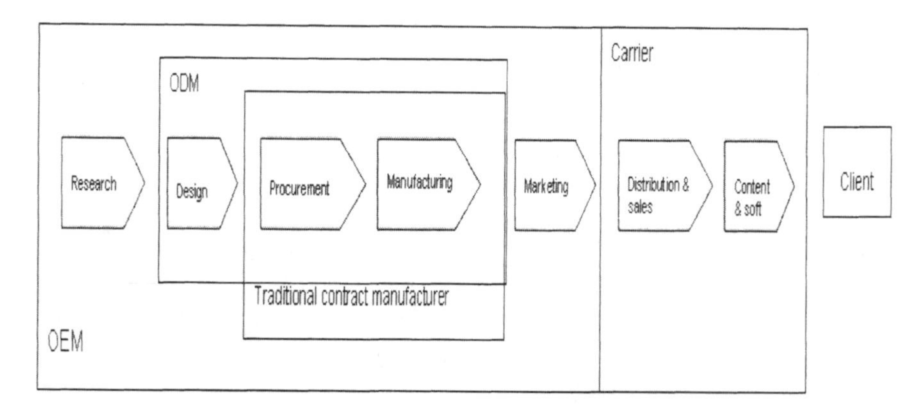

Figure 7.2 Value chain in the telecommunication industry 2005
Source: Adapted from Infineon Technologies 2004.

One factor influencing the supply chains in the electronics industry is that several firms use shared technology platforms between different products. They use similar components and technologies for different products. For example, there is substantial technological proximity between computer monitors made with LCD (liquefied crystal display) technology and LCD TV sets. Companies strong in LCD technology employ it in several applications (TVs, monitors, video cameras, mobile phones).

Electronics firms are also participating in supply chains of other industries, such as the automotive, medical and military. Sometimes they are parts of integrated and sometimes independent entities. Concentration in car manufacturing, aerospace and military industries has led to the reduction of suppliers. The growing medical and healthcare sector has benefited new entrants into medical electronics.

Choice of location for different activities depends basically on supply and demand side factors. Among the supply factors are prices such as general labour cost, availability and cost of specific labour, that is engineering graduates, materials cost, transport cost, environmental cost and land cost. Choice of location is influenced also by proximity to technology-creation regions, technology factors such as software and hardware interaction, and legal issues such as local legislation and standards. A unique role is played by company-specific factors, such as knowledge of a particular market or favourable political relations between countries.

Demand factors consist of consumers with the number of population, purchasing-power and technology preferences. Industrial electronics site location is determined by the location of related industries, such as automobiles, medical equipment and avionics.

During the decade 1995–2005 favourable cost conditions for manufacturing have created new factories in China, Brazil, Mexico and Eastern Europe. Among the fastest growth consumer markets are China, Mexico, Russia and Brazil. Traditionally, the largest electronics products consumer has been the USA and the market of consumers with the most sophisticated demand Japan.

With technical innovations it is possible to reduce size parameters. Decreasing product size has reduced the share of transport cost and allows the use of suppliers from more distant locations. Decreasing transport costs also allow scale-intensive production and concentration of industrial activities. In the period 1980–2000, revenues per ton mile decreased by 30 per cent for air freight and more than 50 per cent for railway (*Economist* 2004).

Global sourcing of components has initiated the consolidation of electronics distribution and logistics firms. Enterprises such as Ingram Micro and Arrow Electronics are present in most of the countries and are able to deliver large quantities of components and offer services. Changes in the supply chain cause concentration of retail outlets for mature products such as home electronics. For example, the USA's biggest retailer, Wal-Mart, is also the number one consumer-electronics retailer (Forbes 2008). With the expansion of supermarkets, the competition between specialized home electronics stores and general supermarkets is expected to increase.

The European electronics industry by geography

By the geographical distinction used by industry experts, Europe could be divided into three regions: Western Europe (EU–15, EFTA and among it the UK, Ireland, France, Germany, Italy, Netherlands, Spain, Switzerland, Nordic Countries), Central Europe (first-wave new members of the EU or 2004 accession members; Poland, Hungary, Czech Republic, Slovenia, Baltic States) and Eastern Europe (Bulgaria, Romania, Ukraine, Russia) (Carbone 2006). Every region has been specializing in certain parts of the value chain and activities. The relative share of employment in the electronics industry in the EU in 2003 is shown in Figure 7.3.

The major headquarters of European TNCs, research centres, design firms and several component and materials suppliers both European and non-European origin are located in Western and Northern Europe. There is also a substantial but declining manufacturing base. Manufacturing units in Western Europe remain as producers of top models of products and components (Philips' wide LCD panels, new semiconductors, medical electronics products and defence electronics products), transform into logistics and service centres, or cease to exist. Furthermore, firms using a substantial volume of electronics such as car factories and aircraft producers are located in Western Europe. The largest electronics manufacturers are Germany and France. Switzerland occupies a niche of top semiconductors. The UK is the location of several design firms and basic research institutions. The factories of non-European electronics firms are primarily located in Spain and Ireland.

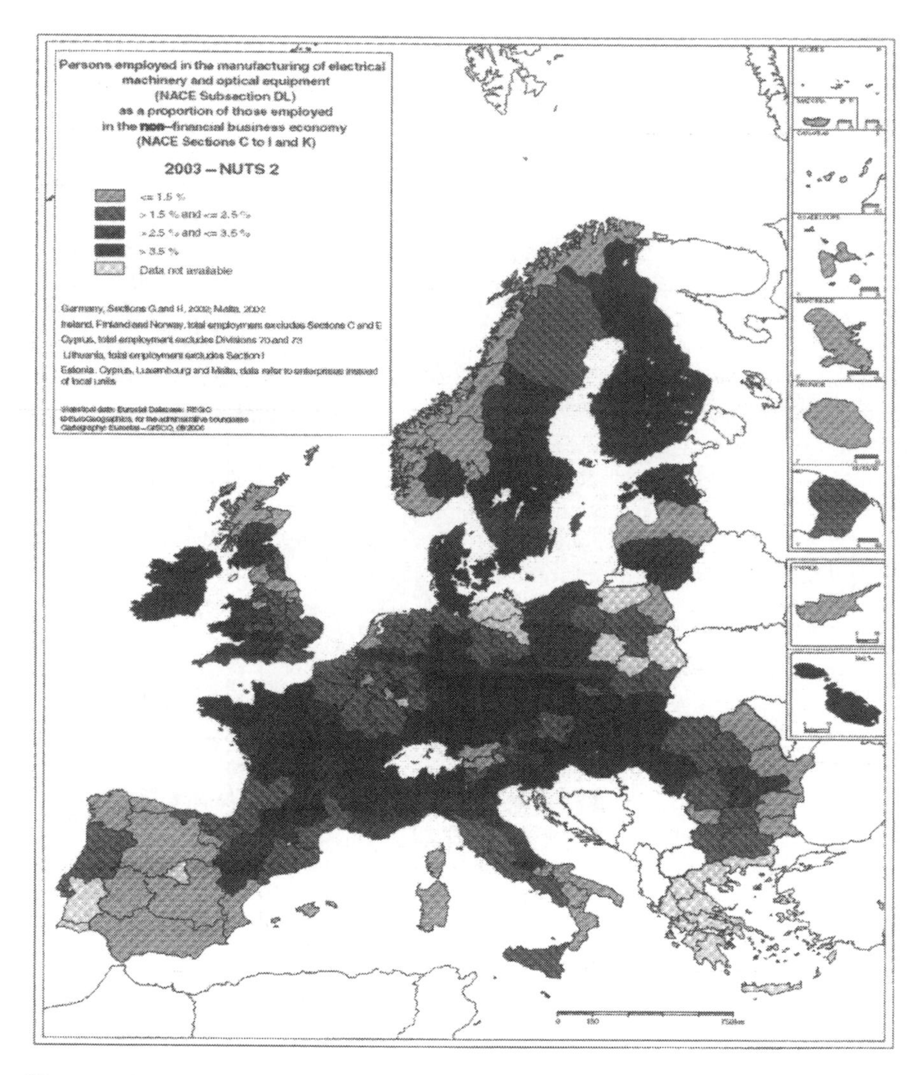

Figure 7.3 Employees in the electronics industry
Source: European Commission 2006:170, © European Communities 1995–2008.

The Enterprise Survey conducted during the MOVE Project in five European countries[1] and 190 electronics firms also showed different functions in the supply chain, change of role in the supply chain and different background and history of firms. The UK is the birthplace of several electronics-industry major inventions, products and technologies. However, active delocalization of firms and operations has caused a substantial decrease in the physical production numbers. Both local and international electronics companies no longer see the UK as a location for the manufacture of labour-intensive mass-products. The role of UK firms is product development and design. Unlike SME firms in Eastern Europe and other emerging locations, several British SMEs have proprietary knowledge, patents and well-known trademarks. Patent portfolio and intellectual property management ensure that inventors also have future resources for research and product development. A different picture of the development path of the electronics industry is presented by Greece. Greece, with a relatively limited tradition in electronics, has as its main function the servicing of other economic sectors such as telecommunications, medicine and maritime. Single-entrepreneur engineering firms constitute the second layer of enterprises supplying large firms and public institutions. A relatively significant share of electronics firms in Western Europe are suppliers for military contractors such as Thales, BAE Systems and Finmeccanica (Labrianidis et al. 2007).

The electronics industry in CEE faced a tough transition period from the end of the 1980s. Enterprises faced the decision to reprofile their activities or cease to exist. This period was particularly difficult for firms producing outdated final products (Labrianidis et al. 2007). The growth of the electronics industry started from the offering of manufacturing services mainly for Western European firms (Radosevic 2003).

Central Europe is now a region of manufacturing units with some design capabilities. Major international contract manufacturers (EMS), European OEM firms and local suppliers are located in Central Europe. Major engineering and design operations are committed to foreign-owned firms. Contacts of Central European firms are based on logistical proximity and trade ties. For example, Polish firms tend to communicate more with German firms while those from Baltic States deal with Finnish and Swedish firms. For example, due to transport costs, previous traditions and the availability of a labour force, Poland has specialized as a mass consumer electronics producer, Estonia as a small batch industrial electronics and telecommunication producer and Hungary as a car electronics and telecommunication producer with a growing electronics industry services sector (Labrianidis et al. 2007).

Eastern Europe is a new opportunity for the next wave of electronics industry siting. In Eastern Europe there are similar types of contract manufacturers and firms that missed the first wave of relocation from Western to Central Europe. Russia, Romania and Ukraine have substantially large internal markets and growing

1 UK, Greece, Poland, Bulgaria and Estonia (Labrianidis et al. 2007).

purchasing power and the need for customization could be a motivation for transfer of manufacturing operations to these locations. Taking into consideration the lower level of labour costs and similar level of qualifications, there could be substantial relocation of manufacturing operations from CEE in the period between 2007 and 2015.

Forms of Delocalization

Trade of electronics goods

Trade of electronics products between countries and trade blocks is influenced by duty rates. The highest duty rates on the import of electronics products are in China and India (between 15 and 30 per cent for different products) and the lowest duty rates are in Japan and the USA (between 0 and 5 per cent). The general trend is towards the lowering of tariffs via the WTO IT Agreement (Borrus and Cohen 1997). Tariffs for electronics imports have two dimensions – geographical and technological. Inside the trading block, countries have lower or zero duty rates, and for outsiders the rates are higher. The EU has low rates for electronics goods for free-trade-agreement neighbouring countries: Turkey, Switzerland, Norway and the Ukraine. The United States has lower rates for Brazil, Israel and NAFTA partners: Mexico and Canada. ASEAN countries have low duty rates among themselves (Fukase and Martin 2001).

Another dimension of the electronics trade is the different duty levels for components and end-products. In 2006, the EU taxed TV sets and DVD players at the rate of 14 per cent and main components at a 4.9 per cent rate (European Commission Taxation and Customs Union). Different rates for components and finished products act as incentives for establishing local assembly units (Borrus and Cohen 1997).

Trade in computers and their parts have grown almost tenfold during the last 15 years. The fastest period of growth was in the beginning of 1990s with computerization and the adoption of personal computers by a wide group of users. The earliest leader in computer trades was the USA, but relocation of factories into East Asia and the growth of indigenous Asian producers has shifted the main focus of trade to Asia. DCs rely on the imports from developing East Asian countries.

Main product groups and volume of trade are presented in Table 7.1. Traditionally the EU has been stronger with positive trade balance in optical goods and medical instruments. The EU has been weaker in the trade of data processing goods such as computers and peripherals.

Table 7.1 Volume of world trade of electronics products* (in billion dollars)

	1990	1995	2000	2005
Data processing goods				
Export	27.5	123.4	198.8	259.3
Import	29.8	137.5	218.4	272.4
Electrical equipment, electrical components, electronics components, telecommunication, car electronics				
Export	162.6	631.6	983.6	1338
Import	125.2	610.3	992.9	1365
Trade of optical, photographic, measuring and medical instruments				
Export	39.9	133.9	194.2	320
Import	35	131.6	191.9	316

Note: * Different figures are the result of excluding reexport and reimport.
Source: United Nations Comtrade database.

Mergers and acquisitions

Delocalization in the electronics industry has different forms, and often firms in different geographical locations are involved. Motivations for delocalization can be divided into marketing, economic and technological. The main forms of delocalization are creation of JVs, selling shares to foreign investors, buying foreign firms, outsourcing different activities and taking subcontracting from other firms. Mainly for marketing reasons, in developed markets Taiwanese and Chinese firms have acquired entities and rights to use well-known brand names in Europe and the United States. In 2005, Taiwan- based BenQ acquired Siemens, AG's mobile phone business and rights to use Siemens's trademark for mobile phones for a certain period. Similar deals have been conducted between Lenovo (China) with IBM (USA) and TCL (China) with Alcatel and Thomson (both of France).

The motivation behind the creation of JVs could be the adjustment of supply and technological partnership. Adjustment of supply of LCD panels was the motivation for the creation of a JV between Sony and Samsung in Tangjung in South Korea. The motivation to develop and launch new products rapidly is behind the technological partnership contracts. New products use a number of different technologies, and for the swift launching of products it is the optimal way to match with the companies whose core competence lies in needed technologies.

Becoming a subsidiary is an option for firms in consolidating industries such as computer manufacturing, telecommunication equipment and electronics component logistics. The last large British electronics firm Marconi-GEC was forced to sell its shares to Ericsson after the loss of competition for modernizing the UK telecommunication infrastructure (21st Century Network project). Major acquisitions and mergers in electronics have been related to consolidation of industries such as Hewlett-Packard with Compaq in the computer sector and Alcatel with Lucent in the telecommunications equipment sector.

Having foreign shareholders is also a common practice in the electronics sector. Major electronics firms in Europe and the USA are listed companies with a large number of shareholders, among whom financial institutions play a major role. Asian firms have more consolidated ownership structure and family holdings in the case of Japan, Korea and Taiwan. Government participation is common in China.

Subcontracting is the process of using other firms in contract bases in one's own supply chain for an enterprise. Short delivery time and the need for capital drive firms towards the use of different subcontractors. Over time several shorter-term partnerships have developed into longer-term outsourcing processes.

Outsourcing is the delegation of operations that are considered as non-core for firms. Outsourcing of manufacturing activities has created whole new industries such as contract manufacturing, third party logistics (3PL) and semiconductor manufacturing. Larger electronics firms have outsourced, in the first case, part of their manufacturing activities, and in the second case, logistics and sourcing activities. Semiconductor production has been split into firms committed to design and sales activities called fabless[2] firms and semiconductor foundries called fabs,[3] whose main task is production. Most contract manufacturing firms have close ties with one to three major customers (Labrianidis et el. 2007).

Factors Affecting Delocalization

Production cost and markets

An initial reason for the transfer of operations from one country to another was price difference for labour. Taiwan, Singapore and Korea were good places for relocation of USA and Japanese manufacturing firms in the 1980s and 1990s. Price differences are still one major reason for the relocation of manufacturing operations, but are not the only one.

Salaries in mainland China are very low compared to DCs. According to statistics, in Shenzen (near Hong Kong) the minimum wage is $72.50 per month. In addition, workers receive housing in dormitories and food from the company. At the same time, farmers in the central and western provinces earn on average $400 per year or less, while workers in large urban areas such as Beijing and Shanghai typically make four times as much, according to PRC government

2 Semiconductor design and sales firms that do not own a manufacturing unit.

3 Semiconductor manufacturers without a design unit.

statistics. According to labour-relations observers and journalists, real salaries are sometimes lower than announced, and a deduction of approximately $30 is made from a $75–100 salary. Employees at a factory typically work 11 hours a day, six days a week, and rack up to 70 hours of overtime a month (McLaughlin 2006).

Salaries are part of the general cost structure. In addition to labour costs, there are other factors that substantially increase the price of bringing the product to market. Operating in different time zones, compressed time frames, visa requirements and higher transport costs are factors limiting the contracting into distant regions. Outsourcing has become more and more process-driven by strategic opportunities (Industry Directions 2005).

Time to market, asset reduction and specialization are becoming more and more important for outsourcer firms. A survey conducted among European firms (MOVE survey; Labrianidis et al. 2007) showed that among the major reason for receiving orders by firms were felt-expertise, reliability and appropriate technology.

Outsourcing has not only helped to save the production cost of international enterprises. It has also provided jobs, helping people to start new lives in urban environments and, in general, raising living standards. In CEE, outsourcing has created new industrial goods markets and in East Asia new consumer markets. Outsource-providing enterprises, which in the beginning delivered only to markets in DCs, have also started to sell products in domestic markets such as China and East Asia. Mass manufacturing has driven prices down and consumers in China, India and the rest of the markets previously considered as undeveloped are able to buy mobile phones and TV sets. Governments in Asia are trying to improve infrastructure and are therefore becoming important customers for telecommunications and other infrastructure-related electronics products. The importance of Asian markets is estimated to grow further (see Table 7.2).

Market factors and government policies are giving an impetus to manufacturing. The fastest market growth is happening in East Asia and the Pacific region. It is expected that in a few years China will become the major manufacturing region for electronics production (see Table 7.3).

Table 7.2 Total electronics, world market share by region consumption

	World (%) 2005	World (%) 2010	Billion €		Growth* (%)	Difference between production and consumption (2005)
			2005	2010		
Total world	100	100	1,070	1,428	6.0	
Europe	27	25	292	358	4.2	-66
North America	31	28	329	404	4.2	-93
Japan	11	10	117	136	3.1	49
China	10	12	106	176	10.7	136
Other Asia-Pacific	11	13	117	187	9.8	39
Rest of the World	10	12	109	167	8.9	-66

Note: * Compound annual growth rate.
Source: Adapted from Decision Etudes and Conseil 2006.

Table 7.3 Total electronics, world production by region

	World (%) 2005	World (%) 2010	Billion €		Growth* (%)
			2005	2010	
Total world	100	100	1,069	1,428	6.0
Europe	21	19	226	272	3.8
North America	22	20	236	284	3.7
Japan	16	14	166	193	3.0
China	23	28	242	394	10.3
Other Asia-Pacific	15	15	156	219	7.0
Rest of the World	4	5	43	65	8.8

Note: * Compound Annual Growth Rate.
Source: Adapted from Decision Etudes and Conseil 2006.

Technology and education

The most important single factor determining the drop in electronics products prices is the development of technology. Among the factors determining the development of technology that might be mentioned is the favourable environment of financial, technological and support institutions and a workforce with appropriate knowledge, skills and attitudes (Radosevic 2003).

Electronics industry localization choice is dependent on technology factors such as the proximity to technology creation centres. Several countries have tried to copy the Silicon Valley model, where technological knowledge is combined with entrepreneurship and venture capital financing, but the level of success has not been equal in Japan or the EU.

Probably the most important factor determining the volume of introduction of new products is high-technology entrepreneurial culture. Very often in Europe, the attitude is that the main supporter of new product introduction should be the public sector with public funding. Mainstream financial institutions are also hesitant to finance new spin-offs. A curious case from 2005 should be mentioned: a USA venture capital fund was behind the success of such innovations as Skype, which was initially developed by Swedes and Estonians (Labrianidis et al. 2007).

The first and most visible policy from promoting electronics and software industries in emerging economies is the education of engineers and scientists. Educating engineers in most of the technology areas requires substantial investments from universities and personal efforts by students. The number and quality of engineers are the concern of governments in highly developed and emerging economies (see Table 7.4). The USA and UK have used simplified visa requirements for programmers, engineers and other highly qualified workers. Eighty per cent of Asian-born foreign students (around 650,000) are studying in the USA and Europe (UNESCO 2004). Two-thirds of them remain in the USA.

Table 7.4 Number of university graduates and science and engineering graduates

	University graduates (million)	Science and engineering graduates 2003 (thousands)
India	3.1 (2003) -6 (2010)	316
China	2.8 (2004) -3.5 (2005)	337
Russia		216
EU–15	2.0	290 (2001)
Japan	1.1 (2001)	250 (2001)
US	2.2 (2001)	380 (2001)

Source: Eurostat 2005.

Countries such as India and China are definitely facing quality problems in the expansion of the university system. By estimation, in India only 25 per cent of engineering graduates and 10–15 per cent of other university graduates are suitable for work in the IT sector. In China, only one in 10 graduates is suitable for work at a TNC (Trinh 2006). Old civilization traditions and talented diasporas are a large contributory factor for this state of affairs. Despite this situation, it is only a matter of time before quantity changes into quality. The rising number of graduates offers new opportunities for several social groups of young people.

Patenting and maintaining intellectual property in general is a vital part in the electronics business. As stated by Fujitsu Inc. (2004), patents are essential for differentiating from partners, assuring strength in alliances and getting revenue from licenses. Patent portfolio management is a part of the standard annual reports' technology section.

Patenting in the United States Patents and Trademark Office is used as a commonly-recognized indicator of technology creation. The distribution of patenting patterns among firms and continents has been very stable during the last years. Specifically, in the top-10 list, five to seven firms come from Japan, three to four from the USA and one (Samsung) from South Korea (Table 7.5).

Patenting is not only the major firm issue. A survey (Labrianidis et al. 2007) among the 34 Greek and UK electronics SMEs firms showed that five firms with European patents had considerably higher turnover per employee. Invention and patenting of invention can ensure long term competitiveness for firms.

Table 7.5 Largest patent recipients in United States

	1997		2005
IBM	1,747	IBM	2,941
Canon	1,499	Canon	1,828
NEC	1,144	Hewlett-Packard	1,797
Motorola	1,192	Matsushita Electric	1,688
Fujitsu	925	Samsung Electronics	1,641
Hitachi	1,152	Micron Technology	1,561
Mitsubishi Denki K.K.	918	Intel	1,549
Toshiba	962	Hitachi	1,271
Sony	964	Toshiba	1,258
Eastman Kodak	799	Fujitsu	1,154

Source: US Patent and Trademark Office.

Social Consequences

Outsourcing influences almost every actor in the supply chain. Those most affected have been manufacturing workers in DCs. Blue collar workers in DCs, who have enjoyed relative work security and high living standards, are facing problems. Loss of job security is a threat not only for blue collar workers. Efforts to graduate more engineers in China and India have produced results and caused the transfer of engineering jobs to offshore.

Offshoring doesn't hurt all labour groups equally. Service people, such as those involved in installing, maintenance and fitting, are less influenced. The decline in workplaces and closing of firms have had a negative social impact on regions. People have to retrain and sometimes also to relocate. Especially hurt are people in distant peripheral areas of regional economies or national economies such as Wales, Scotland, rural Finland and other locations where alternative employment opportunities are limited. The decline in rural industries could lead to further urbanization and concentration of people. It is hard to believe that simpler manufacturing could stay in high-cost regions.

Outsourcing of all or part of manufacturing operations has been a common practice for many electronics firms since 2000. Fierce competition has pressed OEM firms to use design capabilities of manufacturing service firms. In the period 2002–2005, R&D budgets of most electronics TNCs decreased. This decline of research budgets forced the OEM firms to look for outside development capabilities. At the same time manufacturing service firms established design bureaus, and hired hardware and software development engineers. Often the new designers and researchers are situated in emerging economies such as India, Taiwan, Russia, China and the Ukraine. ODM firms are more actively using engineers in emerging countries than OEM firms (Labrianidis et al. 2007). The process of innovation outsourcing has caused conflicts between brand owners and manufacturing service firms (Engardio and Einhorn 2005). Managing intellectual property has also become more and more important for EMS firms.

The unemployment level in Europe and the desire to new create modern manufacturing workplaces have put electronics TNCs in a more favourable negotiating position vis-à-vis national and regional governments. Countries and regions trying to attract foreign investment offer investors different types of financial and non-financial aid. It should be noted that some firms receiving financial aid in Central Europe received approximately the same type of aid that was received a decade ago in locations such as Wales and Scotland. A case of wide public interest is that of LG Philips Display, which received £220 million aid for the creation of jobs in Newport, Wales and closed the factory in 2003. In 2006 a company with similar ownership applied for aid of 206 million Euros (EUR) in Poland and received approval from the EC (European Commission State Aid Register 2006). Similarly, in February 2007 the Finnish Government discussed requesting the return of research money granted to the EMS firm of Finnish origin Elcoteq (Olsson 2007), which in January 2007 decided to abandon Finnish manufacturing operations.

A large number of jobs created by manufacturing service firms in the member states of the EU 2004 and the EU 2007 are relatively simple, consisting of several manual assembly operations and are therefore low pay. Another indicator is that international companies in Poland, such as LG, Jabil or Thomson, have high seasonality of TV set manufacturing with appropriate labour contracts – in some companies only one-third are permanent contracts, whereas the remaining two-thirds are temporary workers or employment agency workers.

Government support money raises several questions. If companies relocated their activities from Western Europe to Central Europe relatively easily, why not move further when there are business grants? The Ukrainian and Turkish governments are eager to develop national economies and willing to negotiate with foreign investors.

The electronics industry is strongly influenced by environmental regulations and laws. This is due to the fact that several production processes in electronics are hazardous for the environment; electronics products contain parts made from toxic materials and the recycling of electronics products is complicated and costly. Life cycles of electronics products are getting shorter and shorter, and consumers create substantial waste. A large driver towards more sustainable production and consumption of electronics is public opinion. The purchase of energy-saving home appliances and the collection of used batteries/accumulators are examples of such behaviour.

The recycling of electronics products is a technologically complicated and labour-intensive procedure. The UN Environmental Programme has warned about the dumping of e-waste in poor African countries (BBC 2006). India and China are affected equally by the problem of electronics waste. It could be also expected that environmentally-conscious consumers and international organizations will put more pressure on China to cope with environmental problems.

Results of the Empirical Survey in the Electronics Industry

Two aspects related to the delocalization are of particular interest: enterprise structure with localization and knowledge creation within enterprise (knowledge management).

The empirical research for this chapter was conducted in the period between November 2005 and May 2006 by five universities in Greece, the UK, Poland, Bulgaria and Estonia. The quantitative survey was conducted among 189 firms in the same five countries. In addition to the quantitative survey with all enterprises, in-depth qualitative interviews were carried out. The respondents represented most of the major sectors of the electronics industry.

Field of activities

Respondents to the survey represented broad sectors of the electronics supply chain including component manufacturers (printed circuit boards, transformers, cable, diodes, sensors, plastic parts), electronics contract manufacturers (EMS, vertically and horizontally integrated firms), contract manufacturers with good design capabilities (ODM type firms), subassembly and system OEMs, as well as service firms (logistics, sales, maintenance) (see Figure 7.4).

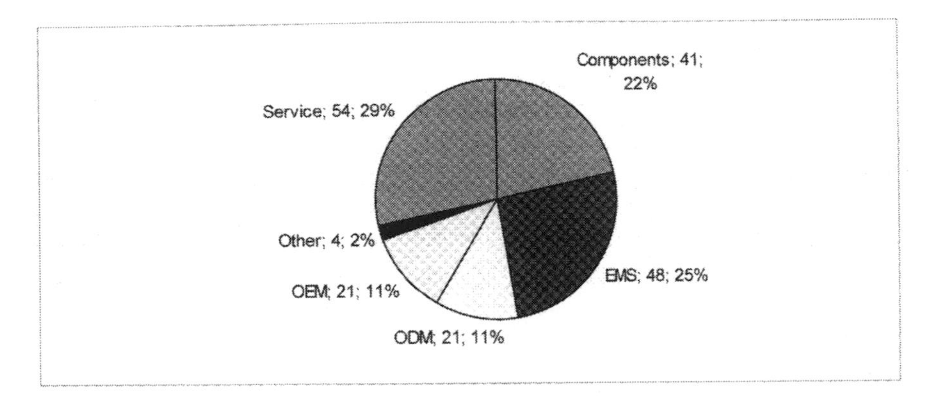

Figure 7.4 Respondents by the position in the supply chain (per cent)
Source: Enterprise Survey.

Among the survey firms, 23 UK firms were mainly from northeast England. Twenty four Polish firms were located in major urban centres, mainly in southern Poland (Wroclaw, Krakow), northern Poland and the Warszawa area. Several Polish firms were located in Special Economic Zones (SEZ) near major industrial centres. Twenty one firms from Greece were situated in Athens and Thessaloniki. Forty four firms from Bulgaria were mainly located in major industrial centres. Seventy seven Estonian firms had a very high concentration of firms around national capital, Tallinn.

Core competence

The electronics industry is moving towards a shared competency model, where certain firms concentrate on knowledge creation and certain firms on efficient manufacturing. Use of the shared competency model can be caused both by a decision of the entrepreneur or an order from corporate headquarters, and it creates a clear distinction between different firms in the supply chain. Seventy eight per cent of firms took subcontracting from abroad and 50 per cent of firms contracted certain activities to foreign firms (Table 7.6). Companies evaluated their core competences before the delocalization process and afterwards.

Table 7.6 Competitive advantages considered by firms before/after delocalization (per cent)

	UK	Greece	Bulgaria	Poland	Estonia	Total
Manufacturing of skill-intensive products	69.6/56.5	36.5/45.0	79.5/90.9	43.5/69.6	50.0/59.2	58.2/66.1
Labour-intensive products	57.1/13.0	36.4/30.0	52.3/36.4	65.2/73.9	60.5/51.3	57.1/43.5
Design and product development	56.5/87.0	36.4/60.0	31.8/38.6	26.1/47.8	18.4/26.3	28.8/26.4
R&D activities	52.2/65.2	45.5/55.0	36.4/38.6	17.4/13.0	11.8/14.5	26.0/30.6
Inputs supply	0/4.3	0/15.0	9.1/6.8	8.7/21.7	50.0/48.7	24.9/32.0
Capital-intensive products	13.0/8.7	9.1/5.0	27.3/38.6	13.0/26.1	11.8/17.1	15.8/21.0
Distribution and marketing	8.7/39.1	0/5.0	6.9/6.9	13.0/26.1	19.7/23.7	13.0/19.9
Other activities	4.3/17.4	0/0	2.3/4.5	4.3/8.7	3.9/2.6	3.4/5.4

Source: Enterprise Survey.

Enterprise self evaluations show growth of knowledge intensity and more sophisticated production both in Western European and Central European countries. Especially strong is growth of knowledge-related activities in the UK sample firms. The period of 2001–2006 was a period of active exit of mass-production electronics from the UK. In this period, the UK production capacity of several consumer electronics producers such as Sony, Celestica and LG Philips was reduced substantially. Delocalization in the UK is symptomatic of not only large enterprises but also SMEs and single entrepreneurs. Delocalization of medium-sized firms happens quite often in the form of outsourcing of less profitable activities. Activities that remain are considered to be competitive. However, Enterprise Survey answers should be assessed carefully. In some cases, undoubtedly world-class British engineering firms with international patents did not consider development their core strength, and at the same time more technologically modest Central European firms considered themselves strong in product development and design. A second paradox is the factor that competitiveness, as such, could be relative towards, for example, other firms in the same region or absolute on the world level.

Most of the firms declared that the workforce is the basis of their major strength. Among the sample, 48 firms had more than half their workforce with tertiary education. One hundred and thirty-one firms had less than 50 per cent of their workforce with tertiary education. The largest employers of highly educated people were small service and OEM firms. In the UK, the employment of educated people, mainly in service-oriented and high-tech components electronics firms, increased proportionally. In general, companies with design activities and service orientation tended to increase the recruitment of educated people both in absolute numbers and proportionally to total workforce.

Company strategy

Company strategy shows how a company sees its current situation and plans its future. Michael E. Porter notes three main business strategies as the following: segmentation strategy to business niche, cost leadership strategy and differentiation (Porter 1985). Due to factors such as trade barriers, new legislation or tough competition, enterprises could have different acute problems. At a certain time-period business strategy could cover aspects such as sales, technology, environmental management and problems coping with labour cost.

Business strategy depends on several factors, such as the perception of markets and development of both distribution and supplier networks. Strategy is also determined by status, position in the supply chain, availability of different resources and several other factors. In general, strategy planning is the joint task of owners and top management. The role of middle management and unit leaders is performing according to a planned strategy.

On the basis of in-depth interviews, we determined that position in supply chain and market orientations are the main parameters determining business strategy (see Table 7.7). Entrepreneurs had a more holistic view of enterprise and the market than managers did.

The main goals of enterprises during the delocalization process were the creation and introduction of new products and extension of product lines. For the Estonian and Bulgarian firms, the main goals were related to the modernization of manufacturing. For the UK and Polish firms, extension of the product line was the main goal (see Figure 7.5).

Table 7.7 Main strategies of electronics firms based on ownership and position in supply chain

	Local origin and locally oriented	Internationally owned and globally oriented
OEM	Creation of niche products for local consumers (often corporate consumers)	Creation of top products for global consumers. In general have strong brand and patent portfolio
ODM	Mixed strategy. Sometimes trying to create 'own products' but at the same time having strong manufacturing skills	Offering both design and manufacturing services for international firms
Service	Satisfying local customer needs. At the same time trying to extend service network to neighbouring regions	Serving particular area reserved by mother firm. Changes in activities are possible in the case of changed corporate policy
Component	Supplying for local firms and firms in neighbouring countries. Smallest of them could be called 'one controller' companies	Supplying for global OEMs. Very often have profound knowledge about component produced
EMS	Serving local customers	Serving international customers in the same way in different geographical destinations

Source: Enterprise Survey.

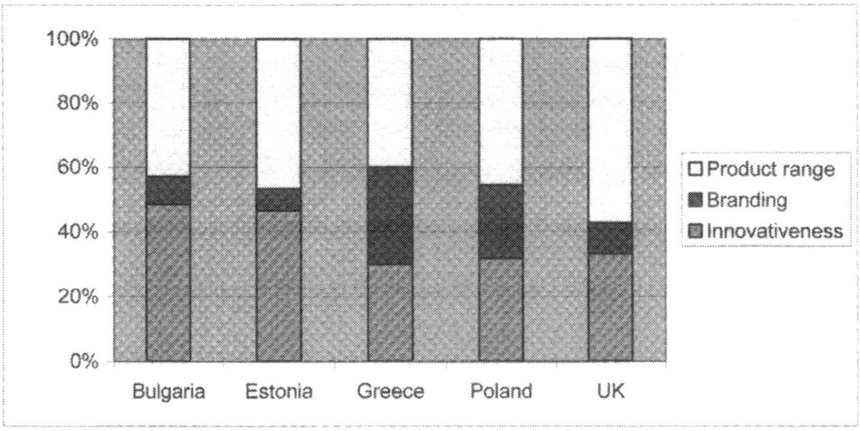

Figure 7.5 Goals during the last five years
Source: Enterprise Survey.

For component manufacturers and EMS firms, extension of the product line and services was the primary goal. OEM and service firms' main goals were linked to the introduction of novelties and strengthening of trademarks (see Figure 7.6).

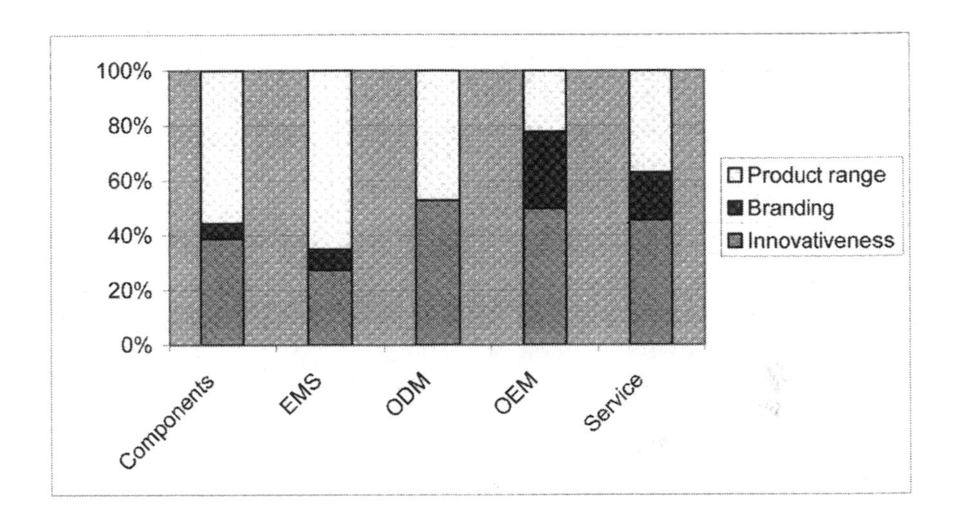

Figure 7.6 Goals during the last five years
Source: Enterprise Survey.

The future strategy of Estonian and Polish firms in our sample was relatively similar regarding high quality manufacturing. The Greek and UK firms' strategy direction is relatively aimed more towards design and quality oriented than in other countries in the Enterprise Survey. The relative absence of cost strategies among the Bulgarian electronics firms could be caused by the fact that the cost level of Bulgarian firms is still lower than in other countries (see Figure 7.7 and Figure 7.8).

Companies at the beginning of the supply chain tend to have more cost- and delivery- related goals. Enterprises closer to the end of supply chain tend relatively more to goals related to after-sales services and product design (see Figure 7.8).

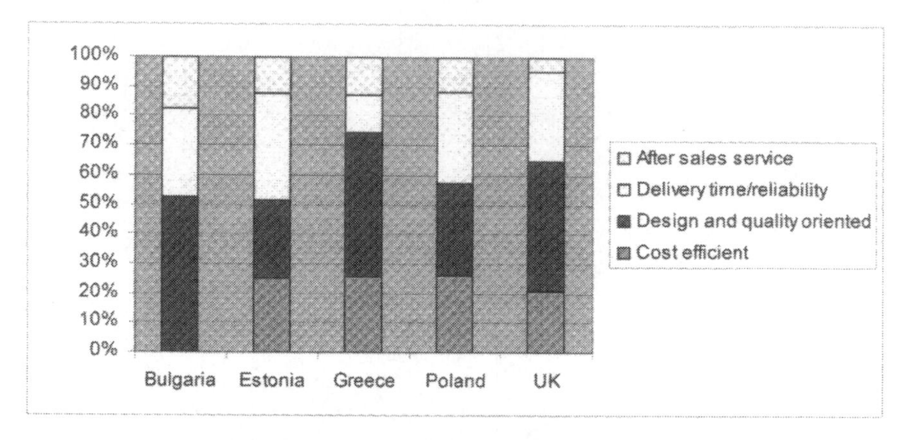

Figure 7.7 Proportions among the different future strategies based on country of origin

Source: Enterprise Survey.

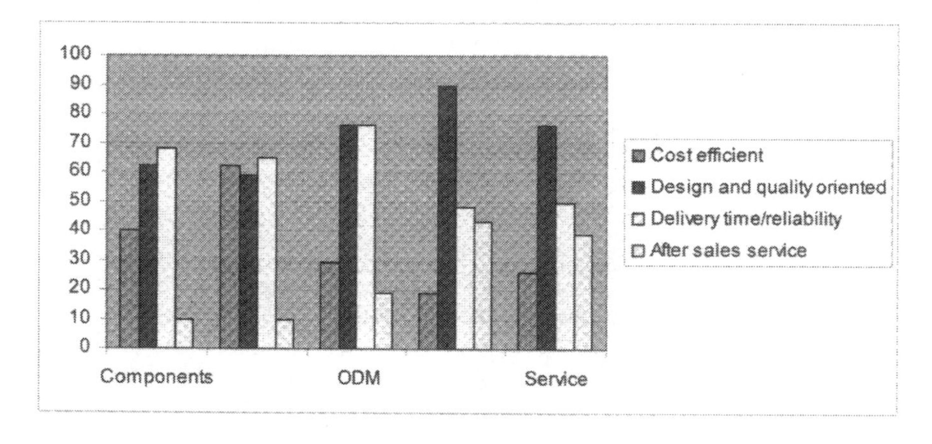

Figure 7.8 Direction of business strategy based on the position in supply chain (per cent of all enterprises)

Source: Enterprise Survey.

What is outsourced?

Outsourcing is one of the key activities for the firm. Typically the cost of purchasing goods is the largest cost item for electronics firms. To a great extent, outsourcing strategy also determines the success of the whole supply chain. Such successful products as Apple's iPod and iPhone or Microsoft's Xbox 360 are also successful due to well functioning supply chains. It is a mutual success for outsourcers and manufacturing firms.

No single firm in the electronics industry is able to rely on its own resources and capabilities alone. Only a few firms, like 'Hon Hai Precision' (Taiwan) and 'Flextronics' (Singapore), are able to produce a wide range of components and have a large number of different technologies in their portfolio, but such large firms are rather exceptional.

The enterprise sample was divided into 3 groups: large international electronics firms (with more than 1000 employees worldwide), small international firms (with fewer than 1000 employees) and local firms operating in a single country. After the delocalization process, large international firms tended to increase their orders to local subcontractors. This trend was especially visible in the case of EMS, ODM and OEM firms. Subcontracting by service firms tended to remain on the same level. Orders for local subcontractors by smaller international firms and local firms tended to remain the same or decrease slightly after delocalization (see Table 7.8). The main reason for outsourcing was the lack of needed technology contained in-house and lack of time. This fact shows that the supply chain of the electronics industry has become more and more sophisticated.

Table 7.8 Subcontracting by enterprises

		Reasons from subcontracting from abroad (per cent of firms)						
	Increase/decrease of using local subcontractors after delocalization*	**Lack of skilled labour**	**Higher labour cost**	**Not enough time**	**Lack of appropriate technology, equipment**	**Lack of capital**	**No access to natural resources**	**Other**
Big internationals	3.5	7	10	16	33	7	0	33
Small internationals	2.9	11	4	15	33	11	8	15
Local firms	2.9	4	12	8	27	4	1	17

Note: * 2: small decrease; 3: same level; 4: small increase.

Source: Enterprise Survey.

Challenges for the European Electronics Industry: Conclusions

The electronics industry in Developed Countries and especially in Europe is facing several challenges in the coming years. Price pressure from both high technology level countries and low-cost countries is becoming stronger. Several manufacturing and design firms must change their business model. The first round of changes in the electronics industry happened during the telecommunication industry decline in 2002–2004. This caused several larger and smaller firms to consolidate and to focus on core competences. Less productive and profitable parts of firms were sold. To a great extent, mass market consumer items such as home appliances were among the divisions or units sold.

The Enterprise Survey also showed that TNCs tend to have extensive negotiating power vis-à-vis firms in the supply chain. Price pressure was felt most strongly by electronics manufacturing service firms.

The survey also showed that service activities play a substantial and growing role in the supply chain of electronics. There is more and more room for maintenance, installation, sales, logistics and consulting type of activities. Firms that could successfully add to their products' additional services gained a great deal.

With the trimmed capacities, European electronics could find new solutions to the problems they face. Europe has several advantages such as a strong capital base, industrial traditions and an educated workforce. High labour costs and living standards should also not be seen as an obstacle but as an opportunity. High labour costs encourage the adoption of new automation technologies and labour-saving technologies. Europe also offers a good platform for design and development activities. The sophisticated market is a good platform for development of different niche products in embedded electronics.

European industry in general, and electronics in particular, also faces several problems. An ageing workforce, low-skill immigrant workforce and meltdown of industrial skills are only a few of them. The popularity of the sciences such as physics and biology as well as mathematics has been in decline for years. This makes a new intellectual ascent difficult. To a large extent, Europe lacks entrepreneurial spirit and willingness to compete and win. Maintaining a high living standard is not possible without innovation, at least in such a geographically small, resource-poor continent as Europe. Therefore, people should be creative, entrepreneurial and well-trained.

In response to price pressure from mass manufacturing firms, several Western European enterprises were forced to move into the military electronics sector, where pressure from foreign firms is lower. In general, public sector contracts tend to be profitable for firms, but too great a reliance on some large contracts puts firms at additional risk. Being reliant on large projects is, in the long term, too risky for medium- and even large-size firms.

The MOVE survey among enterprises showed that electronics firms with strong 'own' products and know-how can survive and prosper in the long term. Continuous participation in the development process, careful management and attention to the sales network offer potential for success in the twenty-first century.

References

BBC (2006), UN Warning on E-waste 'Mountain' [website] (updated 27 November 2006) <http://news.bbc.co.uk/2/hi/technology/6187358.stm> accessed 30 November 2006.

Belderbos, R.A. and Zou, J. (2005), 'Foreign Investment, Divestment and Relocation by Japanese Electronics Firms in East Asia', *Asian Economic Journal* 20, 1–27.

Blinder, A.S. (2005), 'Fear of Offshoring', Princeton University, CEPS Working Paper No. 119, December 2005.

Blomström, M. and Kokko, A. (1998), 'The Determinants of Host Country Spillovers from Foreign Direct Investment: Review and Synthesis of the Literature', Working Paper Series in Economics and Finance 339 (Stockholm: Stockholm School of Economics).

Booz Allen Hamilton and IFC (2003), 'Electronics Manufacturing in Emerging Market' (published online 2 June 2003) <http://www.boozallen.com/media/file/133484.pdf >, accessed 23 May 2007.

Borrus, M. and Cohen, S. (1997), 'Building China's Information Technology Industry: Tariff Policy and China's Accession to the WTO', Berkeley Roundtable on the International Economy [website] (updated November 1997) <http://repositories.cdlib.org/brie/BRIEWP105/>, accessed 19 May 2007.

Carbone, J. (2006), *Electronics Industry Eyes Eastern Europe*, Online Purchasing Journal (published online 16 February 2006) <http://www.purchasing.com/article/CA6305320.html>, accessed 20 March 2008.

Caves, R.E. (1996), *Multinational Enterprise and Economic Analysis* (Cambridge: Cambridge University Press).

Chang, J.-S. and Rosenzweig, P.M. (1995), 'A Process Model of MNC Evolution: The Case of Sony Corporation in the United States', Carnegie Bosch Institutes, Working Paper 1995–09, <http://cbi.tepper.cmu.edu/papers/cbi_workingpaper-1995-09.html>, accessed 30 March 2008.

Decision Etudes and Conseil (2006), *World Electronic Industries: 2005–2010* (updated July 2006) <http://www.decision-consult.com/doc/brochures/exec_toc_wei_current.pdf>, accessed 20 May 2007.

Economist (2004), *A World of Work* (published online 11 November 2004) <http://www.economist.com/displaystory.cfm?story_id=3351416>, accessed 11 January 2005.

Engardio, P. and Einhorn, B. (2005), 'Outsourcing Innovation', *Business Week* (published online 21 March 2005) <http://www.businessweek.com/magazine/content/05_12/b3925601.htm>, accessed 30 May 2005.

European Commission, Taxation and Customs Union website, <http://ec.europa.eu/taxation_customs/dds/cgi-bin/tarchap?Lang=EN>, accessed 1 June 2008.

European Commission State Aid Register (2006) <http://ec.europa.eu/comm/competition/state_aid/register/ii/by_case_nr_n2006_240.html#257>, accessed 5 June 2008.

European Communities (2006), *European Business: Facts and Figures* (Luxemburg: Office for Official Publications of the European Communities).

Fujitsu (2004), *Annual Report 2004* (Japan: Fujitsu Limited).

Forbes (2008), *The Global 2000*, Forbes Magazine Special Report (published online 4 February 2008) <http://www.forbes.com/lists/2008/18/biz_2000global08_The-Global-2000_Rank.html>, accessed 1 June 2008.

Fukase, E. and Martin, W. (2001), 'Free Trade Area Membership as a Stepping Stone to Development: The Case of ASEAN', World Bank Discussion Paper No. 421 (Washington, DC: The World Bank).

Hennart, J.F., Kim, D.J. and Zeng, M. (1998), 'The Impact of Joint Venture Status on the Longevity of Japanese Stakes in US Manufacturing Affiliates', *Organization Science: A Journal of the Institute of Management Sciences* 9:3, 382–95.

Hewitt-Dundas, N., Andréosso-O'Callaghan, B., Crone, M. and Roper, M. (2005), 'Knowledge Transfers from Multinational Plants in Ireland: A Cross-Border Comparison of Supply-Chain Linkages', *European Urban and Regional Studies* 12:1, 23–43.

Hirschmann, A.O. (1981), 'The Rise and Decline of Development Economics', in Hirschmann, A.O. (ed.), *Essays in Trespassing: Economics to Politics and Beyond* (Cambridge: Cambridge University Press), 1–24.

Industry Directions (2005), 'Regaining Control Managing Risk', European Supply Chain Management Association [website] <http://www.industrydirections.com/pdf/ESCARegainingControlManagingRisk.pdf>.

Infineon Technologies (2004), 'Annual Report 2004', <http://www.infineon.com/boerse/jahresbericht2004/download/english/AR2004_complete.pdf>, accessed 23 May 2007.

Labrianidis, L. and Kalantaridis, K. (2004), 'The Delocalization of Production in Labour Intensive Industries: Instances of Triangular Manufacturing between Germany, Greece, and FYROM', *European Planning Studies* 12:8, 1157–73.

Labrianidis, L., Domanski, B., Kalantaridis, C., Kilvits, K. and Roukova, P. (2007), 'The Moving Frontier: The Changing Geography of Production in Labour Intensive Industries: Final Report', Financed by European Commission 6th Framework Programme, MOVE Project [website] (updated 20 November 2007) <http://econlab.uom.gr/~move/index.php?lang=en>, accessed 5 June 2008.

Lall, V., Koo, J. and Chakravorty, S. (2003), *Diversity Matters: The Economic Geography of Industry Location in India*, World Bank Policy Research Working Paper No. 3072 (published online June 2003) <http://wwwwds.worldbank.org/servlet/WDSContentServer/WDSP/IB/2003/07/12/000094946_03070104102643/Rendered/PDF/multi0page.pdf>, accessed 20 March 2005.

Liagouras, G., Zambarloukos, S. and Constantelou, A. (2006), 'When Blueprints are Not of Big Help: Impasses and Challenges of Technology Policy in the Intermediate Economy of Greece', paper presented to Conference in Galatasaray University, Istanbul, Turkey, 2–4 November.

Malecki, E.J. (1997), *Technology and Economic Development. The Dynamics of Local, Regional and National Competitiveness*, 2nd edition (Harlow: Longman).

Mata, J. and Portugal, P. (2000), 'Closure and Divestiture by Foreign Entrants: The Impact of Entry and Post-entry Strategies', *Strategic Management Journal* 21, 549–62.

McLaughlin, K. (2006), *IPOD'S CHILDREN: China's Young Make Them, America's Youth Buy Them — Apple Probes Work Conditions*, San Fransisco Chronicle (published online 16 July 2006) <http://www.sfgate.com/cgibin/article.cgi?f=/c/a/2006/07/16/MNGAFK06MI1.DTL&hw=IPOD+CHILDREN&sn=010&sc=296>, accessed 22 August 2006.

Olsson, J. (2007), 'Finnish Government May Force Elcoteq to Pay back Governmental Funding', *Electronics Industry Portal* [website] <www.evertiq.com>, accessed 9 February 2007.

Porter, M.E. (1985), *Competitive Advantage: Creating and Sustaining Superior Performance* (New York: Free Press).

Radosevic, S. (2003), *The Electronics Industry in Central and Eastern Europe: An Emerging Production Location in the Alignment of Networks Perspective*, Centre for the Study of Economic and Social Change in Europe, School of Slavonic and East, Working Paper No. 21 (published online March 2002) <http://www.ssees.ucl.ac.uk/publications/working_papers/wp21.pdf>, accessed 22 September 2005.

Trinh, T. (2006), 'China and India vs. Europe: 1:0?', *Deutsche Bank Research*, <http://www.dbresearch.com/PROD/DBR_INTERNET_EN-PROD/PROD0000000000202465.pdf>, accessed 5 June 2008.

von Tunzelmann, G.N. (2004), 'Integrating Economic Policy and Technology Policy in the EU', *Revue d'Economie Industrielle* 105, 85–104.

UNESCO (2004), 'The Courier of UNESCO 2004', [website] (updated 10 March 2005) <http://portal.unesco.org/en/ev.php-URL_ID=19155&URL_DO=DO_PRINTPAGE&URL_SECTION=-465.html>, accessed 23 May 2007.

Databases

European Patent Office Database: <http://www.epo.org/>.

United Nations Comtrade Database: <http://comtrade.un.org/db/>.

US Patent and Trademark Office Database: <http://www.uspto.gov/>.

Chapter 8

Footwear Industry: Delocalization and Europeanization

Poli Roukova, Spartak Keremidchiev, Margarita Ilieva, Evgeni Evgeniev[1]

Introduction

Footwear, as one of the most labour-intensive industries, has been among the first sectors to be exposed to the processes of delocalization and global restructuring. The process of internationalization, however, occurred with different intensity in different regions, thus leading to a wide diversity of forms of delocalization, where the links between firms and regions were embedded in different historical, political, institutional and socio-economic environments. Nevertheless, we can say that delocalization followed a general pattern, in which the European footwear industry initially, in the 1970s–1980s, shifted production from the more developed, northern European, to the less developed, southern European, countries. More distant locations such as Brazil, China, Vietnam, India and Mediterranean African countries constituted the next wave of delocalization destinations. As far as the CEECs were concerned, their significance grew markedly, mainly through OPT in the 1980s after the political changes and the establishment of market economies in the 1990s. As a result, currently the largest share of footwear production of these countries is exported to the EU market.

In this chapter, we argue that it is mainly industry- and country-specific factors that are shaping the regional map of the European footwear industry. The most important among those factors are a liberalized trade policy, low labour costs and organizational flexibility. Among the different forms of delocalization of the footwear industry, international subcontracting is the most widespread, while there are also a limited number of JVs and FDI as well as relations based on spontaneous market exchange.

The aim of this chapter is to contribute to a better understanding of the recent delocalization trends in the European footwear industry. Our main focus is on: a) forms and networks of Europeanization of national footwear industries; b) company strategies and c) delocalization effects. The presented outcomes of the research analysis are based on detailed data sets gathered from the Enterprise Survey and

1 The authors are grateful to the collaborators from the Institute of Geography for their valuable assistance on the project implementation and to Mrs. Alexandra Ravnachka especially.

key informant interviews in five EU countries: Bulgaria, Estonia, Greece, Poland and the UK, carried out under the FP6 MOVE Project (Labrianidis et al. 2007).

Overview[2]

Outward processing trade

The leading role that international trade policies played in drawing and redrawing the global commodity map is undisputable, and one of its outcomes is the so-called 'quota geography' of production. The pattern and intensity of the delocalization process in labour-intensive industries is therefore strongly dependent on the implementation of trade policy measures; hence policy issues have been central to delocalization analyses.

The period after 2005 witnessed the introduction and enforcement of important tariff and non-tariff regulations concerning environmental, social and health standards. Currently, new anti-dumping measures are being elaborated in order to protect the EU footwear industry. These measures have been initiated by large footwear producers such as those in Italy in spite of the strong negative attitudes to the enforcement of such measures in half the EU countries. Special attention is given to labour-related issues such as working conditions, social movements and so on. These are considered to be important background factors for the spatial division of labour.

The delocalization of production activities from the EU to the CEEC was enforced under Outward Processing Trade (OPT) agreements in the 1980s. It flourished in the 1990s after political and economic changes in the post-socialist countries. OPT contracting continues to be in force although the agreement formally expired in 1998. OPT has undergone changes in the course of time, not only in terms of quota volume, but also concerning the countries, regions and industries involved. OPT aimed at providing support to the EU manufacturers and retailers, particularly in terms of overcoming import quota restrictions under the MFA and improving their competitiveness (Graziani 1997; Dunford et al. 2002). Prolongation of the OPT might also have a negative impact, as it can undermine the competitiveness of home firms by forcing them to adopt defensive strategies. Some of the negative consequences are loss of jobs and closure of enterprises.

While OPT was very important for the host CEE firms as it provided an opportunity to keep relatively stable levels of production and employment during the period of economic transition, the positive effects were short-lived. The OPT agreements have played a key role in establishing and deepening the

2 The footwear industry is grouped with textile and clothing industries in accordance with their common globalization features as labour-intensive industries. In the literature, the three industries are often examined as one sector designated as TCF (Textile, Clothing and Footwear).

existing linkages between companies from the old member states and the new member states (Pellegrin 1999; Begg et al. 2002). These linkages have been the basis for the establishment of triangular subcontracting networks later (Begg et al. 1999; Kalantaridis et al. 2003). Most of the recent JVs and FDI in LII in CEEC are based on previous OPT subcontracting relations (the case Italy-Bulgaria is a good example of OPT evolution in the footwear industry). In this connection, Pellegrin differentiates the 'footloose offshoring' in the LDCs, such as Mexico, from outsourcing to CEEC (1999).

In the European footwear industry OPT is still in force through the implementation of international subcontracting as a predominant form of organization of footwear production between old and new member states (Rabellotti 2003). Many researchers have highlighted the asymmetric character of OPT contracts and their negative consequences for the development of companies and regions in CEEC. This is in the sense that subcontracting of basic manufacturing activities defines the low position of host firms in the value chain. Being locked in a position of dependence on foreign contractors, host firms only have very limited prospects for upgrading, and most of them actually shift to downgrading (Graziani 1997; Pellegrin 1999; Smith et al. 2005). CEE host firms that manage to upgrade are usually large companies with long-standing OPT relations, but considering the huge number of firms involved in OPT such cases of upgrading are very rare (Smith et al. 2005). OPT intensifies competition within the CEEC and their regions, 'where cost pressure dominates and 'undermines' local firms' positions' (Smith at al. 2005). It leads to the fragmentation of the local industry, decrease of wages, de-skilling of the labour force, and other negative aspects (Begg et al. 2003).

Analysing the competitiveness of the European TCF industry in relation to the EU enlargement, Hanzl-Weis (2004) pointed out that Hungary, Slovenia and Romania were the main outward producers of footwear in the beginning of the 1990s. In the second half of the 1990s, Romania's export of footwear to the EU 'skyrocketed' between 1995 and 2001, while other CEEC market shares were stagnating or slightly decreasing. Currently, the footwear trade data shows that 90 per cent of the shoes and the intermediate products manufactured in new member states are exported to old member states (SEC report 2005).

Research background

The delocalization of production is a dynamic process with high complexity leading to a great diversity of organizational forms, network configuration and changes of functions ensuing from the distribution of power-control and rent distribution. The delocalization forms are structured temporally, spatially and by sector, and the diversity of organizational forms and production networks is a result of changing patterns of competition and governance in global contracting (Pickles et al. 2006). A range of external and internal factors creates development opportunities and constraints for the firms and regions involved.

The most widely-applied research approaches for studying the globalization of labour-intensive industries are the GVC approach and the cluster approach, as well as the GPN approach. The last includes elements of the GCC and elements of the GVC analysis and actor network theory. Currently, different combinations of the above-listed approaches are considered to be crucial for overcoming the limitations of any single research method.

Our survey is based on the main concepts of the GVC as a network-centred analysis. From the GVC perspective, the footware industry is integrated into global networks of a buyer-driven commodity chain, and the value comes from relational rents and from design, marketing and branding (Gereffi and Memedovic 2003, 3). The top position is occupied by the lead firm, which controls the access to major resources and rent distribution. Firms operate in different ways 'combining various production models' within one commodity chain (Bair 2006). This is relevant to a higher degree for companies that operate mainly as subcontractors. Humphrey (2003:17) emphasises that 'a diagnosis of value chain linkages and the particular requirements for competitiveness that they create' are important to be analysed.

It is envisaged that participation in global networks creates development opportunities and advantages for improving company competences and for the development of new capabilities based on learning from foreign buyers. In this process, the role of the lead firm (marketers, branded manufacturers and retailers) is of key importance (Gereffi 1999). Research on the footwear chain suggests that in some cases global buyers discourage, if not obstruct, the development of high value-added activities by local producers, and the local upgrading opportunities depend on the way chains are governed (Schmitz and Knorringa 2000; Humphrey and Schmitz 2002).

CEE firms operating as subcontractors under OPT are more often involved in regional rather than global chains, which lessens the learning effect (Pellegrin 1999; Pickles et al. 2006). Producer-producer OPT relations prevail over retailer-producer relations in the European labour-intensive industries (Bair 2006), and this fact has a negative impact on the learning process.

Once a company has acceded to the chain, it needs to improve competitiveness in order to maintain its position. Industrial upgrading, which is a key concept in GVC studies, is envisaged as crucial in this process (Gereffi 1999; Gereffi and Memedovic 2003), that is, the firm has to move up to higher value-added activities. There are four types of upgrading that are discussed by GVC scholars. These are product upgrading, process upgrading, functional upgrading and inter-sectoral upgrading (the last is defined in terms of clusters, but in terms of GVC it is considered as organizational upgrading) (Humphrey and Schmitz 2000; Yoruk 2001). From a GVC perspective, Rabellotti has summarized the processing stages in the footware industry as pre-assembly, assembly and post-assembly (2003). Upgrading of the networks' functions improves company competitiveness. The most important among these is the development of backward and forward linkages as well as performing key organizational functions in triangular configurations. However, CEE companies working as subcontractors (SMEs especially) have very limited ability to take key positions in the triangular production (Smith et al. 2005). Recent studies on GVC focus on the impact of the international trade policy and the regional context of upgrading (Pickles et al. 2006).

Company strategies depend on the company's access to resources, knowledge and freedom of decision-making, as well as on its capacity in terms of capabilities. The issue is strongly related to the ability of the host firm to first, *lock-out* from a dependent position and second, on the existing opportunities for upgrading (Evgeniev and Roukova 2007). Humphrey (2003, 19) defines that 'the main strategy options for combating lock-in are: market diversification, excellence in manufacturing and effective use of knowledge accrued from within the value chain. Neidik and Gereffi (2006) argue that company strategies 'were devised in a particular national context', and other authors analysing the topic in depth find that there are regional aspects to strategies (Pickles et al. 2006).

A successful strategy is committed to the company's ability and capability to adapt to the dynamic global economic environment, to create and maintain appropriate vulnerability and to respond quickly to the changes in both the international and domestic markets. Industrial upgrading strategies in terms of shift to higher value-added activities cannot be regarded as a panacea for successful economic performance. There are cases of redirection of upgrading or replacement of functional upgrading by process and product upgrading, that is, a shift to lower value-added activities. The latter strategy could in some cases generate better performance in terms of company sales and profits both in old and new member states (Amighini and Rabellotti 2003; Pickles et al. 2006).

The EU footwear industry

In 2003 more than 27,000 companies operated in the footware industry of the EU–25, employing about 361,000 workers, with their turnover reaching 26.7 billion EUR. If the figures for the new members, Romania and Bulgaria, are added, the above figures will change significantly. Romania occupies second place after Italy in the footware industry (in terms of number of industry employees) and together with Bulgaria ranks among the top 10 footwear suppliers of the EU–25 (in terms of value).

In recent EC reports (SEC 2001, 2005) concerning changes in the EU footware industry, the focus is on industry competitiveness, particularly in relation to the processes of globalization and EU enlargement. The main features and trends outlined for the European footwear industry are as follows:

- The footwear industry has marginal position in manufacturing with 0.5 per cent of the total value-added generated in the manufacturing sector and about 1 per cent of employment in manufacturing.

- The decline of firms, employment and production from recent decades has not changed in recent years. During the period 1995–2003, 160,000 jobs have been lost in the EU–15 footwear manufacturing amounting to a 32 per cent drop.
- The footwear industry is highly labour-cost sensitive: labour costs account for 67 per cent of value-added.
- The productivity is low, about 40 per cent of the average for manufacturing. This is due to the manual operations, which cannot be automated yet.
- The increase of labour productivity measured as value-added per employee is indicated as a competitive advantage of the EU footwear industry, but it is mainly due to the decline in employment. The average labour productivity in new member states is equivalent to 30 per cent of that at the EU–15 level, but labour productivity was considered to be less significant than labour cost in influencing company decisions to delocalize.
- The footwear industry is represented mainly by SMEs: 45 per cent of value-added is produced in small- and 25 per cent in medium-sized enterprizes.
- The industry distribution within European countries shows high concentration by country: Italy alone produces 50 per cent of all EU footwear, and together with Portugal and Spain accounts for a two-thirds share. The new member states have significant contributions to the EU total accommodating 30 per cent of employment and producing 9 per cent of the industrial value-added in the sector.
- The increase of exports in 2005 by 33 per cent as compared to 2004 came after years of grave decline. These figures, however, are far below the 1995 export data. In contrast, the import growth trend has been constant for the last decade. For 2002–2005 alone, its rate was 57.3 per cent. New member states export out of the EU less than 10 per cent of their production output and the rest of the export goes to the old member states.

In 2005 the pressure of cheap imports from China and Vietnam drastically increased competition in the EU market. Imported shoes accounted for less than 50 per cent of the sales in 1995, but in 2003 three out of any four pairs of shoes purchased on the EU market were imported from third countries. The European footwear market is evaluated as one of the most open markets. The trade liberalization will support EU imports of raw materials and export of shoes for the higher-price segment of the market on the one hand, but on the other hand, market penetration will increase competition.

The shift to low labour-cost countries has led to a significant decrease in production volumes in the EU, but the data for the industrial value-added indicates a decline twice as slow as that of production volume. This fact is due to a range of factors, but the most important is that value-added activities remain in the home countries, that is, EU companies preserve higher positions in the footwear value chain.

It is also noteworthy that the specialized distribution and niche markets play a particularly important role in the European footwear sector and account for half its turnover. A special relationship of reliable service and trust has been developed between retailers and consumers, in which children's footwear occupies a special place.[3]

Research Objectives

The research questions put forward are grouped as follows:

- *Forms and networks*: Who delocalizes activities and what activities are delocalized? In what kind of networks do the companies participate and what functions do they have within the network? What kinds of relationship are established within the networks? What is the distribution of power and dependence within the networks?
- *Company strategy:* What delocalization challenges do the companies involved face? What actions do the companies undertake to cope with them? What kinds of company strategies are employed and which of them can be deemed 'successful'?
- *Delocalization effects:* What is the delocalization impact on the company's economic performance? What are the social consequences?

The created database combines the findings of a detailed enterprise survey and key informant interviews. The survey of 119 footwear firms is based on in-depth interviews with managers, owners or other managerial staff of randomly selected enterprises affected by delocalization in five EU countries. Two-thirds of the firms investigated are located in Bulgaria, Poland and the UK. Estonia and Greece have quite a small weight in the sample. The causes and effects of delocalization of footwear production were further discussed with 26 key informants. These were national and regional experts, representatives of business associations, trade unionists and researchers.

The Survey

General characteristics

The survey was carried out at the end of 2005 and the beginning of 2006. A total of 119 footwear firms (with 17,056 employees) were interviewed. The average number of employees per company is the highest in Bulgaria (193) and in Poland (142). In the UK and Estonia this number is about 100, whereas in Greece it is

3 SEC 2001, 2005.

44. Large firms with more than 250 employees are well represented in Poland and Bulgaria, which could be due to the legacy of the industrial structure during the socialist period (Figure 8.1).

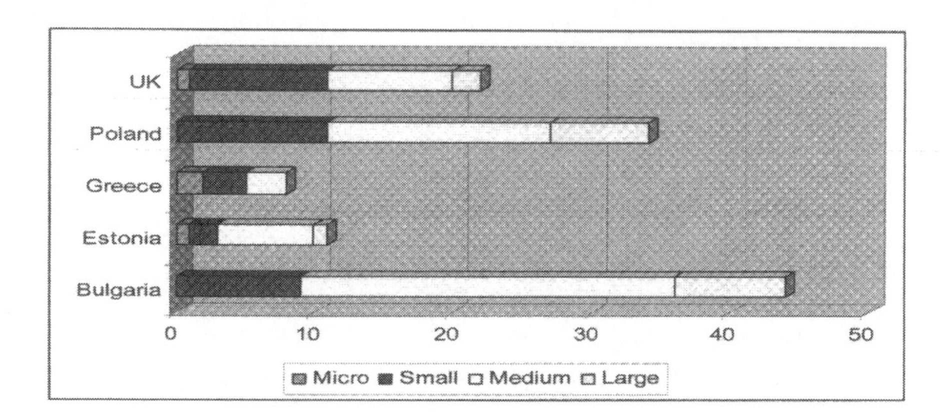

Figure 8.1 Distribution of companies by size (number)
Source: Enterprise Survey.

The Greek and UK companies surveyed were established before the end of the 1980s (87 and 95 per cent respectively). The greatest share of Polish and Estonian firms were established in the first half of the 1990s, whereas half the Bulgarian companies were founded in the second half of the 1990s. The emergence of the newly-established CEEC enterprises is especially linked to the post-socialist restructuring of CEE economies and the rapid increase of OPT with old member states.

The product specialization of the surveyed firms does not differ from that of the EU footwear industry. More than half the firms surveyed produce footwear with uppers of leather. The second place is taken by upper parts of shoes. A few firms produce footwear with uppers of textile (Bulgaria and Poland), high fashion sports footwear (Bulgaria and UK) and children's shoes.

Forms and networks

In terms of ownership structure, in more than 90 per cent of the footwear companies the equity capital is of national origin. The ownership structure in CEEC confirms the significance of international subcontracting or insourcing in the sector. Bulgaria has the highest share of FDI and JVs – 36 per cent (Figure 8.2).

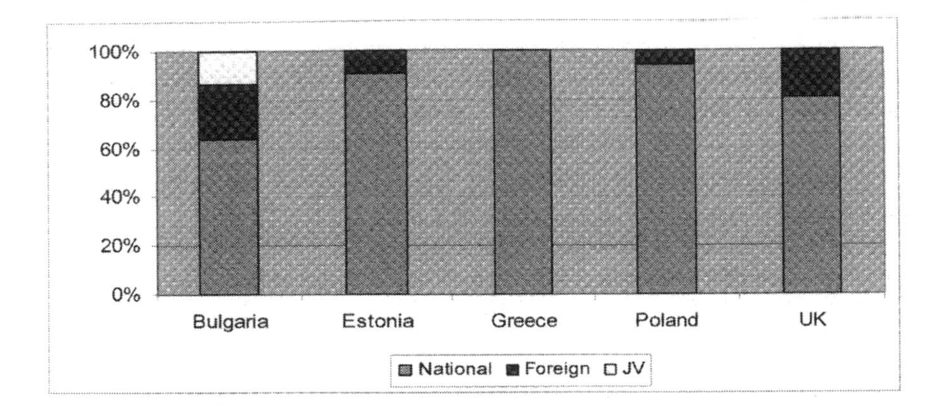

Figure 8.2 Ownership structure of companies
Source: Enterprise Survey.

The distribution of companies by forms of delocalization highlights the predominance of insourcing and outsourcing. Footwear firms in the UK (68 per cent) and Greece (75 per cent) outsource production and have subsidiaries abroad (22 and 25 per cent respectively). UK enterprises have joined the international networks with other activities such as R&D. Insourcing predominates among 88 per cent of the companies in Poland, 98 per cent in Bulgaria and 91 per cent in Estonia. Bulgaria has the highest share of companies (23 per cent) that are affiliates of foreign, mainly Italian, companies. One-third of the Polish and almost half the Estonian firms outsource production. There are companies in Greece that work under international subcontracting, but not a single one in the UK does so according to the survey.

The results in Table 8.1 show significant differences between insourcing and outsourcing companies along a set of criteria. Having a subsidiary abroad shows a certain form of upgrading for the domestic firm, since it tries to achieve lower prime-cost production and be close to its export market. Our data shows that only a limited number of footwear enterprises have subsidiaries abroad (25 per cent of the Greek firms and 23 per cent of the UK firms). However, an important outcome from our study is that there is a significant and strong negative relationship between companies that do subcontracting and have subsidiaries abroad (r=.39**). Moreover, the insourcing firms usually do not have subsidiaries abroad, whereas outsourcing firms do not show a pattern of determinant focus on having a subsidiary abroad. It also becomes clear that insourcing companies are predominantly not involved in outsourcing to firms abroad, which means that triangular manufacturing in Bulgaria, Poland and Estonia (mostly being insourcing firms, according to our sample) is present, but is very limited.

The companies from the sample were asked several questions relating to their upgrading features when they first began delocalizing. Thus, from the correlation analysis we detect two important characteristics of insourcing firms that differentiate them from outsourcing firms. Firstly, most of the insourcing firms indicated that design and product development were not among their competitive advantages at the beginning of delocalization, which is somewhat different compared to outsourcing firms, which have less significant, but a strong, positive relationship with this indicator. Secondly, most of the insourcing firms stated that distribution and marketing were not among their competitive advantages at the beginning of delocalization, as there is a highly significant and very strong relationship between the variables, whereas in the case of outsourcing firms the relationship is not significant. In other words, insourcing firms are somewhat weak in forward channels, whereas we could not identify what the case for the outsourcing firms is. Furthermore, the dependency on buyers is explicit, as we analyse the competitive advantage of the footwear firms based on labour-intensive activities. We found a highly significant and strong positive correlation (r=.69**) between the competitive advantage of labour-intensive activity at the beginning of the companies' involvement in delocalization and at present. Moreover, the relationship is even stronger (r=.85**) when we take into account footwear firms that consider capital-intensive production as their competitive advantage.

Table 8.1 Comparison of two main delocalization firms

Variables	Insourcing	Outsourcing
Firms with subsidiary abroad	–.39**	NS
Firms outsourcing abroad	–.62**	NA
Design and product development as company advantages at the beginning of delocalization	–.31**	.18*
Distribution and marketing as company advantages at the beginning of delocalization	–.26**	NS
After delocalization, do you produce more complicated (high value-added) goods?	.37**	NS
After delocalization, do you have services such as design, marketing, distribution, etc.?	–.26**	.25**
Your company is: part of a cluster/industrial district?	–.41**	.25**
part of a national subcontracting network	.32**	–.32**
part of a national network	.20*	–.20*
part of inter-firm trade (hierarchy type)	.16*	NS
Importance of intensity of competition	.26**	NS

Note: **correlation is significant at the 0.01 level (2-tailed); *correlation is significant at the 0.05 level (2-tailed); NS – non-significant; NA – non-applicable.
Source: Enterprise Survey.

Many scholars consider that working under international subcontracting creates a possibility to learn from doing, and one important effect of this practice is to move from assembly of simple goods to producing more complicated goods. Our findings support this claim, since insourcing firms have a significant and very strong and positive relationship with the variable which explores the change of company's production from simple to more complicated goods. However, the learning opportunities have constraints. The survey demonstrates that moving to higher value-added activities such as design, branding, distribution and so on is very difficult. If outsourcing, footwear firms have managed to move to offering these services after delocalizing of production (r=.25**), the insourcing firms are found on the opposite side (-.26**). In other words, once the footwear firm begins outsourcing, it has far more chances to upgrade and to assume a higher position in the value chain compared to the firm that is only insourcing.

The companies are involved in different types and numbers of networks presented in Figure 8.3.[4] The insufficient sample of the Greek firms impedes the identification of this issue.

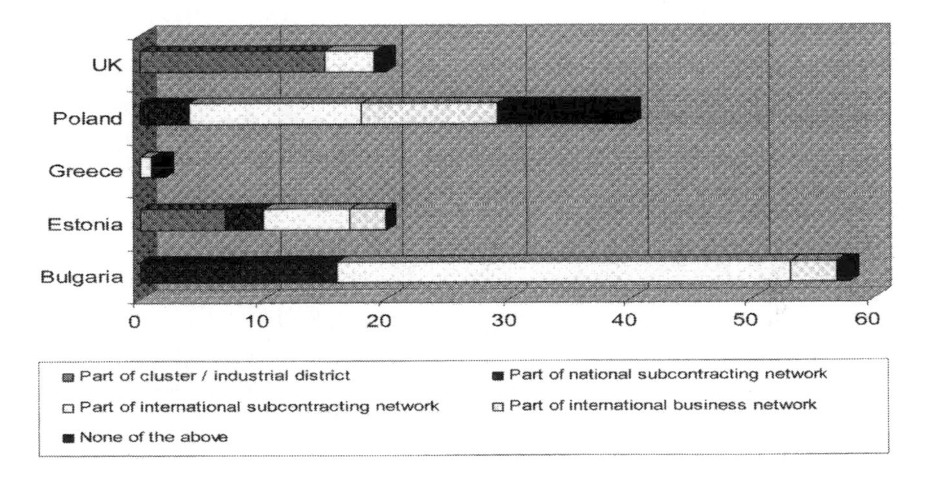

Figure 8.3 Companies' affiliation by type of networks
Source: Enterprise Survey.

The majority of the insourcing firms are not part of clusters/industrial districts, which is quite different for outsourcing firms (Table 8.1). Another important finding is that there is an opposite relationship of the comparison between insourcing and outsourcing firms in respect to the company participating in a national subcontracting network. If participation of insourcing firms in this type

4 The number of answers is higher than the number of firms.

of network is important, the participation of outsourcing firms is irrelevant, as our results demonstrate. The national network concerns the second- and third-layer subcontractors. Our findings indicate that there is a less significant, but medium to positive, relationship between insourcing firms and participation in national networks, whereas outsourcing firms are outward oriented, as in their case there is the opposite relationship with respect to the same variable. Moreover, the type of inter-company trade characteristics is also important. The correlation analysis shows that there is a less significant, but positive and medium, relationship between hierarchy type inter-company trade and insourcing firms, whereas this relationship is not significant in the case of outsourcing firms. It is likely that intense market competition affects the insourcing firms more severely compared to outsourcing firms. A highly significant and strong positive relationship is designated in the group of the former, while there is no significant relationship in the group of the latter, according to our results. The firms in Bulgaria, Estonia and Poland more often participate in more than one network.

The survey provides information about the *changes of network functions* of the surveyed firms. Triangular manufacturing is a good option for firms from developing economies to upgrade their network functions, which yields higher value-added. However, key functions in triangular manufacturing are still very limited in Bulgaria, Poland and Estonia.

The largest share of Bulgarian and Polish companies does not have any orders with local subcontractors and does not purchase intermediate products. Most of the Greek companies do not develop their contacts with local subcontractors, and some of them reduce the existing linkages. Similar practice is applied by two-thirds of the UK firms.

In terms of export, on average 57 per cent of the total sales of the surveyed footwear enterprises was directed to international markets (Table 8.2). Bulgarian and Estonian firms are largely export oriented. The average share of subcontracting is 60 per cent for the whole sample, but it varies significantly – from 91.5 per cent for Bulgaria to 0.0 per cent for UK and Greece. The companies' ranking of the most important factors for receiving orders from foreign firms is as follows: labour costs, expertise, reliability and geographical proximity.

Bulgaria takes a leading position in terms of the share of export in the total sales. The largest share is attributed to subcontracting-based export and almost half of it to export of intermediate products.

Table 8.2 Export and subcontracting in 2004

Country		Exports share of total sales	Share of total exports on subcontracting basis	Share of intermediate products in total exports
Bulgaria	Mean	82.2	91.5	46.5
	N	43	43	43
Estonia	Mean	65.9	64.6	33.5
	N	11	11	11
Greece	Mean	8.7	0	0
	N	6	6	0
Poland	Mean	42	71.0	34.4
	N	31	29	30
UK	Mean	38.8	0	0
	N	22	22	1
Total	Mean	57.2	60.4	39.6
	N	113	111	85

Source: Enterprise Survey.

The share of export in total sales has not changed for half the companies since the beginning of delocalization and for more than two-thirds of UK and Bulgarian firms. It has increased to 35 per cent for all firms, and for some of them this increase is considerable. These positive trends are registered by 72 per cent of the Estonian firms, 60 per cent of the Polish firms and by less than 20 per cent of the Bulgarian, Greek and UK firms. A decrease of the export share has been noted for about 10 per cent of the UK and Polish firms.

The weight of subcontracting indicates the degree of dependence on foreign firms. The export of more than 80 per cent of the Bulgarian and Estonian companies is fully (100 per cent of their export) on subcontracting basis versus only one-third for the Polish firms. Over 80 per cent of Bulgarian and Polish firms sell their production under a foreign company's brand name.

The companies surveyed specify three main markets (accounting for more than 50 per cent of the company's export). In Bulgaria, 73 per cent of the companies export to Italy and the rest to Germany, Greece and UK. In Estonia, 87 per cent of the companies export to Finland. Greek firms have various external markets: Germany, the UK and Russia. Poland's main markets are the old member states.

Almost 71 per cent of the Polish and 84 per cent of the Bulgarian firms have up to five customers. The established contacts may be considered optimal for the company's stability in this case. Higher diversity by number of customers is observed for the Estonian firms, because wholesalers are their main clients. In the case of Bulgaria and Poland, manufacturers have the leading position among the types of customers, followed by wholesalers and large retailers. Therefore, the production networks are more important than the buyers' networks. Half the Bulgarian firms and two-thirds of the Polish firms practice splitting the risk of loss of orders by using local, second-level subcontractors.

The nature of relationships between partners characterizes the distribution of power and access to resources, knowledge and freedom of decision-making. The selected variables are presented in Table 8.3. The answers were given by companies that insource production, and therefore UK firms were excluded. The results obtained allow formulation of the following findings:

- The relationships are of high stability and contacts are signed regularly rather than whenever necessary.
- The control of the production process by the main contractor is often flexible rather than tight.
- Almost all Bulgarian and Estonian firms have formal contracts with both their contractors and their subcontractors, which is not a common practice in Poland.
- Personal contacts in establishing business links are important for all Estonian companies and for the half the Bulgarian and Polish firms.
- The most frequent manner in reconciling differences of opinions in the Estonian, Polish and Bulgarian companies is by striking a balance (in this case answers account for the lowest share for the three countries). In Bulgaria the decision of the main contractor is dominant for 44 per cent of the companies.
- Interruption of relations with customers will lead to moderate or slightly negative effects, according to 60 per cent of the Bulgarian and the Polish firms, but it will be severely negative for 70 per cent of the Estonian firms. For the past three years, half the Bulgarian firms and 73 per cent of the Polish firms have interrupted relations with customers. The most frequently cited reason is financial.

Table 8.3 Main features of relationships in case of insourcing (per cent)

		Bulgaria	Estonia	Greece*	Poland
Stability of links	High stability	93.0	80.0	100.0	73.3
Frequency and type of contacts	Regularly	65.2	70.0	0.0	80.0
	Whenever necessary	30.2	30.0	100.0	13.3
	When possible	4.6	0.0	0.0	6.7
Nature of contractual relationship	Formal	88.4	90.0	0.0	46.7
Do you have contracts with your subcontractors?	Yes	58.5	90.0	0.0	46.7
When there are differences of opinion, who wins?	You	0.0	0.0	0.0	0.0
	Main customer	44.2	20.0	100.0	33.3
	We find balanced decision	55.8	80.0	0.0	66.7
What would be the implications of breaking down of the relationship on you and your partner?	Severely negative effects	34.9	70.0	0.0	36.7
	Moderate negative effects	44.2	30.0	1	10.0
	Slight neg. effects	20.9	0.0	0.0	53.3
How important are personal relations as opposed to formal ones for you and for your partners?	Very important	23.3	40.0	0.0	30.0
	Medium importance	18.6	50.0	0.0	13.3
	Slightly important	25.6	10.0	0.0	13.3
	Not more important	32.5	0.0	0.0	43.4
Have there been cases when relationships have collapsed?	Yes	51.2	50.0	0.0	73.3
How does the main contractor control the process of the work?	Tight Control	34.9	10.0	0.0	25.0
	Flexible Control	65.1	90.0	100.0	75.0

Note: * Greek firms which insource footwear production.
Source: Enterprise Survey.

Company strategy

Our survey investigated company objectives for the period 2000–2005. The responses of the managers of the companies may be divided into five groups. They differ mainly according to the position of company in the life-cycle curve, the value chain and the commodity chain.

Operational improvement, product development and capacity expansion are expressed in the following manner: to increase or keep the level of production output and orders from customers; to increase the volume of production of uppers;

to modernize the machine park, improve the quality; to start cutting uppers, not only sewing them; to introduce children's footwear into the company product mix; to implement information technologies that enable better production control, management and supervision of distribution.

Market development as a priority objective is structured in several subgroups: to set up a retail network (distribution network and trade contract), to develop the company's own chain of retail shops, to open and develop an Internet site in order to have broader access to clients, to create the company's own trademark and brand for securing the position (brand building), to penetrate and become established in the European market (Western markets, that is, France and Germany, among others) and the markets of Russia, to sell directly to big companies, to penetrate markets demanding higher quality and price, and to produce shoes for uniformed services.

The most popular financial objectives are as follows: to enhance financial liquidity, to improve collection of receivables from sales, to keep the profit margin at a stable level, to cut down costs, and to organize production of high value-added goods.

Social objectives: to improve working conditions for workers, to limit employment reduction or stabilize the number of employed workers and to provide training for employees to understand and use newly implemented technologies.

The corporate restructuring and strategic alliances development objectives, which are very rarely articulated, are the following: to split the company into several divisions (office department, sewing workroom, production department), to get rid of unnecessary property and to develop strategic alliances with other firms from the sector in order to compete better with other rivals.

The companies take a range of upgrading actions towards these objectives, but two are of utmost concern: product range and functions.

Product range The decisions about selection of the product range are entirely determined outside the companies in the cases of outsourcing, subcontracting and subsidiaries of foreign companies. In these cases, the contractors' headquarters assign orders for production to the companies that are merely producers. The functions regarding market studies, creation of new models and marketing of the production are not developed in these companies.

In the companies producing their own brands of shoes, the top management level takes the decisions about production variety, either alone or in cooperation with designers and marketing specialists. The surveys of the market, the fashion trends, the customers' taste and dealers' requirements are everyday activities in the company.

Such decisions revolve around two poles. The first is associated with narrower specialization and response to the needs of specific client groups, for example, dancers, uniformed staff, athletes and others. The second most often exploited possibility is to scrutinize the market trends and offer quick, and what is deemed the most adequate possible response to the demands in several broader segments such as male or female, or casual or high-quality shoes. Purchasing of new cutting-edge technologies is another possibility for launching new models of shoes. A mixed approach for choosing a product range is also used where part of the production is ordered by an external customer and the remaining part of the production capacity is used to produce an own-brand of shoe products.

Function (changes in the firm's position in the chain) The majority of the companies investigated thought that they were moving up the value chain. The most frequent statements were: 'We have moved up the commodity chain from manufacturing to brand management, marketing and design.' These companies take activities that demand special knowledge and skills, changes in production activity into designing. Another way of upgrading is by using the possibilities for outsourcing of the production functions and activities. Therefore some of the companies stated: 'The company upgraded its position thanks to higher level of outsourcing to factories in Poland and India'; they cited, 'development of the company's own distribution network', and 'subcontracting part of the footwear production to China'.

Some of the producers in the UK have entirely ceased their production activities, as they have outsourced production abroad and are only dealing with distribution at present. ('[We have] move[d] away from manufacturing to 100 per cent outsourcing. We emphasize wholesaling and retailing', the firm Stead and Simpson pointed out.)

Only a small number of the surveyed companies reported upgrading in the value chain, while a large part of the companies in the sample did not report any change in their chain position. This trend is mainly because the nature of the orders from their contractors has remained unchanged. Even in these cases, the companies claim that they are dealing with the production of more complicated models of shoes or parts thereof. Others report that they have obtained a licence for the factory from big companies such as Reebok and Asics.

Delocalization effects All the interviewed experts have given prominence to the strong impact of globalization and internationalization on the configuration of national footwear industries. The respondents have estimated that outsourcing of production activities allows the home firms to cope with the global market challenges and increase profitability. The role of a competent firm management is underlined as crucial. The positive effects, indicated in key informant interviews, are structural changes in national economies, shift of firms from low value-added activities to higher value-added activities, including trade (Greece), and retailing and design (the UK). The negative effects cited are decrease of the production volume and number of employees involved in processing activities (the UK) and decline of the company's competitiveness (Greece), among others. Experts argue that host firms gain from delocalization through learning. The introduction of know-how into new organizational technologies and training in international production and marketing practices has lead to an increase in branch competitiveness, employees' skills and qualifications and aspiration for better education and training. Other effects are improvement of labour productivity and quality, development of additional supporting activities and incentive for development of own product and own design parallel to subcontracting. The negative effects for host firms have been indicated as follows: full dependence on foreign or national contractors (owners and middlemen), decapitalization of the companies, short-term development prospects, worsening of labour conditions and low wages.

The *trends of economic performance*, as consequences of involvement in delocalization, indicate that one-fifth of the firms interviewed did not experience any changes in terms of profit and turnover. More than half the firms report an increase, and some of them a significant increase of turnover and profits. Companies that have faced difficulties in managing new challenges constitute 24 per cent of the sample, and most of them are in Poland. The UK and Bulgarian firms have succeeded in maintaining and increasing employment, turnover and profits after delocalization, although different factors stand behind these two cases. The UK firms cited the shift to the market niche production and development of high value-added functions, such as design, marketing and distribution. Bulgarian firms gain from labour-intensive low-cost activities such as production under orders from abroad and production of upper parts of shoes. The Estonian and Polish firms shift to upgraded production processes. They produce 'more complicated goods' ordered by foreign firms. The Greek firms gain from cost reduction, because they do not develop high value-added functions like the UK firms, and most of them do not diversify their production activities.

The labour cost has grown in the CEEC and its share as a percentage of the total costs increased considerably for more than 70 per cent of the companies in each country (Figure 8.4). This fact may lead to a shift of production to locations with cheaper labour costs in coming years. In the UK, half the firms have reduced their labour costs, because they outsourced labour-intensive activities. Two-thirds of the Greek firms have not reported any changes.

Delocalization processes also have significant *social consequences*. The shift of jobs has different social, quantitative and qualitative dimensions in the home and host countries. The changes in employment when comparing the situation before and after delocalization show considerable reduction in the number of employees in the UK firms and much less in the Greek firms (Figure 8.5). Within the new member states, Bulgarian and Polish firms display a considerable increase of the number of employees, while two-thirds of the Estonian firms report a decrease in the number of employees.

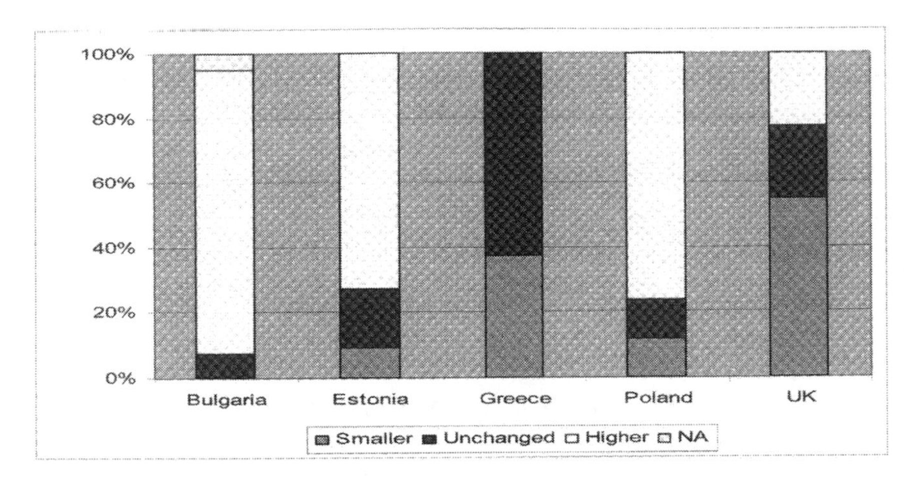

Figure 8.4 Changes of labour costs as a share of total costs
Source: Enterprise Survey.

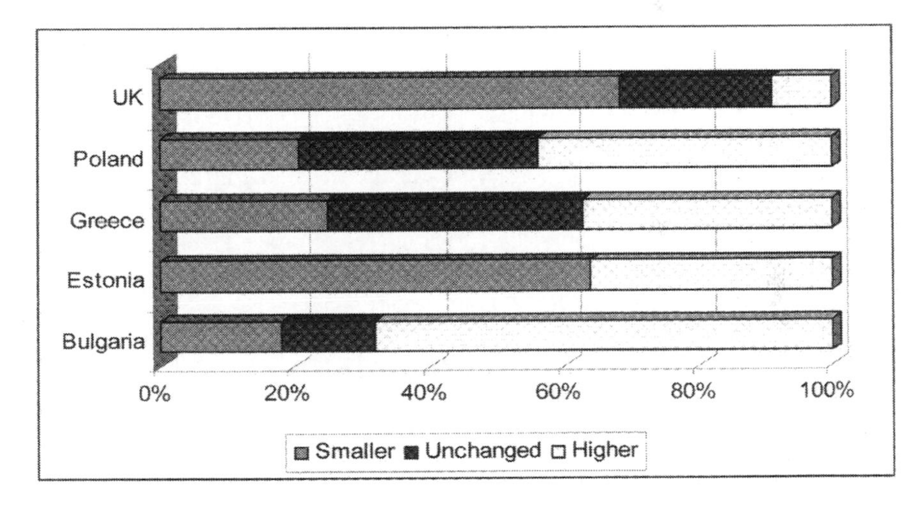

Figure 8.5 Employment changes after delocalization
Source: Enterprise Survey.

The quantitative changes in employment are accompanied by qualitative ones – an increase in the number of employees with tertiary education and of 'white collar' workers. The changes in the quality of companies' personnel by country disclose the type of activities developed during the delocalization and the changes in position within the value chain. The increase in the share of highly educated personnel in half the UK firms confirms that they have managed to retain higher value-added activities within the UK. The development trend with respect to the white-collar personnel hints at keeping administrative functions connected with production organization enlargement. In contrast, the employment rate of both groups in the rest of the countries under investigation has changed slightly. This fact shows their capability to respond to delocalization challenges.

Experts interviewed from the countries investigated point out that *wages* in the footwear industry are lower than in the other labour-intensive industries and in manufacturing as a whole. The surveyed companies have compared the wages they pay to the national average wages in the footwear industry. Average and higher wages are paid in UK firms (46 and 36 per cent of the firms respectively) and in Bulgarian firms (45 and 50), but the wages in Estonian and Polish firms are either average or lower than average. The wages correspond to the labour-market potential of footwear industry employees. Scarcity of labour is reported by 80 per cent of the Bulgarian firms and 73 of the UK firms. In Poland, this share is 65 per cent.

Delocalization influences the changes in labour contracting. Half the Estonian companies report a decrease in the number of temporary and part-time employees. One-third of the Polish firms have increased this type of labour contracting. According to the key informant interviews, a certain part of employment is seasonal, which has led to an increase in the level of temporary unemployment in Poland and Greece. In Bulgaria, labour contracts are mainly permanent. When firms have large orders, they employ part-time workers or extend the working time of the permanent personnel. On the other hand, low wages, unattractive (that is, monotonous) work and bad working conditions are in some cases factors in frequent personnel changes in this industry in all countries under investigation.

The experts interviewed in Poland and Bulgaria are concerned about the low degree of observance of the labour regulations related to contracts and working conditions. There are cases of informal employment in terms of contracting and tax payment in small companies.

Conclusion

Outsourcing and insourcing have linked *home* and *host* firms in the European footwear value chains. A high degree of industry consolidation is observed in all countries except Greece. The footwear firms are mostly incorporated in regional production networks. The UK firms are an exception to the above case, since their networks are global and buyer-driven.

Classic patterns of upgrading, which lead to the top of the value chain, have been achieved by *UK firms*. UK footwear export indicates a movement towards production of specialized and designer footwear. Vertically integrated companies with an overseas network of suppliers dominate the UK footwear industry. One of the specificities of the UK footwear sector is the complex distribution and retail structure.

After the sector readjusted by declining significantly, *Greek firms* have again gained a momentum to preserve their position somewhere in the middle level of the value chain. Production under subcontracting occupies a significant place in the export of footwear. More recently some enterprises have tried to develop in the field of design, but the majority of the companies prefer outsourcing of production to suppliers from neighbouring low labour-cost countries.

German and Italian footwear firms have been outsourcing production to *Poland* under the OPT agreements since the beginning of the 1990s. Recently, the share of subcontracting production in the total output has decreased, mainly because of the changes in the exchange rates and the growing labour costs in comparison with other Eastern European countries. Some Polish firms subcontract production to other countries, aiming to implement new organizational functions as network organizer of triangular manufacturing. A few Polish firms are trying to locate factories in the Ukraine, Russia and Belarus and to gain access to these markets.

Most of the *Estonian* footwear enterprises are involved in cross-border production networks as subcontractors to Finnish and Swedish firms.

The *Bulgarian* footwear industry occupies a low position, from a GVC perspective, producing parts for shoes, semi-finished products and shoes on a subcontracting basis. Some of the large former state enterprises have succeeded in keeping their contacts with the UK and other EU firms, dating back to the period before the end of the 1980s. In the second half of the 1990s, many Italian firms outsourced production to Bulgarian firms. Some of them have established JVs or bought large ex-state enterprises. Many firms exporting Bulgarian footwear have taken advantage of the demand on the domestic market for which they produce own-brand production and develop company retail chains.

The experience of Greek and Polish companies in taking the position of intermediates in the chains (functional upgrading) did not bring the expected success. Their strategies comprised targeted stabilization of their current position rather than rising to a higher position in the European footwear chain.

The long-standing insourcing by firms from CEEC strongly affects their ability to upgrade. The large shares of their export on a subcontracting basis indicate the high degree of dependence on the main contractor. The effect of learning is one-sided and leads to manufacturing process improvements only, thus 'pinning firms down' to their current low position in the chain. There is no access to a broader set of knowledge, which may be beneficial for the firms in their attempt to become internationally competitive. A very small number of cases were found where the companies have demonstrated capabilities to absorb and implement knowledge about international markets. Industrial upgrading is considered by managers as an improvement of existing activities rather than as a shift to activities implemented at the top of the value chains. Some managers have realized the need for such a shift, but they estimated the constraints on this. They have found a solution in increasing the number of contractors and the size of orders instead of moving up the chains. In fact, market diversification is considered in quantitative and not in qualitative terms. The cases of a strategic mix of market diversification, excellence in manufacturing and effective use of knowledge are rare. Here, the companies' competence level confines them in the best case to the domestic markets.

The further trade liberalization and the EU enlargement will continue to reconfigure the production linkages between member states. The role of other European non-member states will increase in European footwear production.

References

Amighini, A. and Rabellotti, R. (2003), 'The Effects of Globalization on Italian Industrial Districts: Evidence from the Footwear Sector', Paper presented at the Conference on Clusters, Industrial Districts and Firms: The Challenge of Globalization (Modena, Italy).

Bair, J. (2006), 'Regional Trade and Production Blocs in a Global Industry: Towards a Comparative Framework for Research', *Environment and Planning A* 38, 2233–52.

Begg R., Pickles, J. and Roukova, P. (1999), 'A New Participant in the Global Apparel Industry: The Case of Southern Bulgaria', *Journal of Problems of Geography* 3–4, 121–35.

Begg, R., Pickles, J. and Smith, A. (2002), 'The Wrong Trousers: Complexity, Product Specifity and Division of Labour in Pan-European Apparel Production Systems', paper presented at the 4th EURS Conference on '(Re)placing Europe: Economies, Territories and Identities', Spain, 4–7 July (Autonomous University of Barcelona).

Begg, R., Smith, A. and Pickles, J. (2003), 'Cutting It: European Integration, Trade Regimes, and the Reconfiguration of Eastern and Central European Apparel Production', *Environment and Planning A* 35, 2191–2207.

Dunford, M., Hudson, R. and Smith, A. (2002), *Restructuring the Geographies of the European Clothing Industry* (unpublished manuscript).

Evgeniev, E. and Roukova, P. (2007), 'Foreign Dependency and Upgrading of Bulgarian and Turkish Textile and Clothing Firms: A Comparative Analysis', *Journal of Problems of Geography* 1–2,106–119.

Gereffi, G. (1999), 'International Trade and Industrial Upgrading in the Apparel Commodity Chain', *Journal of International Economics* 48, 37–70.

Gereffi, G. and Memedovic, O. (2003), 'The Global Apparel Chain: What Prospects for Upgrading by Developing Countries', *Sectoral Studies* (Vienna: UNIDO).

Graziani, G. (1997), 'Globalisation of Production in the Textile and Clothing Sector: The Case of the Italian Foreign Direct Investment and Outward Processing Traffic', paper presented at the BRIE Conference on 'International Production Networks, Foreign Direct Investment and Trade in Eastern Europe: Will there be a Unified Economy?' Vienna, 5–6 June.

Hanzl-Weiss, D. (2004), 'Enlargement and the Textiles, Clothing and Footwear Industry', *World Economy* 27:6, 923–45.

Humphrey, J. (2003), 'Opportunities for SMEs in Developing Countries to Upgrade in a Global Economy', SEED Working Paper 43, ILO.

Humphrey, J. and Schmitz, H. (2000), 'Governance and Upgrading: Linking Industrial Cluster and Global Value Chain Research', IDS Working Paper No. 120 (Brighton: Institute of Development Studies, University of Sussex).

Humphrey, J. and Schmitz, H. (2002), 'How Does Insertion in Global Value Chains Affect Upgrading in Industrial Clusters?', *Regional Studies* 36:9, 1017–27.

Kalantaridis, C., Slava, S. and Sochka, K. (2003), 'Globalization Processes in the Clothing Industry of Transcarpathis, Western Ukraine', *Regional Studies* 37, 173–86.

Labrianidis L., Domanski B., Kalantaridis C., Kilvits K. and Roukova, P. (2007), 'The Moving Frontier: The Changing Geography of Production in Labour Intensive Industries: Final Report', Financed by European Commission 6th Framework Programme, MOVE Project [website] (updated 20 November 2007) <http://econlab.uom.gr/~move/index.php?lang=en>, accessed 5 June 2008.

Neidik, B. and Gereffi, G. (2006), 'Explaining Turkey's Emergence and Sustained Competitiveness as a Full-package Supplier of Apparel', *Environment and Planning A* 38, 2285–303.

Pellegrin, J. (1999), 'German Production Networks in Central/Eastern Europe between Dependency and Globalisation', Discussion Paper FSI 99–304, Wissenschaftszentrum Berlin für Sozialforschung.

Pickles, J., Smith, A., Bucek, M., Roukova, P. and Begg, R. (2006), 'Upgrading, Changing Competitive Pressures, and Diverse Practices in the East and Central European Apparel Industry', *Environment and Planning A* 38, 2305–2324.

Rabellotti, R. (2003), 'How Globalisation Affects Italian Industrial Districts: The Case of Brenta', paper presented at the RSA International Conference Pisa 12–15 April.

Schmitz, H. and Knorringa, P. (2000), 'Learning from Global Buyers', *Journal of Development Studies* 37, 177–205.

SEC (2001), 'Report on the Promotion of Competitiveness and Employment in the European Footwear Industry' (updated 21 December 2001) <http://europa.eu.int/comm/enterprise/footwear>, accessed 14 April 2005.

SEC (2005), 'Economic and Competitive Analysis of the Footwear Sector in the EU 25', Commission (updated 12 July 2005) <http://europa.eu.int/comm/enterprise/footwear>, accessed 12 September 2006.

Smith, A., Pickles, J., Begg, R., Roukova, P. and Bucek, M. (2005), 'Outward Processing, EU Enlargement and Regional Relocation in the European Textiles and Clothing Industry: Reflections on the European Commission's Communication on "The Future of the Textiles and Clothing Sector in the Enlarged European Union"', *European Urban and Regional Studies* 12, 83–91.

Yoruk, D. (2001), 'Patterns of Industrial Upgrading in the Clothing Industry in Poland and Romania', *Centre for the Study of Economic and Social Change in Europe*, Working Paper No. 19.

Chapter 9

The Impact of Delocalization on the European Software Industry

Robert Guzik, Grzegorz Micek

Introduction

The IT sector is a fascinating case for research on delocalization not only because the information sector has been the most rapidly developing part of the economy for the last few years, but also because it was the first where offshore outsourcing was for well-paid white-collar jobs. It was also the first of services to be delocalized in spite of the traditional view that services are characterized by unity of production and place of consumption. This condition no longer applies, as the place has become virtual due to advances in telecommunications and the Internet. Unlike workers in other labour-intensive sectors, those who work in IT have to be highly skilled and well educated (Arora and Gambardella 2005). Competitiveness in the software industry is not based on productivity or even quality, but on ideas and design; therefore, the software industry is sometimes called the 'industry of the mind' (Florida et al. 2003).

The aim of this chapter is to analyse the delocalization of the European software industry in the context of subcontracting and FDI. It draws from the results of a three-year research project based on extensive field work (primary data) and analyses of secondary data. Based on the latter source and literature, the third section of this chapter has been developed as a review of ongoing processes in the software industry. In the fourth section we try to cover four specific aims: to analyse forms of delocalization and their extent in the European software industry, to examine reasons behind delocalization from both perspectives of host and home countries, to describe briefly linkages and cooperation networks between companies from various countries, and to analyse prospects of further delocalization to locations outside Europe, especially in the context of India's success in IT outsourcing services.

Definitions and Data Sources

The term 'software industry' is interchangeably used with 'IT sector' and is defined here as NACE 72 Group[1] (computer and related activities). It must be remembered that such a definition of IT is only a part of the broader ICT sector,[2] which, as a whole, is not discussed here. However, because of data availability, most reports produced by international organizations (the UN, World Bank and OECD) are devoted to the ICT sector. In this report, we sometimes employ data for ICT in order to show a general setting, but it should not be interpreted as equal to data for the IT sector. Another important point to bear in mind is that IT activities are generally larger than the IT sector portrayed in official statistics. Many enterprises, governmental agencies and the like posses their own IT departments – classified elsewhere. A final observation we would like to note here is that many software companies deal with hardware (for example, IBM and Hewlett-Packard), and therefore, some data for the IT sector are exaggerated by hardware sale revenues.

Most of the secondary data used in the following section comes from Eurostat publications and databases. The conclusions in section four are based on two main sources: 27 key-informant interviews with selected representatives of software companies and organizations (for example, IT chambers of information technology and information processing societies) and 191 interviews conducted with senior managers and executives of software companies in five countries (Bulgaria, Estonia, Greece, Poland and the UK).

Whenever we use the term CEE we mean former socialist countries – now members of the EU (for example Poland, the Czech Republic, Hungary, Slovakia, Slovenia and the Baltic States). Eastern Europe is a broader category and includes, apart from CEE countries, other countries of the former Soviet Union and the former Yugoslavia.

IT Industry – Trends

The software sector is among the most rapidly growing sectors in OECD countries, with a strong increase in value-added, employment and R&D investment. According to ICT Outlook 2006 (OECD 2006), the rapid growth of the sector, especially in CEE and some non-OECD countries in the developing world, for example, India, deserves recognition as a new wave of globalization in global

1 This includes hardware consultancy (NACE 72.10), publishing of software (72.21), other software consultancy and supply (72.22), data processing (72.30), database activities (72.40), maintenance and repair of office, accounting and computing machinery (72.50); and other computer related activities (72.60).

2 The ICT sector consists of manufacturing (electronics, office equipment, telecommunications equipment) as well as services (IT, telecommunication, postal services, radio and TV broadcasting).

ICT. This trend occurs not only because of the rapid growth of producers in these countries, but also because of the huge growth of the ICT market. Thanks to advances in ICT, more services are now tradeable and may be provided from remote locations. Therefore, simple software code writing is now accompanied by a full range of consulting, R&D and other services previously reserved to limited locations in DCs.

The European software industry

The EU (EU–25)[3] software industry employed 2.5 million people in 2004 and generated EUR 153 billion of value-added with a turnover exceeding EUR 308 billion (Table 9.1). The UK is the country with the highest share in the EU software industry. The turnover of the UK software sector accounted for 26 per cent of the EU, and the generated value-added was 30 per cent of the number for the EU. Other countries according to their share in the EU software sector turnover are: Germany (18.4 per cent), France (15.1 per cent) and Italy (11.6 per cent). Country ranking according to contribution to employment does not differ much (the UK, Germany, Italy, France). The share of the largest new member country, Poland, in the EU software industry is 1.1 per cent of the turnover, only 0.7 per cent of value-added and with a relatively high share of 2.9 per cent in employment.

Table 9.1 Main indicators of the software industry for selected European countries, 2004

Variable	EU–25	UK	Germany	Poland	Bulgaria
Value-added at factor cost (in million EUR)	152,337	47,006	28,375	1,140*	77
Turnover (in million EUR)	308,209	80,365	56,840	3,281*	204
Persons employed	2,483,170	573,424	370,346	71,280*	12,183
Average personnel costs (in Thousand EUR)	49.0	55.9	57.7	16.0*	4.6
Value-added per person employed	61.3	82.0	76.6	28.9*	6.3
Wage adjusted labour productivity in per cent	123	146	133	125*	137

Note: * Data for 2003.
Source: Eurostat.

3 No data for Greece; for some countries data for 2003.

The larger share in employment than in turnover or value-added is a consequence of low apparent labour productivity. Value-added per employee is the highest in Ireland (EUR 98,200) and very high in the UK, Denmark and Germany (above EUR 70,000). In Italy and Spain, such productivity slightly exceeds the level of EUR 40,000. In new member countries, the highest productivity is found in Slovenia (EUR 31,000), the Czech Republic (EUR 20,400) and Poland (EUR 17,000) and is the lowest in Lithuania (EUR 9,800), only slightly higher than for the newest members, Romania (EUR 9,000) and Bulgaria (EUR 6,300). Such a measure of productivity has to be adjusted by differences in labour-costs in these countries. The average value-added per employee calculated as a percentage of personnel cost was 123 per cent for the whole of the EU and was highest among the 'old members' in Ireland (206 per cent), the UK (146 per cent) and Germany (133 per cent). Such wage-adjusted productivity for new member states was also quite high (above 150 per cent in Latvia, Lithuania, Slovakia and Romania) and around the EU average in the rest of the countries. It clearly shows that, thanks to lower wages, the new member countries are very attractive locations for different forms of nearshoring from Western European countries.

The software industry is very sensitive to the general economic situation, and its growth is well correlated to changes in GDP. For example, in the period from 1995–2004, the most rapid growth of employment in NACE 72, as well as in turnover or value-added in this sector, was observed in Ireland and Portugal. In new member countries, employment growth in this period was higher than 200 per cent, whereas in Germany and the UK, the level of employment in 2004 was similar to 1995. Similarly, value-added grew by less than one-third in Germany or the UK, whilst in Ireland, it grew twelve-fold and in Slovakia and Poland six-fold. Another reason for such a difference is the fact that new EU member countries are lagging behind better developed, Western European economies in terms of computer usage and Internet accessibility among others.[4] Therefore, the internal market is still far from saturation there.

International trade

Ireland is the world's biggest exporter of IT and computer services (more than $18 billion in 2004) followed by the UK ($10.5 billion), the United States and Germany (Table 9.2). The advantageous position of Ireland can be accounted for due to the corporate tax environment, making this country a location for exporting activities, especially for American TNCs. What is to be observed is the enormous growth in scale of trade, much quicker than the growth of the sector, which proves the thesis of growing globalization in the IT sector. A second important observation is the diminishing role of the United States in the computer services trade, although we have to remember that a significant part of trade accounted to other countries is generated by local affiliates of USA-based TNCs. Although

4 With the exception of Estonia and Slovenia.

exports from expanding economies such as Poland have risen more sharply than the OECD average, local markets have expanded faster and imports still exceed exports. Last but not least, the diminishing position of Greece is observed, where exports dropped almost two-fold with a simultaneous four-fold import growth.

However, Ireland lost its leadership in exports of software goods (Table 9.2). Spectacular growth of exports is to be noticed from Germany, among others, thanks to the great success of the German company SAP and its ERP systems software, placing it as the biggest exporter of software packages in the world.

Table 9.2 International trade in computer services and software goods, 1996–2004, in million USD

Industry	Computer services				Software			
	Exports		Imports		Exports		Imports	
Country/ year	1996	2004	1996	2004	1996	2004	1996	2004
Germany	1,603	7,810	2,379	7,906	734	3,210	946	1,813
Greece	362	197	55	221	24	41	43	140
Ireland	105	18,484	306	362	3,567	2,029	636	246
Poland	28	195	135	419	38	151	16	133
UK	1,706	10, 469	519	3,536	1,102	1,523	1,137	1,754
USA	2,775	8,501	422	5,804	3,087	3,030	714	1,244

Source: OECD International Trade in Services database.

The analysis of international trade in software goods and services is undermined, however, by the quality of available data. In fact, the real scale of outsourcing seems to be much higher than reported in the statistics. The OECD (2006) report gives an example of ambiguity between Indian and OECD statistics – India reported $9.6 billion worth of exports of computer services to OECD countries in 2002, whereas OECD countries reported only $294 million imports from India. Among foreign-owned software companies surveyed in Poland, we have found a mechanism of functioning as cost centres, which have reported no sales here (sales were reported in tax-friendly Ireland). Another source of difficulty in capturing trade in software and ICT services is the diversity of delivery channels. Often software is counted as hardware trade when computers with pre-installed software are sold. Trade statistics capture less than one-third of the real trade value of the software sector. Additionally, this data may be distorted by the fact that for tax/duty purposes it is often not the value of software but the value of physical supports (CD-ROM, diskettes) (OECD 2002).

Value chain in the software industry – Delocalized stages

The most value is added in the software value chain during product development, sales and services (Competitive Alternatives 2008). The latter two stages are usually carried out in host countries. The position of a particular foreign software company within the value chain is different for mass-production typical for a majority of companies and customized solutions. The growing emphasis on localization and product development requires a higher level of software engineering skills and is more reliant on outsourcing and indigenous supply chains, including translation, packaging, manual printing, transport and technical support.

A minority of software companies located in Asian and CEE countries develop original software, while a vast majority operate as software distributors and sales offices for large international companies. Subcontracting for Western Europe (Germany in the case of Poland) and Northern European (Finland and Sweden in the case of Estonia) software companies is also important. Most CEE software companies operate in a low segment of the value chain. Although basic software maintenance remains the most popular software-related activity, CEE companies seem to be moving up the value chain to software development strategy and systems design (Figure 9.1). Many overseas companies are set up initially to carry out a basic function, such as relatively low-skilled software manufacturing. Within a very short time, corporate management recognizes the quality and skill of the indigenous staff and moves other functions there, such as software localization and eventually high-skilled product development, technical support and marketing (Coe 1997a; *The Software Industry in Ireland* 1992).

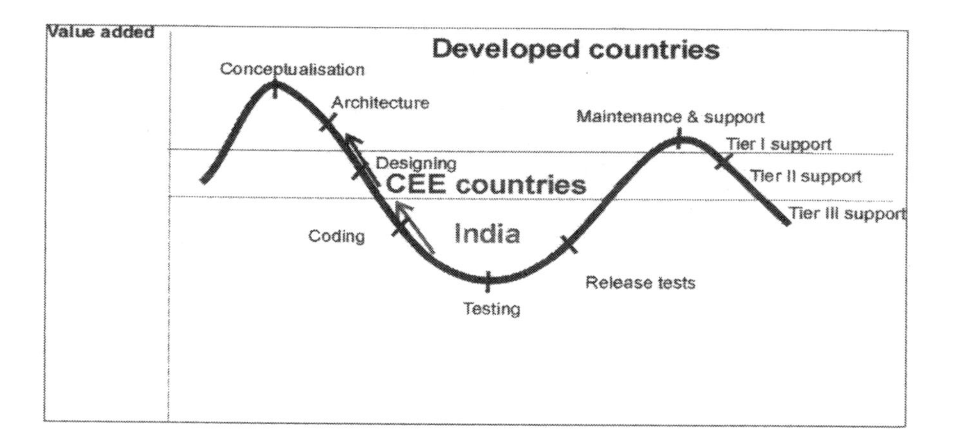

Figure 9.1 Division of work in outsourcing software development
Source: Partly based on Ali-Yrkkö and Jain 2005.

The division of labour is also important. Consultants estimate that, in an offshore outsourcing arrangement, 15 per cent of the client's IT staff are retained in a home country, 15 per cent of the supplier's IT staff are onshore, and 70 per cent of personnel are offshore (Overby 2003).

Forms of delocalization

The main types of delocalization in the software industry identified in the literature are as follows: FDI, subcontracting and offshoring (offshore outsourcing). Non-market based types of FDI appear not only in 'old' EU member countries (Coe 1997b), but also in CEE countries. There are three main forms of footloose or non-market based FDI in the software industry: offshore data processing (business process outsourcing – BPO), offshore programming and software packaged production. While the emergence of the first two of these developments has been well chronicled (Coe 1997b), the emerging manufacturing-style international division of labour in the software products industry has received less attention. Software package production and localization are among the key activities conducted by foreign companies in LDCs and DCs (Coe 1997b). Mass-produced software is very different from many other computer services in that it is essentially like any other product and need not be manufactured in proximity to its consumption.

The extent to which process is subcontracted varies among software companies. Many of them actually replicate disks and assemble the products using their own staff, but some outsource the entire process. The most important type of activities subcontracted in Europe are application development, followed by program and architecture planning and testing (Ali-Yrkkö and Jain 2005).

Offshore outsourcing,[5] a type of BPO, is the exporting of IT-related work from DCs to areas of the world where there is both political stability and lower labour costs or tax savings. Broadly, there are three types of offshore contract supply of IT:

- Supply of application development and support functions.
- Supply of systems design and integration; and support for networks and infrastructure.
- Supply of IT-enabled services, such as accounting, records management, claims administration and call centres, among others (*Trends in the Offshoring of IT Jobs* 2004).

5 Outsourcing and offshoring represent two different dimensions. By outsourcing we mean that other firms take over operations that were previously conducted within the firm (note that relocation is not a requirement for outsourcing). Offshoring, in turn, means relocating activities from one country to another, but not necessarily from one firm to another (see Labrianidis 2008, 4–10).

India's position in the offshore outsourcing market is dominant, with an estimated market share of approximately 80 per cent (Sahay et al. 2003). A recent study by the Frost and Sullivan consulting company (*Global Offshore Outsourcing* 2004) revealed that offshore outsourcing goes far beyond what has been occurring in India for the last several years. Centres of offshore supply are found in Ireland and Israel, and interestingly, in China, the Philippines, Central and Western Europe and Russia.

Factors behind delocalization

Delocalization factors from the point of view of the host country are noted here. A favourable tax regime was one of the main factors in the 1980s. Nowadays, the low cost of labour input and the availability of skilled staff are among the most important reasons. Other factors are low set-up costs, infrastructural investments, often made by the public sector, and significant, prior investments in educational services. A strong external effect associated with a large outward shift in demand for education, resulting in the entry of private educational providers, also enhances the range and scope of IT skills. Organizational changes and internal restructuring may lead to the outsourcing of IT-related non-core activities (Coe 1997a). However, the software industry is a clear example of an industry where the flow of ideas has been as important as the flow of physical capital (Commander 2004; Crone 2003). Therefore contingent events may also be significant in attracting foreign investment.

The main reason for delocalization may be explained in terms of classical location factors. Production moves to where it is cheaper. Many papers support the view that the most important motive for offshore outsourcing is lower costs (for example, Carmel and Agawar 2000; Girma and Görg 2002; Ali-Yrkkö and Jain 2005). However, additional costs, such as management and communication costs, make the cost difference clearly smaller than the wage difference. Relative to USA costs, typical cost savings from offshoring fall between 20 and 40 per cent depending on the type of work (*Trends in the Offshoring of IT Jobs* 2004). For programming, the cost savings are closer to 20 per cent, while for BPO the current savings are in the range of 40 per cent. Savings on maintenance and support for legacy systems are around 25–30 per cent. Relative to Canadian costs, savings would be about 10–15 per cent lower (Huws et al. 1999). CIO Magazine (Overby 2003) estimates 'the hidden cost' of moving IT work offshore at 15–57 per cent of the contract's value. In reality, most IT organizations save 15–25 per cent during the first year; by the third year, cost savings often reach 35–40 per cent as companies 'go up the learning curve' and modify operations to align to an offshore model (Davison 2007).

The rise of the export software industry in such countries as India, the Philippines, Russia and Bulgaria has drawn attention to the delocalization of more highly-skilled information processing work. The new global distribution of work in this sector follows yet another pattern. Here, a good supply of highly skilled work, especially IT professionals, constitutes an important attraction (Huws et al. 1999). Both product specialist subsidiaries (developing and producing a limited product line for the global market) and strategic independent subsidiaries (developing lines of business for local, continental or global markets) can have substantial benefits for economies compared to the traditional marketing or manufacturing sites (Coe 1997b). Crucially, the emergence of such new, higher-performance overseas plants has created employment opportunities for the less-developed areas of the EU such as CEE, Spain, Portugal and Ireland (Coe 1997b), which rely heavily on FDI to provide employment.

The disadvantages of delocalization from a host country perspective are presented below. Growing labour cost may constitute a danger for offshoring. There is an increase in software specialists' wages in India, which over the past few years have been increasing at an annual rate of somewhere between 15 and 25 per cent, according to one of the vendors (Ali-Yrkkö and Jain 2005). The lack of internal quality-control procedures is another disadvantage. Apart from well-paid jobs, employees experience some disadvantages of delocalization. Seasonal and temporary work is typical for IT project-oriented companies, for example, system or application integrators (Coe 1997b). The level of embeddedness of software companies still seems to be relatively low. Some foreign software development firms do not seem to be establishing links with indigenous firms. The majority of large foreign firms are part of international value chains with limited local clustering, which is shown in the case of Flanders (Larosse et al. 2001). Most of the large IT companies have to align their alliance strategies with headquarters abroad, thus limiting the scope of local cluster development. However, a few of the biggest players (for example, Microsoft and Symantec in Ireland) purchase a majority of their raw materials locally.

There are also remarkable social (language difficulties) and cultural barriers to delocalization. Delocalization decisions are hampered by a lack of trust and a perception of risk among clients who are uncertain of the skills, capabilities and credibility of potential foreign subcontractors. Potential information leakage and data security are a concern for investors (Lai et al. 2004). Lack of continuous client-developer interaction is another delocalization barrier. Despite good communications links, interaction sometimes needs to be face-to-face.

Delocalization of the European Software Industry: Company and Key Informant Survey Results

The general characteristics of the poll are presented in Table 9.3. For the purposes of the study, the survey focused primarily on CEECs which are viewed as the main beneficiaries of the process of global integration in the industry. Thus, 52 enterprises were surveyed in Estonia, 51 in Bulgaria, 50 in Poland, 20 in Greece and 18 in the UK. The enterprises surveyed were not randomly selected. The methods of selection have led to small peculiarities of the enterprises surveyed in terms of size. There appears to be a slight over-representation of medium and large-scale firms (especially in the UK) at the expense of micro-enterprises (apart from Estonia). Age distribution seems to be typical with an average 28 per cent of young companies within the survey. Young companies' involvement rate is higher in countries where the software industry constitutes a new wave in the economy (Estonia and Bulgaria). Poland is in the middle of the history of the software industry, with the first foreign companies entering the market in the first half of the 1990s. Then come Greece and the UK, both with at least 70 per cent of surveyed enterprises established before 1995.

Table 9.3 General characteristics of the enterprise survey

Type of delocalization (share of total in country, per cent)	UK	Poland	Greece	Estonia	Bulgaria	Total no. of firms	Share in total
Foreign companies	22	66	5	29	16	61	32
Subsidiaries abroad	50	18	25	8	4	36	15
Subcontracting in/outsourcing from companies abroad	61	68	75	54	98	135	72
Subcontracting out/outsourcing to companies abroad	50	20	25	25	4	36	20
Total number of firms	18	50	20	52	51	191	100

Source: Enterprise Survey.

The forms of delocalization in the software industry are diversified. Out of 190 interviewed companies, 72 per cent undertake subcontracting or outsourcing from a company abroad. It is very common in Bulgaria and Greece (75–98 per cent of companies) and slightly less popular in Poland, the UK and Estonia (54–68 per cent). Almost one-third of interviewed companies were affiliates of foreign entities. There are two less popular forms of delocalization: having a subsidiary abroad and giving subcontracting or outsourcing to a company abroad. These forms are most popular in the UK, although in fact they have become more common in Poland. The Bulgarian software industry ranks the lowest (four companies out of 51 involved in these two forms of delocalization), but based on key informant interviews, it must be said that Bulgarian companies recognize such opportunities and will take advantage of them in the near future. The above-mentioned significant role of FDI leads to a conclusion about the expansion-based nature of internationalization of the software industry. Foreign companies do not employ a large number of people

(19 per cent of total enterprises), but they represent a high financial turnover (42 per cent of total) (Table 9.4).

Table 9.4 Share of employment and turnover of companies involved in different types of delocalization (per cent)

Type of delocalization	Share of total employment	Share of total turnover
Foreign companies	19	42
Subsidiaries abroad	60	35
Subcontracting in/ outsourcing from companies abroad	73	65
Subcontracting out/ outsourcing to companies abroad	56	52

Note: A single company may be included in different types of delocalization.
Source: Enterprise Survey.

The extent of delocalization largely differs in analysed countries. It is very limited in the software industry in Greece where wages are relatively high, so it is almost impossible to compete with companies from LDCs. Additionally, the domestic market is of a small size. The largest scale of delocalization (caused mainly through expansion) is in the UK and Poland, with Estonia and Bulgaria rapidly achieving the same level.

The interviewed companies largely benefited from delocalization and reported increased sales after being involved in the process. Turnover has risen in about two-thirds of the interviewed companies. The number of companies that reported a turnover increase is between 69 per cent and 74 per cent in Bulgaria, Estonia and Poland. Profits have also increased in almost a majority of companies interviewed, especially in Polish and Bulgarian firms (see Table 9.5).

Table 9.5 Changes of turnover and profits after delocalization

	Bulgaria	Estonia	Greece	Poland	UK	Total
Turnover	●	●/●●	–/●	●●/●	●/–	●
Profits	●	●/–	–/●	●/●●	●/–	●/–

Note: ●● – strong increase, ● – slight increase, – – no change, ○ – slight decrease, ○○ – strong decrease.
Source: Enterprise Survey.

Forms of delocalization and reasons behind the process

Seven variables were chosen in order to identify modes of delocalization and clusters of similar companies. These indicators include: exports and subcontracting share, number of companies serviced in 2004, share of foreign capital, year of first establishment and employment in the company and in its foreign subsidiaries. The companies analysed were compared to species of birds based on their size, level of travelling (the level of engagement in exports and subcontracting) and number of friends and partners (companies serviced). Enterprises were clustered by the k-average method. The distribution of companies within clusters is not balanced (Table 9.6) – cluster III includes 118 companies. However, this group of companies is very homogeneous and resistant while increasing the number of clusters.

Table 9.6 Typology of companies based on involvement in delocalization

Number of cluster	I	II	III	IV	V	VI	Total (average)
Type	Hawks	Woodpeckers	Sparrows	Swallows	Owls	Parrots	
Indicator	**Final centres of clusters**						
Average export share (per cent of total sales)	50.0	62.7	28.3	72.6	14.5	28.6	39.8
Average share of subcontracting activities (per cent of total exports)	96.7	70.3	48.9	74.0	30.0	36.6	55.5
Average number of foreign companies serviced	9.7	6.9	3.9	2.7	10.0	5.6	4.2
Average foreign share (per cent of share capital)	20.0	0.0	4.5	97.0	0.0	73.1	25.8
Average year of first establishment	1995	1992	1995	1997	1982	1987	1995
Average employment	402	93	25	33	2,486	212	105
Average employment in foreign affiliates	15	15	1	0	6,075	0	130
Number of companies	5	17	118	36	4	11	191

Source: Enterprise Survey.

Insourcing/subcontracting in The majority of companies interviewed are involved in subcontracting in. Such companies may be found within the first three identified groups of enterprises (clusters I-III). The group of 'hawks' consists of five relatively young, usually indigenous and large companies. In comparison to the mean of 40 per cent of all turnover from exports hawk-like enterprises are export-oriented and highly dependent on subcontracting activities (Figure 9.2), usually working for numerous customers. Seventeen companies classified among 'woodpeckers' are very-active, medium-aged and medium-sized indigenous firms. This group includes, among others, seven Bulgarian and four British companies. The most common type ('sparrows') consists of 118 small and micro firms. About three-quarters of Greek, Estonian and Bulgarian companies interviewed belong to this group. Subcontracting in is the domain of Greek companies, where over 80 per cent of export value comes from subcontracting from a limited number of partners – almost two-thirds of firms have up to three partners. It is worth mentioning that sparrows are involved in export to a relatively small extent.

Subcontracting is of medium significance in export activities: in the survey, 56 per cent of the total exports of companies are on a subcontracting basis. However, only 21 per cent of the total exports of software companies are intermediate products/solutions (sold to other firms abroad for further processing). In other words, products developed in host countries are commonly final solutions.

The reasons behind insourcing (subcontracting abroad) reported by key informants are quite striking. According to the Enterprise Survey, 78 per cent of companies have orders due to representing a suitable level of expertise. Among other important reasons are reliability and appropriate technology. Low cost mattered only in the case of 35 per cent of companies interviewed. There are many companies in Poland and Estonia that have similar costs to Western European competitors and compete largely by dedication to work and the resulting high quality. Also companies from home countries (see the paragraphs below) claim low cost is not of the highest importance when choosing a subcontractor. It seems to be that factors behind the Europeanization of the IT sector are less cost-efficiency driven than it seems at a first glance.

> The rule is to provide a quality product. In bigger contracts we prefer not to sell than sell cheaper. (Polish small-sized subcontractor)

Low-cost reasons behind getting orders are important only in Bulgaria. In other countries, this reason was mentioned more rarely (less than one-third of companies). Once more, this supports the thesis about the over-exaggerations of the dominant role of low-cost reasons in decision making about choosing a subcontractor. Many key informants representing IT organizations also claim that innovation and skills are more important in delocalization growth than low cost. The quality of software development in CEE companies complies with European requirements.

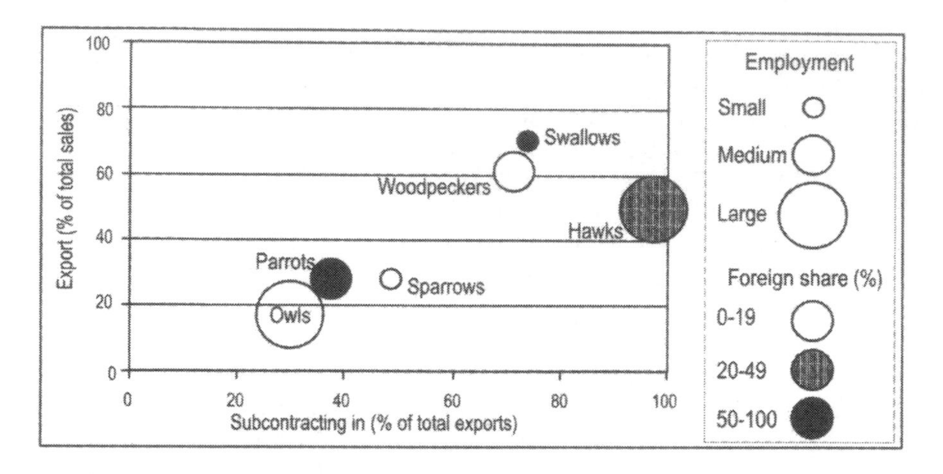

Figure 9.2 Typology of companies based on employment, ownership involvement in subcontracting and exports
Source: Enterprise Survey.

Table 9.7 Reasons for receiving orders from subcontractors

Reason	Total (per cent of companies involved)	Countries (per cent of companies involved)	Share of exports (mean, per cent)	Average employment
Low cost	35	Bulgaria (62), Poland (29), Greece (27), Estonia (14)	49.5	72
Appropriate technology	52	Bulgaria (68), Estonia (68), Greece (40), Poland (38)	41.9	63
Expertise	78	Poland (91), Estonia (86), Bulgaria (84), Greece (73)	43.9	63
Geographical proximity	20	Estonia (32), Bulgaria (24), Poland (21)	43.9	54
Reliability	57	Greece (80), Bulgaria (78), Estonia (46), Poland (44)	41.7	64

Source: Enterprise Survey.

Table 9.7 shows that companies that have received orders due to low cost can be found among the largest enterprises. Low-cost companies are also more oriented towards exports. What appears to be one of the important reasons behind the search for foreign customers in several countries is the significant delay of public spending in ICT (Greece, Bulgaria Poland). Also, a general decline of the economy was a significant factor for some CEE companies to search for customers abroad.

Foreign companies FDI is the second form of delocalization in terms of frequency of occurrence in the sample. Foreign companies are younger and smaller, but more export-oriented than indigenous firms (Table 9.8). The latter most often have contacts with subcontractors and are more dependent on them. They represent the lowest number of subcontracting partners and sometimes depend on one foreign partner supplying them with software.

In the survey, FDIs are represented by two types of firms: young companies (cluster IV: 20 firms in Poland, eight in Bulgaria, six in Estonia, one in the UK and one in Greece) that report the highest level of exports and a large involvement in subcontracting, and relatively old firms focused on the internal market with a low level of exports and subcontracting in (cluster VI). The first type may be called swallows as they find nourishment and come back home to feed their offspring. Swallow-like companies predict a spring of foreign investments in the software sector. The latter type will be described in reference to subcontracting out.

Table 9.8 Characteristics of foreign companies

Indicator	Is your firm an affiliate of a foreign company?	
	No	Yes
	Average	
Exports (per cent of total sales)	32.7	54.8
Number of face-to-face interactions per year	10.9	22.1
Balance of power (1 – customer least powerful; 5 – customer most powerful)	2.6	2.2
Mutual dependence (1 – companies highly dependent on each other; 5 – independent)	2.3	1.9
Average number of years of continuous relationship	5.8	4.8
Total employment	123.9	63.9
Year of first establishment	1994	1996

Source: Enterprise Survey.

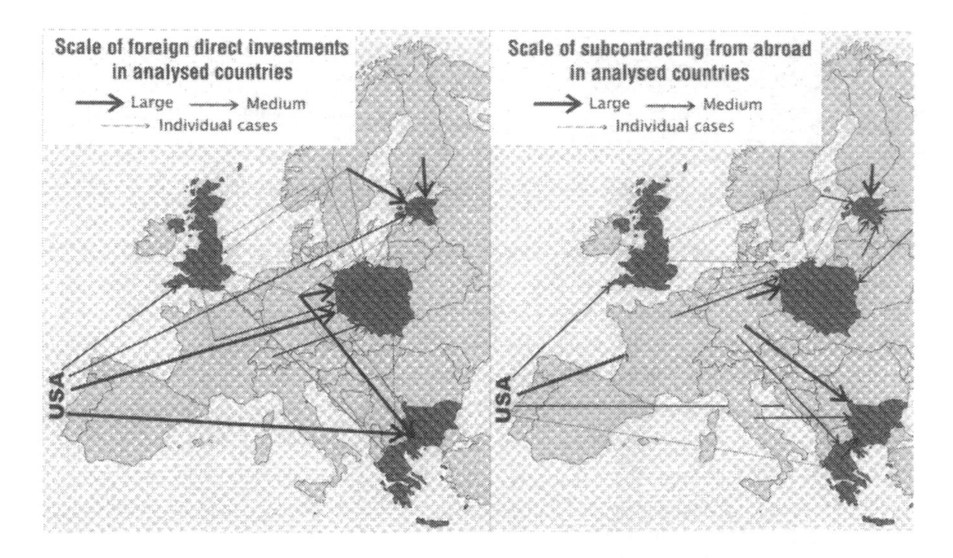

Figure 9.3 Foreign investments and subcontracting from abroad in analysed countries

Source: Enterprise Survey.

> The company has a policy of employing only very good graduates without experience in order not to pay them very much. The owner saves money on everything, for example, computers, furniture, company cars – everything is low-cost and underinvested. (Medium-sized American software development centre in Poland)

The major foreign investors in the European software industry are USA and German companies. Scandinavian enterprises invest in Estonia and northern Poland (Figure 9.3) – the most interesting examples are traditional media enterprises that tend to have stakes in IT companies. Many Estonian software companies are controlled by Scandinavian (usually Finnish) companies.

The reasons behind FDI are strictly connected with the need for a company's expansion in order to secure growth. In CEE, both skills and cost are very important, and also those countries are themselves market opportunities. Foreign companies deal with competition less often by reducing costs and more often by improving quality. Obviously, in some cases pure cost also matters. It is typical for some large American software companies to conduct labour-intensive software development and testing in Central Europe. There are usually cost-based locations. Cost considerations mean that some of these companies save money and want to pay as low as possible. On the other hand, Polish managers of some foreign enterprises try to find a way of becoming more independent from their mother-companies.

> Within Polish subsidiary – there is an unwritten strategy – which the Polish team realizes – to be more and more self-reliant – to employ not only software specialists, but also business analysts in order to be wholly responsible for projects. To be seen by HQ as not only of a very good technical quality, but also as wholly capable for innovation. To be better and better (self-improvement). (Medium-sized American software development centre in Poland)

There are three types of supply chains in which foreign subsidiaries are included. The most common supply chains are limited to foreign partners and closed within TNC. Additionally, they are not embedded locally. A second, rare type is represented by the local embeddedness of supply chains from the beginning. It usually takes place when the director of a foreign company is of CEE-origin and his/her friends become business partners there. Thirdly, foreign enterprises may become embedded in a region after several years of operating in CEE – surprisingly, this is not a common example.

Subsidiaries abroad

> What we tend to do is to set up a small team in Europe that analyses the market there to see if there are sufficient companies with potential ... We are trying to set up a partnership with them; we would work with them to try to make the first one or two sales and then try and make them sufficiently skilled to be able to do that on their own. (Large multi-office Scandinavian company operating in UK)

The fifth type of companies that are largely involved in subsidiaries abroad is not common (cluster V with four companies) and embraces old, very large Polish and British indigenous companies. A distinction between 'owls' and other types of companies can be made based on employment in foreign affiliates that are very large (UK) or grow rapidly (Poland). In comparison to other enterprises, the involvement in subcontracting in and exports among 'owls' is the smallest in relative terms, but very large in absolute numbers. Typical for this type is the wise choice of partners and consequently the highest number of partner companies that are allowed to be relatively independent from one subcontractor.

> Company strategy is to be very close to clients. When clients internationalize or make international expansion, we follow them or acquire a firm in a new region. (Finnish company operating in Estonia)

Managers of companies that have subsidiaries abroad claim that among the most important factors in decisions to invest abroad are the size of the market and its growth and access to regional and global markets. Access to a skilled labour force also matters. What motivated companies to invest abroad is usually the high cost of domestic labour and high social contributions. Greek firms went to the Balkans (Romania and Bulgaria) mostly because they were pushed away by the negative conditions (high wage expectations and desired high posts since the mid–1990s) of the labour market in Greece. They went there in order to find trained and cheap employees.

Search for new markets is sometimes necessary in order to secure the growth of a company. One of the reasons behind setting up subsidiaries abroad by some large CEE companies is to be able to create long-lasting relationships with the main customers by supporting their indigenous customers in expansion to old EU member countries.

Companies that subcontract out The last group (cluster VI) consists of old, usually foreign companies operating largely in host markets. For these large- and medium-size enterprises, cooperation with foreign or mother companies is not as important as penetration into large markets. The majority of parrot-like companies give subcontracting to companies abroad. Some of them mainly sell foreign software. Just as parrots tend to mimic, these companies use what was developed abroad and create a new context providing the host country market with localization and customization services.

Among European countries, Germany is to a large extent involved in outsourcing (nearshoring). German enterprises are the largest customers for Polish, Bulgarian and Greek companies (Figure 9.3).

What is most striking are the reasons listed behind this form of delocalization. Companies subcontract from abroad because they feel a lack of specific skilled labour at home (the UK and Poland). Shortages of IT skills in a home country were also reported in many papers to be one of the most important types of lack of in-house resources (*Trends in the Offshoring of IT Jobs* 2004).

Enterprises also have to deal with a lack of appropriate technology or equipment (Poland, Greece, Estonia, Bulgaria). Higher labour cost ranks third (the UK, Poland, Greece). Greek companies tend to focus on Bulgarian subcontractors, while a few Estonian companies focus on Finnish suppliers. Polish companies use both Russian and British enterprises to provide necessary solutions. Almost all of the largest Polish enterprises have invested abroad (chiefly between 2003 and 2006), mainly in Russia or the Ukraine. However, cases of delocalization from new EU member countries are rare, since companies do not possess the required financial capacity to invest abroad. Some examples in this respect are the Bulgarian Scient (Vietnam), Polish Sygnity (Russia), Prokom (Czech Republic), ComArch (Ukraine) and Asseco (Slovakia). There are three modes of activities abroad presented by CEE companies; most common is establishing subsidiaries and arranging partnerships with foreign representatives, while quite rare are acquisitions of medium-size foreign companies (in the cases of Polish Prokom and Asseco).

Linkages and cooperation networks A significant number of companies (over 38 per cent of entities) are involved in supply chains that are a part of international networks. However, there are only nine cases of enterprises that are only included in international networks and do not operate in the domestic market (over 90 per cent of turnover from export). This means that the level of real involvement in international networks is relatively low.

The position of a majority of companies (56 per cent) within supply chains has been upgraded in the software industry. Downgrading cases are rare (four cases reported). Surprisingly, there are no statistically significant differences between different modes of delocalization and the upgrading/downgrading processes. Different types of delocalization represent a similar share of upgrading (from 45 to 60 per cent of firms). Qualitative change is observed in the activities conducted by software firms. There are cases of upgrading from simple subcontracting to more arm's-length relations (simple exports). The positive process of a shift from software development to consulting services is also observed. There is a transition from simple codewriting to implementation of whole projects.

A relatively sophisticated mode of involvement in subcontracting from foreign companies may be shown by a high number of foreign companies serviced by an individual company. On average, it gives 4.2 customer companies per one subcontractor. Surprisingly, the average number falls to 3.5 for affiliates of foreign companies. Almost two-thirds of subsidiaries of foreign companies work for up to two customers. The limited number of foreign subcontractors observed in foreign subsidiaries leads to the conclusion of the greater dependence of foreign subsidiaries than of foreign subcontractors. About one-third of foreign subsidiaries stressed that they represent the same level of power as their owner. Foreign subcontractors more often find a balanced decision with their partners.

In general, only 34 per cent of subcontractors exclusively service one company. It must be concluded that some companies have already established a network of foreign subcontractors and are not dependent on one of them. The same number of companies is dependent and independent on their subcontractors. Almost half of the managers feel that there will be moderate consequences for breaking off the relationship with the partner.

Existing subcontracting relations have been continuing on average for 5.5 years: for the longest time in the UK (9.2 years) and for the shortest in Estonia (4.1 years). This average points to a relatively long and strong relationships between subcontractors.

Conclusions: Delocalization or Expansion?

It must be concluded that geography matters. There are not many cases of companies that invested in a distant country. For some German, French and Scandinavian subcontractors, both geographical and cultural proximity matters and that is why they choose Poland or Estonia instead of India.

Many key informants have pointed out that whereas large, routine projects (for example, two- or three-year contracts, 1,000 employees involved) often come to India or China as well as the Philippines and South Africa, CEE companies are relatively small and are not able to execute larger tasks. However, there is a possibility to focus on more innovative, flexible projects that require smaller teams. Based on company interviews and the Enterprise Survey, it must be said that CEE companies recognize this possibility and enter into rather smaller projects.

In reference to the first aim of the chapter, there are two simultaneous processes taking place in the European software industry. The most visible form of delocalization is expansion conducted by foreign companies, mostly USA- or German-owned, but also more often by companies from former host countries (for example Poland). The average value-added per employee shows that new EU members are not lagging behind older EU countries. Similar productivity with lower wages constitutes a key element that attracts foreign partners to CEE. There are two types of foreign companies. One is a remarkable group of old, usually foreign companies that are not largely oriented towards home markets and mainly operate in host markets. For these large- and medium-size enterprises, subcontracting is not as important as penetration into large markets. The second type of foreign company is represented by young firms that report the highest level of exports and a large involvement in subcontracting. They report a relatively low number of subcontracting partners.

The second major form is subcontracting abroad, which has been rapidly growing in recent years but started in Europe at the beginning of the 1990s at a very low level. Many indigenous companies have a very limited number of partners and represent a medium level of involvement in subcontracting. There is also another specific type of company: very large British and Polish enterprises which have invested abroad. They have implemented a wise policy of signing contracts with many partners and remaining independent from them.

Despite the dynamic development of offshoring activities thanks to the rapid growth of the entire IT sector, there is no job loss in DCs. Therefore, delocalization in the IT sector may be considered a win-win game so far. It is more the expansion of foreign and domestic companies abroad than delocalization as a sector and market growth is observed. Even if delocalization is observed, it is, as Arora et al. (2002) argue, due to a shortage of IT skilled professionals in the USA and Western Europe as a main factor. An important reason for the delocalization of IT and BPO services is the option of round-the-clock operation. In general, the global expansion of IT firms is driven, firstly, by the need for market access and growth, secondly, by economies of scale and costs savings and, lastly, by access to skills and technology.

Considering the second aim of the chapter, the factors behind delocalization of the IT sector are less cost-efficiency driven than it seems at first glance. There are many companies in Poland and Estonia that have similar costs to Western European competitors and compete largely by dedication to work and the resulting high quality. Also, companies from home countries claim low cost is not of the highest importance when choosing a subcontractor. It is clear that going abroad is more closely connected to growth strategy than as a means of seeking higher profits. The shortage of human capital in home countries and the promising markets in host countries are reported as the main reasons for delocalization. Thanks to expansion, FDI and subcontracting, the entire IT industry grows, skills and knowledge are spread and, as IT is auxiliary to other economy sectors, general economic growth is fuelled (OECD 2006).

When it comes to cooperation networks, changes in scale and types of activities being outsourced over time in comparison to the 1990s are reported. Nowadays, a full range of activities is outsourced instead of low-value, labour-intensive software codewriting. A move up the value chain among companies who were subcontractors or subsidiaries of foreign companies is apparent.

In reference to the fourth aim of this chapter, the success of many CEE IT companies is accounted for by the quality of human capital (dedication and commitment to work), flexibility, involvement and level of expertise rather than to its low cost. According to most of the managers and key informants interviewed, all this, together with a cultural proximity to the USA or Western Europe, gives an advantageous position in relation to India-based competitors, where cultural differences are a barrier to successful development of more sophisticated tasks or problems in India. Other factors lowering the competitive position of India are wages, time-zone incompatibility, drainage and the poor creativity of IT specialists.

Further delocalization of IT sector activities to India or to another low-cost country is not perceived as a danger to the European software industry. The cultural barrier, geographical distance and high profitability only for huge projects are factors responsible for eroding India's competitive position as a location for European outsourcing.

According to a vast majority of interviewed managers, delocalization does not need to be controlled: only political and fiscal stability is of high importance. Therefore, EU membership was reported as an important factor behind investment or choosing a partner. Entrance into the EU by Bulgaria and Romania should further catalyse internationalization of the IT sector in these countries as was earlier observed in Poland and Estonia.

Future research should cover several fields that include, for example, the in-depth analysis of the cost structure of delocalized companies. It is necessary to conduct such an analysis in order to establish the role of wage differences in the growth of delocalization. A more detailed assessment of the scale of neighbourhood effects in various countries would help to understand the role of a cultural and geographical proximity-driven cooperation in software industry growth

References

Ali-Yrkkö, J. and Jain, M. (2005), Offshoring Software Development – Case of Indian Firms in Finland (Keskusteluaiheita: Discussion papers of the Research Institute of the Finnish Economy 971).

Arora, A. and Gambardella, A. (2005), *From Underdogs to Tigers: The Rise and Growth of the Software Industry in Brazil, China, India, Ireland and Israel* (Oxford: Oxford University Press).

Arora, A., Gambardella, A. and Torrisi, S. (2002), *In The Footsteps Of The Silicon Valley? Indian And Irish Software In The International Division Of Labor*, Carnegie Mellon University Software Industry Center Working Paper No. 1. (Pittsburgh, PA: Carnegie Mellon University).

Carmel, E. and Agawar, R. (2000), *Offshore Sourcing of Information Technology Work by America's Largest Firms*, Technical Report (Washington DC: Kogod School, American University).

Coe, N.M. (1997a), 'Internalisation, Diversification and Spatial Restructuring in Transnational Computer Service Firms: Case Studies from the UK Market', *Geoforum* 28:3–4, 253–70.

Coe, N.M. (1997b), 'US Transnationals and the Irish Software Industry: Assessing the Nature, Quality and Stability of a New Wave of Foreign Direct Investment', *European Urban and Regional Studies* 4:3, 211–30.

Commander, S. (2004), *Brains: What Can They Do For Development?* Lecture at the London Business School and EBRD, 4 March 2004 (London: Sussex Development Lecture).

Competitive Alternatives (2008), *KPMG's Guide to International Business Location*, KPMG [website] <http://www.competitivealternatives.com/industries/profiles/CA2008_indprofile_softwaredesign.pdf>, accessed 1 June 2008.

Crone, M. (2003), 'Clustering and Cluster Development in Knowledge-intensive Industries: A "Knowledge and Learning" Perspective on New Firm Formation and Firm-building/Firm Growth in Ireland's Indigenous Software Industry' (St Andrews: Regional Science Association International: British and Irish Section, 33rd Annual Conference, 20–22 August 2003).

Davison, D. 'Top 10 Risks of Offshore Outsourcing', Tech Update IT Management, Meta Group [website] (updated 17 December 2007) <http://techupdate.zdnet.com/techupdate/stories/main/Top_10_Risks_Offshore_Outsourcing.html>, accessed 17 June 2008.

Florida, R., Knudsen, B., Stolarick, K. and Lee, S.Y. (2003), 'Software, Creativity and Economic Geography', Carnegie Mellon University Software Industry Center Working Paper No. 2 (Pitsburgh, PA: Carnegie Mellon University).

Girma, S. and Görg, H. (2002), 'Outsourcing, Foreign Ownership and Productivity: Evidence from UK Establishment Level Data', The University of Nottingham Research Papers 16 (Nottingham: University of Nottingham).

Global Offshore Outsourcing and Offshoring of IT Labor Markets. Offshore Outsourcing – Looking Beyond India (2004) (San Antonio: Frost Sullivan Group).

Huws, U., Jagger, N. and O'Regan, S. (1999), *Teleworking and Globalisation*, IES Report 358 (Brighton: Institute for Employment Studies).

Lai, E., Riezman, R. and Wang, P. (2004), *Outsourcing of Innovation* (Nottingham: ETSG, 9–11 September seminar).

Larosse, J., Slaets, P., Wauters, J., Bruninx, S., Simkens, P. and Wintjes, J. (2001), 'ICT Clusters Organisation in Flanders: Co-operation in Innovation in the New Networked Economy', in *Innovative Clusters. Drivers of National Innovation Systems* (Paris: OECD), 112–31.

OECD (2002), *OECD Information Technology Outlook* (Paris: OECD Publications).

OECD (2006), *OECD Information Technology Outlook* (Paris: OECD Publications).

Overby S. (2003), 'The Hidden Costs of Offshore Outsourcing', *CIO Magazine* (published online 1 September 2003) <http://www.cio.com/article/29654/>, accessed 29 December 2007.

Sahay, S., Nicholson, B. and Krishna, S. (2003), *Global IT Outsourcing: Software Development Across Borders* (Oxford: Oxford University Press).

The Software Industry In Ireland – A Strategic Review (1992) (Dublin: Industrial Development Authority of Ireland, National Software Directorate).

Trends in the Offshoring of IT Jobs (2004) (Ottawa: Prism Economics and Analysis, Software Human Resource Council).

PART 3
The Experience from Elsewhere in the World

Chapter 10
Corporate Strategies for Software Globalization[1]

Alok Aggarwal, Orna Berry, Martin Kenney,
Stefanie Ann Lenway, Valerie Taylor[2]

Introduction

Today, nearly every significant software firm has globalized some portion of its software production process. The goal of this chapter is to provide insight into this globalization process, that many term 'offshoring'. Our interest is in how firms use low-wage environments to undertake software work for their global operations. Our interest is not in the relocation of work from companies in a high-cost nation such as Germany or Japan to firms in another high-cost nation such as the USA or the UK even though this type of relocation is common. This chapter also omits work sent to Canada, which does have somewhat lower wage rates than the USA and is the beneficiary of what some have termed nearshoring from the USA. Also excluded from this presentation are the operations of TNCs or domestic firms that service the local economy of a low-wage nation. In most cases, these are relatively small operations except in the case of China, whose domestic consumption of software is increasing rapidly.

We do not debate the reasons firms offshore to nations with significantly lower wages. The fact that wages are significantly lower is simply accepted. In addition to reducing costs, a company's decision to offshore is often dependent upon two dimensions. The first dimension involves its strategic decision regarding the kind of human capital that it would be able to access when it goes offshore. Put differently, it is an uncontroversial observation that if the nation being considered as a recipient for work relocation did not offer the proper skill set in its workers, firms would not relocate work to that environment. The second dimension is cost. It is uncontroversial to state that, given the right skill set,[3] a sufficiently low cost of

1 The material in this chapter is based on Globalization and Offshoring of Software, W. Aspray, F. Mayadas, and M.Y. Vardi (eds), 2006. Reprinted with permission from the Association of Computing Machinery (ACM). See full report at: <http://www.acm.org/globalizationreport>.

2 O. Berry and M. Kenney took a leadership position in the writing of this Chapter.

3 This chapter focuses on the availability of technical personnel as the attraction for offshoring. It is important to add that capable managers are also extremely important. As

labour and work that can be done remotely, firms will find it attractive to locate to that environment even if there is no market in that locale.[4]

A decision to offshore software work maybe made for a variety of reasons and may take a variety of forms. The decision to offshore work has traditionally been made by the manager with responsibility for a project, including profit and loss responsibility. In cases where the contract involves strategic operations, critical company proprietary information or very large budgets, for example, the decision is often made at a higher level in the organization, sometimes as high as the CIO, CFO, or CEO. In certain cases, the real reason for offshoring might be simply that competitors have already done it or the board of directors is demanding an offshoring initiative to save money. Rarely are the answers so simple, but there are numerous anecdotes about how an executive team in the United States will demand that an operation achieve a certain headcount reduction by a clearly unrealistic date. The responsible executives will achieve the headcount goal regardless of the economic justification. In other words, rationales for action vary. Moreover, similar firms often have different recipes for using offshore resources. One basic decision a company that has decided to offshore must make is whether to undertake the work in its own offshore premises or outsource it.

This chapter considers five kinds of firms that are involved with software production or software services that are provided by one or more LDCs:[5]

- Packaged software firms headquartered in DCs that make and sell software as a product, for example, Adobe, Microsoft and Oracle.
- Software service vendors headquartered in DCs. These companies may also provide packaged software, though not all of them do so. Examples include IBM, Accenture and EDS.

Parthasarathy (2005) points out, the executive management team is critical for the success of an offshore subsidiary.

4 For example, Nike produces athletic shoes in a large number of nations where there are few, if any, customers for its shoes.

5 We do not use the term captive in this chapter, even though it is used elsewhere in the aforementioned report. In keeping with the literature on international business, we use the term subsidiary. It is more accurate and does not suffer from the bias reflected in the term 'captive'. The categories are not divided on the basis of which firms are subsidiaries because numbers 1, 2, 3, and 4 are all subsidiaries. Only the LDC firms' operations are not subsidiaries. The categories developed in this chapter are for the purposes of understanding the impact of globalization on software professionals and thus they may not be useful for other purposes. For example, if one was merely interested in globalization, it might be that the packaged software firms should be combined with software services firms. Or, alternatively, software and software services might be combined. The separation of small start-ups from large software and software services firms is justified only because of the importance they have for the high-technology economy. For other purposes, this separation might not be proper.

- Internal software operations in firms headquartered in DCs that have software operations but are not part of the software industry. This encompasses all the companies producing non-IT goods and services. (The group is eclectic and enormous. The importance of this category is that software is now at the heart of value creation in nearly every firm. This is true of financial firms such as Citibank and HSBC and manufacturing firms such as General Motors and Siemens. Each of these companies has a large staff writing software. To illustrate, it is estimated that by 2010, 40 per cent of the value of a car will be in its electronics, of which embedded software will become an increasingly important component. In 2002, it was estimated that the typical luxury car had 105 microprocessors, up from 70 in 1998 [Tsai 2004].)
- Software-intensive, high-technology startups based in DCs. (This category, though small in numbers of jobs, is important because these firms provide many of the jobs of the future. For these firms, frequently there is no job displacement at all. Rather it is the location of the future employment growth that is in question.)
- Offshore IT service providers headquartered in LDCs that provide services for firms in the DCs. These firms have emerged in a number of countries, though the largest by far are located in India.

This classification of firms is only heuristic. The global Fortune 1000 firms have complicated webs of relationships which might include newly-built facilities, facilities they acquired, contractors from DCs (for example, IBM and Accenture), and contractors from LDCs, for example, Infosys (India), IT United (China), and Softtek (Mexico). Some product firms outsource certain activities to contract R&D firms and even form JVs. There are also intermediate solutions such as the build-operate-transfer option, which lies in between building one's own facility and outsourcing. The tasks being undertaken vary widely and include activities such as low-level software support, product testing, product development and R&D. The options and permutations are numerous, and the case studies in this section are merely overviews; thus they cannot do full justice to the breadth and scope of the software and software service operations of these firms.

For each category, the chapter gives a general discussion of the outsourcing issues faced by a specific kind of firm, followed by several case studies to illustrate the types of operations the firm carries out in LDCs and why those particular countries were chosen. These case studies are intended to be illustrative but not exhaustive. The particular cases were selected in order to provide a balance across sizes of companies and do not constitute a random selection upon which generalizations should be based.

Offshoring Firms

Large, established, DC software firms

Because of the somewhat different dynamics of the packaged software firms and the software services providers, we discuss them separately despite the fact that there is significant overlap between these two categories of firms. For example, Oracle and Cadence, which are usually considered packaged software providers, have large consulting arms to assist with the installation and operation of their software. IBM sold $15 billion in software in 2004, yet it is today more of a software services firm (with revenues of $46 billion in its global services unit). Accenture is a massive consulting firm that provides a variety of services, including software services. Thus the line between the two categories is somewhat difficult to draw, but it is, nevertheless, a worthwhile distinction, because a pure packaged software firm such as Adobe or Microsoft hires programmers almost exclusively, while a firm such as SAP, IBM or Oracle also hires consultants and analysts who are not necessarily working on products but are providing services.

Packaged software firms

The packaged software firms are what most people think of when they think of software. As a general rule, the largest and most successful packaged software firms in the world are headquartered in the USA (the notable exception is SAP in Germany). Successful packaged software firms can be very profitable, because they only need to write an application a single time (although perhaps in several variations) and then reap their revenues from the sales of many copies. One reason for establishing offshore facilities is to localize the package for particular language groups. For example, Ireland has a large industry that specializes in localizing products from USA software firms for the European markets (O'Riain 2004). Localization work characterizes a significant portion of the work by the R&D laboratories of packaged software firms in various nations. This type of work, though important for the global economy, is not of particular interest here.

There are, of course, other motivations for package software firms to locate in developing countries. The most frequently given reason is access to the talented labour force working in these lower-cost locations.[6] One important motivation behind offshoring for these package software producers is that their packages

6 The decision to move to a location for lower cost is a complex one. Lower cost includes not only wages but also the lower cost of benefits, including health care. It also includes issues such as reduced concerns about discipline problems, substance abuse in the workplace, and governmental regulations concerning harassment, racial policies and so on that are part of the protections commonly expected in DCs. This chapter does not place a judgement upon these policies. Quite naturally, in each nation, there are different regulations and standards that channel business activities and create various costs and benefits.

are constantly increasing in size and complexity, driving the cost of writing the software, testing, and debugging it ever higher. Whichever of these causes is the most significant, what is certain is that nearly all major packaged software firms are establishing offshore facilities in lower-cost environments, ranging from Eastern Europe and Russia to India and China. In the following case studies of Adobe and SAP, we examine this new geography of the software industry.

Adobe On account of its Acrobat program, Adobe has a wide global footprint. Its software has applications in digital imaging, design and document technologies. The firm does its product development in the USA, Canada, Germany, Japan and India. In India, Adobe has its largest physical office space outside the USA, and the Indian operation is growing more rapidly than any other location.

In 1997 Adobe established a sales office in New Delhi, India to market its products. In 1998 it established an R&D centre in New Delhi (Noida) to utilize the low-cost R&D talent available in the country. By 2005 Adobe had 3,800 employees worldwide of whom approximately 500 (13 per cent) were located in India. Adobe has invested $10 million in India but plans to increase that to $50 million over the next two years as the R&D centre grows. Adobe was perhaps the first international software company to develop a full-fledged product in India, Page Maker 7 (Rediff 2005). The Indian centre has filed 25 patents in the last four years, an indication of the sophistication of the work it has undertaken.

In 2005 Adobe acquired Macromedia, another Silicon Valley firm. Rather than consolidate Macromedia's Bangalore research operation into its own research operation in New Delhi, it is retaining and expanding the Bangalore facility, which in April 2005 had 150 workers and was expected to grow to 250 by year-end 2005 (Verma 2005).

Adobe's Indian R&D centre works on Adobe Acrobat desktop applications and server-based products as well as products related to digital imaging and video. It develops components for almost the entire range of Adobe's product line. Products it has worked on include PageMaker, FrameMaker, Postscript, Photoshop Album Starter Edition and the Acrobat Reader on Unix and alternate platforms.

From the managerial perspective, the Indian operation is becoming increasingly integrated into Adobe, as is evidenced by the fact that Naresh Gupta, who has managed the Indian operation since its inception, is being relocated to the San Jose headquarters, where he will join the executive management team (Rediff 2005).

There can be little doubt that India has become Adobe's low-cost development centre. To date Adobe has not established development centres in other low-cost countries. In 2002 there were articles in the press stating that Adobe might abandon sales in China because of concerns over software piracy; this was quickly denied by Adobe spokespersons, but the company has not moved to open a development centre there (Sim 2002). Adobe's Indian operations have continued to grow through 2007, and the percentage of the company's employees located in India is likely to increase (given that the total global headcount is growing slowly).

SAP Laboratories SAP, established in Walldorf, Germany, is one of the world's largest software vendors with operations throughout the world. SAP Laboratories is its R&D arm and has units in nine locations worldwide: Walldorf, Germany; Palo Alto, California, USA; Bangalore, India; Tokyo, Japan; Sophia Antipolis, France; Sofia, Bulgaria; Montreal, Canada; Tel Aviv, Israel; Shanghai, China; and (most recently, in 2005, with about 50 employees) Budapest, Hungary. The role of SAP Laboratories is to distribute global development efforts, enable SAP to access the world's best IT experts, support local and global markets, develop first-class solutions and drive innovation and competitive advantage for SAP, its customers, and partners.

Most of the laboratories are relatively small. For example, SAP's fourth largest lab worldwide is in Israel and it employs 500 people (it also has another, smaller R&D operation in Israel located within a firm that it acquired), while Sofia, Bulgaria employs only 200 programmers (SAP 2005). Each laboratory has its own specialties. The Shanghai laboratory has focused on localization work, but is scheduled to grow to 1,500 engineers by 2009 and is expected eventually to do more than just localization (People's Daily Online 2004). In China, SAP is cooperating with the Chinese Linux supplier Red Flag Software to develop corporate applications for Linux (Bishop 2005).

The India SAP Laboratory was established in 1996 and has grown to be the largest lab outside Germany with more than 3000 employees in 2008. According to SAP AG executive board member Shai Agassi, 'Indian developers had contributed substantially to the global success of the NetWeaver, the first appli-structure platform for enterprises across the verticals. The SAP Labs India team is one of our most important development teams for NetWeaver worldwide' (Indo-Asian News Service 2005). Even though the Indian operations for the global economy are categorized as R&D, many of the employees are in the services and consulting operations (Barlas 2004).

SAP has a global R&D and operations strategy with its various laboratories specializing in different areas of software development. The plans for employment growth for both India and China are aggressive. If current plans are realized, India and China will become even greater portions of SAP's total global headcount. Given their economic growth, these countries will also become sizable markets for SAP.

Conclusion The large package software firms are building increasingly global operations. In many cases, their offshore operations are for localization work for the domestic market. However, particularly in the case of India, but also in Russia, the work is for their worldwide software packages. Locating in low-wage countries enables these firms to have access to lower-cost programmers, many of whom are comparable in skill levels to the workers in the DCs. This is not the only benefit. Having operations in other time zones can speed up production by facilitating round-the-clock production. These opportunities are encouraging the rapid expansion of employment by major packaged software firms in India and other lower-cost nations.

Offshoring will have a complicated effect on the packaged software firms and DCs. First, it is likely to put employment pressure on software firms to decrease employment in the DCs. Alternatively, the lower cost and faster production could allow the development of new features in old software and could contribute to the production of lower-priced software products, thereby increasing usage that could result in higher revenues and greater employment. If the trends as described in these case studies continue for the packaged software firms, elements of both of these scenarios may occur.

Software services providers

Software service firms have been among the fastest growing firms in the IT sector, and in general they are far larger than the packaged software firms. This section confines discussion to the software service activities of these firms, but it is important to remember that firms coming from the software service side (such as IBM or Hewlett- Packard) and from the service side (such as Accenture) are converging. In the case of IBM, this has been achieved both through recruitment and its recent acquisition of the Indian service firm Daksh (with its approximately 6000 employees). For service providers, software and various other software-based services (that is, anything done on a computer) may be converging. The software services firms are basically in what might be called a headcount business; they grow by recruiting more workers. Thus they tend to have more employees than most of the packaged software firms.

IBM Established in 1911, IBM has been the global leader in computer hardware and software products and services. In this section, we focus on three different IBM activities, namely, software products, software services and R&D. It is important to understand IBM's scope and scale. In 2004, it had annual revenue of approximately $96 billion. Global headcount at the end of 2004 was expected to be more than 330,000 employees, excluding employees gained from acquisitions and strategic outsourcing contracts (IBM 2004). The company's geography of revenue growth is shifting dramatically. In Brazil, China, India and Russia, IBM's annual revenue growth from 2003 to 2004 was 25 per cent (though from a small base), while growth in the DCs was of the order of 4 per cent. Between 2002 and 2004 IBM increased its workforce in these four nations by 30 per cent (Palmisano 2005).

As of 2005, IBM's Software Group had revenues of $15 billion and contributed one-third of IBM's profit. In the Asia-Pacific region, this group employed 5,000 people, including sales and marketing. In India and China, IBM's software development laboratories employed 1,500 in each country (Smith 2005).[7] Richard Smith, the vice- president of the Asia-Pacific region for IBM Software, stated that 'the Chinese market is internally focused. In India, a lot of the software development activity is mixed – it is focused internally as well as on exports'. As an example of the contributions of offshore centres, 'a significant chunk of the code for its AIX version of the Unix operating system was developed in India' (Smith 2005).

IBM is already well advanced in using global software development teams. Hayward (1997) described a global application development team it created that uses two shifts. The first one is a small group of 25 people in Seattle that would set a daily work specification for a particular application and assign it to offshore teams of 31 programmers each in India, China, Latvia and Belarus (a former Soviet republic). The offshore team in each location would write code to those specifications during their daytime work hours. The code would then be sent back to Seattle, where it would be reviewed and tested. In principle, this process should not only lower labour cost but also accelerate production.[8]

Software development at IBM is now a global process with the offshore, low-cost nations growing rapidly to meet increasing demand. In the illustration by Hayward, the Seattle team was clearly dominant. However, given the increasing capabilities in LDCs, this hierarchical division of labour may no longer be as distinct in the future.

IBM Global Services is the largest service provider in the world with revenues in excess of $46 billion and 175,000 employees spread across 160 nations as of 2004. The services it provides include application development, data storage, infrastructure management, networking, technical support, business consulting and outsourcing services. At the end of 2004, IBM employed 23,000 people in India, and an internal planning document stated that, by the end of 2005, this would increase to 38,000; the bulk of these employees were in Global Services. India now has more IBM employees than any nation except the USA (Hamm 2005).

IBM Global Services is active in providing services to domestic Indian and Chinese firms. For example, in March 2004 it signed a ten-year IT outsourcing deal for $700 million with Bharti Tele-Ventures Ltd., India's leading telecom company, that included the transfer of Bharti's IT-related assets (including workers) to IBM. Not only did IBM acquire a new customer; it also purchased more skilled employees to expand its Indian operations. In 2004 IBM also purchased a leading Indian business process outsourcing firm, Daksh, which though not an IT firm, had 6000 employees. This acquisition illustrates how the IT and non-IT services

---

7 According to the IBM (2005) website, its China Software Development Laboratory employed 2,000 engineers.

8 There continues to be debate regarding the success of such follow-the-sun strategies. Carmel (1999) argues that these global development projects are difficult to manage and often are unsuccessful. On time-shifting, see Carmel and Tjia (2005).

are blurring for the providers. For this reason, the discussion in this section incorporates an overview of IBM's entire range of offshoring service operations not only the software services.

India is becoming IBM's central delivery centre for services. However, like all of the multinational service firms, IBM has also established facilities in a number of other lower-cost nations, including China. IBM Global Services has three centres in China, including one opened in Dalian in 2005 with 600 workers. The Dalian centre is expected to grow rapidly with its main purpose being to serve the Asia Pacific market (ZDNet 2005). IBM Global Services also has a service centre in Mexico.

In August 2005 IBM announced that it was establishing an IT services research centre in Bangalore as an extension of its India Research Laboratory located in New Delhi with an initial staff of 10 researchers. According to P. Gopalakrishnan, the director of IBM's India Research Lab, it would look 'at how technology can improve the capabilities and efficiency of delivery. This would include the whole spectrum of services from infrastructure management, application maintenance, BTO to BPO services' (CIOL 2005). If this pattern continues, India may become the hub not only for doing offshore work but also for developing ways to automate service delivery using software.

India has clearly become a core location for IBM to provide offshore software services, and with the establishment of a research laboratory there to develop methodologies for the automation of service provision it appears as though India may become IBM's global centre of excellence for these functions. However, all of the multinational software service providers have a global footprint so that they can offer their customers a wide variety of services in many different languages. IBM is likely to continue expanding its workforce in software and other services in lower-wage nations, while growth in the DCs is expected to be slow.

With eight laboratories around the world (three in the USA and one each in Switzerland, China, India, Israel and Japan), IBM Research employs approximately 3,050 researchers. The company has steadily increased its R&D expenditures outside the USA, from 28 per cent in 1993 to close to 60 per cent in 2003. In the 1990s, IBM opened three new research labs in Austin (1995), China (1995) and India (1998). For the research laboratories, access to the most creative individuals is the greatest priority, but it is also true that the research centres in China and India have lower operational costs. The main point of these research centres is to attract local talent and to conduct some of the research on problems that are relevant to the local environment using global-class research.

There appear to be some differences in emphasis between the Chinese and Indian laboratories. The India Research Laboratory (IRL) has about 100 researchers and focuses on areas critical to expanding India's technology infrastructure so while IRL researchers work on some local issues such as text mining and speech recognition for Indian languages, they also work on more general research problems in the areas of bioinformatics, natural language processing, grid computing and autonomic computing. The IBM China Research Laboratory (CRL) also has approximately

100 researchers. It has been working on Text-To-Speech systems and can now provide language support for Chinese, Taiwanese Chinese, Cantonese, Korean, Japanese and French. It has also been working on IBM's Websphere Translation Server that provides machine translation between English and Chinese. In this sense, the research profile in the Chinese laboratory is more localized.

All IBM research laboratories actively cultivate relationships with local academic institutions. For example, the India research lab is located on the Indian Institute of Technology (IIT), Delhi, campus, where it has access to a vast pool of talent. In Israel, IBM has built strong relationships with Haifa University and Technion. The R&D laboratories in India and China are still quite small; however, there appears to be a commitment to increase their size rapidly. Their missions are different: in the case of China, much of their work will continue to be on localization and the Chinese language, while the Indian laboratory is more likely to undertake work directly applicable to global business needs.

As the largest software/software services firm in the world in terms of both revenue and headcount, IBM has the most sophisticated global footprint of any firm. Not only is it increasing its employment in LDCs in the more mundane and routine aspects of service delivery, it is also increasing employment in software product development and R&D. In the process, IBM's global posture is changing from being heavily weighted toward the DCs to a more equal weighting globally.

Siemens Business Services Siemens Business Services (SBS) is a Siemens subsidiary that has a global practice in performing software and other outsourced work. It employs approximately 36,000 workers and derives substantial revenue from installing, customizing and maintaining SAP software in businesses. Its 2004 revenues of 4.8 billion EUR were roughly divided between Germany (48 per cent), the rest of Europe (39 per cent), the USA (8 per cent) and the rest of the world (5 per cent) (Siemens Business Services 2004). SBS has been under significant cost pressure and has instituted layoffs to bring its costs under control (Blau 2005). SBS, like many other large service firms, has been globalizing its service delivery operations and in the process has downsized its domestic workforce. Of SBS's 36,100 global employees, only 15,100 are now located in Germany, and 4,000 are located in its rapidly growing Indian subsidiary.

SBS has developed a business strategy that uses a matrix of vertical industry knowledge and sets of general competences to serve its customers. One aspect of the matrix is the industry expertise (vertical knowledge) or competency centres that are scattered in different countries; for example, the paper and pulp vertical is located in Finland (Hallez 2004). The other part of the matrix is the general activities, located in offshore sites in Canada, Ireland and Turkey, which handle stabilized processes. India has two roles: it functions as a back office operation for finance and accounting, and it does general software programming and service and applications development for SAP programs. SBS uses Russia for very labour-intensive and repetitive back office and software application development (Hallez 2004).

Siemens also operates its Siemens Information System Laboratory (SISL) in Pune, India.[9] SISL has been involved in the development of an atmospheric disturbance model for Boeing flight simulators, engine and auto throttle control simulation, modeling and simulation of Weibul clutter, and GPS and INS error modeling for measurement simulation. It has also designed a control system for wind shear control on the Boeing 767, a control system for the flight management system for the Boeing 747, and primary flight control system software, as well as executing the development of Kalman filters for GPS and INS, integrated with GPS in feed-forward and feed-back configurations. SISL has been able to use Indian engineers to design sophisticated software for LDC customers.

SBS and other parts of Siemens are interesting because, in contrast to USA firms, they place a strong emphasis on nearshoring facilities to Eastern Europe, Russia and Turkey. Nevertheless, SBS India is the location with the largest non-German headcount, and it continues to grow rapidly.

Conclusions about DC software services firms Software services is in most respects a headcount and labour-cost business. The multinational software services firms have been experiencing increasing pressure on costs due to competition from LDC producers, particularly the Indian service giants (as described later in this chapter). This has forced the TNCs to secure lower-cost offshore labour. Both IBM and SBS are typical of other service firms such as EDS, ACS and Accenture in that they operate globally, but only in the last five years have they found it necessary to build significant operations in LDCs to decrease their labour costs. Today, the larger firms such as IBM and Accenture are rapidly increasing their headcount in a number of LDCs, particularly India. At the same time, these firms are holding steady on their DC headcount or gradually drawing it down. Given the ferocious competition in software services there is little likelihood that prices will increase substantially. This suggests that, for the large TNCs, the offshoring of services will continue to increase in both absolute numbers and percentages of the global workforce.

Software operations in non-software firms

Today, virtually every firm in every industry sector is dependent on software. These needs range from routine software for personal computers and small servers to more complicated and customized software for complex and proprietary systems. All of these systems require customization, maintenance or updating on a regular basis. IT systems have become an increasingly significant expenditure for businesses in DCs, and firms are actively trying to control these costs. One way to lower them is to offshore the work to nations with lower labour costs.

It is difficult to even estimate the amount of software work that is offshored. Businesses do not provide this information in their reports. If work is transferred

9 This section draws heavily upon Express Computer (2002).

to an overseas subsidiary, this is an internal transfer and may remain unannounced and difficult to trace. It is clearer who does the work. If it is not an overseas subsidiary of the company, then it is likely to be one of two other kinds of firms. The service might be supplied by a large service firm from a DC such as IBM, CapGemini, SBS or Accenture. Alternatively, the work might be outsourced to a firm from a LDC such as TCS or Infosys (India), Luxoft (Russia) or Softtek (Mexico). When a TNC does the software work for its DC facilities itself in one of its LDC locations, it is likely that this is not the only work done at that location. For example, as of April 2005, Dell Computers employed approximately 10,000 people in India in a variety of tasks, one of which was to produce software for Dell's internal operations. The overseas operations undertake many tasks, only one of which is software production. Having a *mélange* of activities can provide the scale needed to make establishing an overseas subsidiary more attractive since the software work may not have been of a sufficient scale to justify a subsidiary.

Agilent Technologies Inc. (ATI)[10] In the technology sector, ATI is a good example of how a firm normally considered a hardware firm also undertakes considerable amounts of software-related work. ATI develops tools and technologies that sense, measure, interpret and communicate data. The company operates in four business areas: test and measurement, automated test, semiconductor products and life sciences and chemical analysis. ATI, which was separated from Hewlett-Packard in 1999, established its first Indian offshoring operation in 2001. By 2005, it had offices in over thirty countries. Manufacturing was located in the USA, China, Germany, Japan, Malaysia, Singapore, Australia and the UK. ATI Laboratories are located in California; Mizonokuchi, Japan; South Queensferry, Scotland; and Beijing, China.

The dot-com crash had a severe effect on ATI. At the end of 2003, revenue was $6.1 billion, down from $9 billion in 2000,[11] and the number of employees had been pared down from 40,000 in 2000 to 29,000 in 2003. In addition to eliminating headcount in the DCs, ATI decided to establish an offshoring centre in India. It was already outsourcing some software work to India. Although the company established a facility in India, it also decided to outsource maintenance and technical work (largely programming) to outside vendors, while retaining strategic control.

ATI introduced what it terms the hybrid model, where outsourcing service providers are required to operate out of its offices. This has proved to be advantageous because it mitigates the perceived security risk of having separate leased lines from non-ATI locations feeding into the VPN (virtual private network). It also allows ATI to induce competition among outsourcers and minimizes transition and operational costs, and it facilitates cross-functional communication between outsourcers and the firm.

10 The material in this case study is taken from Dossani and Manwani (2005).
11 These figures exclude the company's healthcare business, sold to Philips in 2001.

Work transfer has not been simple. For example, in early 2002 ATI established a communications software engineering group in India to automate some software test suites. When the project encountered release delays, there was friction between the USA and Indian engineers. This was exacerbated by the dot-com crash, which resulted in large USA layoffs. These difficulties slowed the transfer of additional work, and, over a period of 18 months, the Indian team experienced a greater than 70 per cent attrition rate. Despite these difficulties, the software test suite project has expanded to include the development of new modules and maintenance and defect correction for the entire product in India.

ATI India began with simple projects. For example, the first technical project was data entry related to engineering services. Other initial tasks assigned to India were similarly simple such as CAD support for engineering and quality assurance. Rather rapidly, however, the work became more sophisticated in both the technical and administrative areas. For example, only three years later, Indian engineers were designing application-specific integrated circuits. The Indian engineers took on more and more R&D work in wireless solution systems, OSS and billing software for telecom service providers. Employment has grown at 20 per cent per year, and total employment in India reached 1,250 in March 2005.

ATI's Indian operation is typical of those established by high-technology firms. It uses both offshore outsourcing and LDC subsidiaries. ATI has established operations and R&D laboratories in a number of nations, but India has become its largest and most important centre. Though it does the more mundane software testing and maintenance, the Indian operation also does more challenging work, for example, developing software that is embedded into ATI's core telecommunications and wireless test equipment products. ATI is an example of a process that is underway in many high-technology and other industrial firms whose core products are becoming more complicated and more software-intensive.

Citicorp There is relatively little information available about offshoring of business or software services in financial firms. What is well known is that the large money-centre banks, insurance firms and financial firms are among the largest IT users in the world. To support their operations, they have large internal staffs and many software service vendors. One of the world leaders in using offshore facilities for global operations is Citicorp. It uses outsourcing both onshore and offshore and was one of the first firms to establish a substantial software service subsidiary in India.[12]

In 1984 Citibank established its Indian software subsidiary, Citibank Overseas Software Limited (COSL). COSL wrote software in India for Citibank's global operations and particularly its effort to computerize its worldwide operations (Arthreye 2003). By the time the global computerization was completed in 1989, COSL had developed a robust banking solution and had approximately

12 For an excellent account of Citicorp's early Indian operations that was drawn upon heavily for this account, see Arthreye (2003).

500 employees (BITSAA 2004). COSL used other domestic companies such as Silverline and Nucleus Software for coding, while it handled the development of the architectural components itself. In 1992, while COSL was being converted into a proprietary subsidiary, two executives convinced 150 employees to follow them to form Citicorp Information Technology Industries Limited (CITIL) which was funded by Citicorp's venture capital arm. CITIL did not sell to Citicorp but rather became a merchant software firm. In 2000, CITIL was renamed I-flex. As of 2005, I-flex had 5,500 employees worldwide and over 500 customers. In August 2005, Oracle purchased a 40 per cent stake in I-flex for $900 million.

The remaining part of COSL continued to work for CitiGroup. Then in 2001, COSL was merged with another arm of Citibank, India (known as Global Support Unit (GSU)) to form OrbiTech Solutions Ltd. which developed a suite of banking products. In 2002, OrbiTech merged with Polaris Software Laboratories (Udani 2001), and, by 2005, Polaris had approximately 6,000 employees, working mainly in the financial arena.

Citicorp pioneered the use of India to lower its cost of software production. From Citibank's initial investment in India, it spun off CITIL and COSL and apparently today does not have large in-house software operations in India. In addition to the software operations, Citibank also had a large service operation that did everything from transaction processing to customer-focused call centres. In 1999, this was spun off as e-Serve and listed on the Bombay Stock Exchange. In 2004, Citi delisted e-Serve and brought it back in-house. As of 2005, e-Serve employed more than 10,000 workers in India. In terms of software services, Citibank was the financial industry's pioneer in using India and has been very important in training Indians in software development for the global market.

Since Citicorp's pioneering establishment of a wholly-owned software services subsidiary in India, many other banks and financial institutions, including Deutsche Bank (Deutsche Software), Bank of America, Barclays, ING and JP Morgan Chase, have established facilities in India to provide software services support for their global operations. Regardless of the ownership configuration, there is ample evidence that the relative amount of software service offshoring by financial institutions to India and possibly other locations will continue to grow. For example, insurance firms, which thus far have been more conservative than banks, have recently begun offshoring their IT operations.

Conclusion It is difficult to be certain that offshoring will lead to a decline in the number of software service employees in the internal IT operations of firms outside the software industry, but it does seem possible. At ATI, there were lay-offs in the IT sector; however, the losses came in the context of massive lay-offs because of the dot-com crash. In the current recovery throughout the IT sector, existing firm headcount in the USA appears to be stagnant. In other sectors, there is very little data available. For example, in financial services, it is unknown whether the increasing headcount in LDCs such as India has had any impact on employment in the DCs. The most that can be said is that non-IT firms are increasing their IT and engineering-related employment in LDCs, and this trend is underway across many different industries, including manufacturing firms such as Caterpillar and Nissan (Kenney and Dossani 2006).

Software-intensive, high-technology startups

For small startups offshoring is often a difficult decision, although recently a number of firms in the USA have been established with the express purpose of leveraging lower-cost engineers offshore. For smaller firms, an offshore facility can be demanding on management time. This is especially true because in India hiring and retaining highly skilled individuals is difficult. In LDCs, particularly China (but also India), the protection of intellectual property, which is usually the only asset that a technology startup has, can be difficult. Despite these obstacles and risks, under pressure from their venture capital backers and due to the need to conserve funds, there is ample anecdotal information suggesting that small startups are establishing subsidiaries abroad, particularly in India, to lower the cost and speed software development.

There is a wide variety of models for utilizing offshore skills, and the following case studies are intended only as examples of what high technology startups are doing abroad. These case studies are by no means exhaustive, and whether they are even representative of current practice is uncertain. However, all of these cases indicate that engineers in lower-wage nations can be an important resource for entrepreneurial firms.

Hellosoft Hellosoft is a private company established in Silicon Valley in 2000 and funded by Venrock Associates, Sofinnova Ventures, Acer Technology Ventures and JumpStartup Venture. It is a growing provider of high-performance communications intellectual property for Internet telephony (VoIP) and wireless devices. The founders are Indian-Americans who had entrepreneurial experience in USA startups, and the company was established with the express purpose of using low-cost Indian engineering talent to create the intellectual property that would be marketed by the USA headquarters team. By plan, nearly all of Hellosoft's R&D is done in Hyderabad, India, where the company employs over 100 digital signal processing engineers (Hellosoft 2005). Marketing and sales operate out of the company's San Jose headquarters.

The Hyderabad centre develops software in areas such as 3G wireless, 802.16 (a broadband technology) and EDGE (advanced data rates for GSM evolution). It has already had significant research success, and in July 2005 Hellosoft raised another $16 million from venture capitalists which will be invested in marketing and further R&D.

Hellosoft's business plan is based on leveraging low-cost engineering talent, and the USA headquarters operates largely as an interface with the market and customers. Nearly all the growth in technical employment will occur in India. Should Hellosoft be successful, the other beneficiaries will be venture capital firms that may garner significant capital gains and further relationships with other Silicon Valley service firms that assisted in the establishment of the firm.

Netscaler[13] Netscaler was founded in 1998 to redesign a specific piece of infrastructure, the load balancer, used in regulating Internet traffic flow. Netscaler aimed to reduce the set-up and tear-down time for each backend server connection. After Netscaler developed a product to demonstrate its more efficient way of handling Internet traffic, the company needed to add other features in order to attract customers who were unsure about moving from legacy products to new hardware that did not have industry backing. Netscaler understood that, as long as it had the ability to see inside a connection, it could offer other on-the-fly services. To create this ability, Netscaler hired an Indian firm known as NodeInfoTech to help develop an on-the-fly SSL encryption engine (NodeInfoTech 2005). With the aid of NodeInfoTech, Netscaler introduced an extension to its product, allowing the backend servers to send unencrypted data to the Netscaler product that encrypted it and forwarded it to the client over a secure connection.

The success with NodeInfoTech convinced Netscaler to establish an Indian subsidiary, Netscaler India. To staff the new operation, Netscaler hired many of the developers from NodeInfoTech (Tillman and Blasgen 2005). In 2004, Netscaler India employed approximately 60 engineers to develop other features such as on-the-fly compression, virtual private networks (VPNs), and integrated cache, and it planned to double the number of Indian employees in 2005 (Kumar 2004). Netscaler had grown to 200 total employees by 2005 when it was purchased for $300 million by Citrix Systems, who retained both the Silicon Valley and Indian operations.

The reason Netscaler formed an Indian subsidiary was to allow the company to increase the types of work it could do and develop tighter engineering integration (Tillman and Blasgen 2005). Netscaler's CEO, B.V. Jagadeesh, found that '[Indian] employees of similar skills are as efficient as they are here. The only handicap they have against their counterparts in the US is that they are not directly exposed to customers and customer challenges as India is not a destination market yet. When Indian companies start to buy our products, even that gap will be reduced pretty dramatically' (quoted in Tillman and Blasgen 2005, 12).

Netscaler continues to both offshore to its subsidiary and outsource to vendors lower-level engineering support. With the aid of both the internal Indian and USA engineering teams, Netscaler can provide all levels of support 24 hours a day. Since the low-level support is fully outsourced, it is hard to learn much more about its operation.

At Netscaler, technical writing is done by in-house technical writers because it is necessary for the writers to work closely with the engineers to provide good documentation. Netscaler originally employed a single technical writer in the

13 This section draws heavily on a case study done by Joshua I. Tillman and Nicholas W. Blasgen (2005).

USA, but in 2003, as the staff in India grew, a technical writer was hired there. The company's main reason for dividing up the writing was that the writer had to work with the engineers in order to correctly document the various product specifications. This allowed Netscaler to divide documentation writing between the two development sites, and the lower wages in India allowed a net reduction in the costs of producing documentation.

As a part of Citrix, it seems likely that Netscaler's future growth will be divided between the USA and India. The exact division is not yet clear, but cost pressures indicate that Indians will become an ever greater portion of the entire workforce.

Ketera[14] Ketera is a venture-capital-financed firm established in 2000 to help firms cut purchasing costs, streamline procurement processes and achieve higher performance from suppliers without the expense and overhead of traditional software applications. The company provides its software as a service. To lower costs, Ketera made a strategic decision to use India for all functional areas in the company. In its first phase of offshoring in 2002 and 2003 it contracted three Indian firms to provide software development, client services, customer support and IT support. In April 2004, the company decided to create a wholly-owned subsidiary in India and to transition from all outsource to mostly in-house offshore operation. In 2005 Ketera has a wholly-owned subsidiary in Bangalore employing about 75 people. The company still outsources a small portion of work to a legacy provider and contracts with new providers for special needs.

Why did Ketera set up a subsidiary? In 2004, the company was offshore outsourcing some software development of its core service product, customer support, IT support and some other functions. However, the company decided that the engineers at the outsourcing firms were not as productive and quality-oriented as Ketera desired. This problem seemed to be due to compensation and attrition issues, and to engineers with no motivation to innovate. There were also difficulties in the USA, where there were too few USA managers to handle the Indian engineers, resulting in significant communication gaps. These issues prompted Ketera to establish its Indian subsidiary. Their first Indian hire was a general manager who had experience working in both a Silicon Valley start-up and in India.

The software-related functions offshored internally were software development, operations IT, marketing and customer support, and portions of product management. In 2005 there was discussion of whether to move certain back office functions and telemarketing to India. According to one report, the centre was tipped to be the product engineering and development site for the company's entire suite of spend management solutions (Times News Network 2004).

Shah (2005) believes that as the Indian teams mature they will be able to perform more sophisticated work and that other functions could be at least partly

14 The source for the discussion of Ketera is Shah (2005).

offshored. Maturation is occurring quickly, and Ketera is already creating a new technology prototype in Bangalore.

Conclusion An increasing number of USA technology startups are utilizing lower-cost workers in LDCs. These case studies indicate that, although startups may initially use outsourcing as a strategy, they often soon opt to establish a subsidiary for a variety of reasons, including concerns about intellectual property protection, workforce control and management efficiency.

According to Shah (2005), the minimum staff size for an offshored operation is about 10 people. If this is accurate, then it may be possible for many more small firms to establish subsidiaries in LDCs. Unfortunately, data on the scale and scope of offshoring by startups is unavailable.

It is tempting to view this offshoring as an unmitigated loss of jobs for USA workers. However, the reality is more complicated. Lowering the cost of undertaking a startup means that the barriers to entry are lowered, and this policy is likely to encourage greater entrepreneurship in the USA. The jobs created by this entrepreneurship should be counted against those lost to offshoring. For example, Rakesh Singh, Netscaler's General Manager of Asia Operations, was quoted as saying, 'The cost savings through outsourcing have helped us become more competitive and experience rapid growth as a company. As a result, we have a lot more employees in the USA today than we did when we set up the India operations' (Tillman and Blasgen 2005, 13). So correctly estimating the employment effect of offshoring in the case of startups is difficult when one takes into consideration jobs created as well as jobs lost.

Offshore IT service providers

The availability of capable software programmers in LDCs provided an opportunity for entrepreneurs and existing firms to hire them and offer their services on the global market. It was in India where this practice first began in a significant way. Initially, in the early 1980s, because telecommunications links were not so sophisticated, the Indian programmers were moved to the USA customer's premises. This practice was profitable and gradually expanded and evolved as both customers and providers became more comfortable.[15] This level of comfort and the lower cost that could be offered through remote provision of services led to a shift wherein a major portion of the contract work was completed in the offshore offices of the contractor.

---

15 Obviously, comfort is a subjective term that refers a person's faith that another person(s) will respond in certain predictable ways or that a set of agreed upon tasks will be discharged according to a set of expected criteria. Cultural, social, economic, legal and other practices and beliefs impact our comfort with a relationship. Comfort is increased through repeated successful interactions. As levels of trust increase due to positive interactions, the client becomes more willing to escalate its commitment.

Indian firms were the pioneers in providing the offshore outsourcing of software production and services. This is a lesson that can be generalized to firms in other nations, as the real explosion of outsourcing came during the dot-com boom of the late 1990s, when there was great concern about a shortage of programmers. USA firms in particular were concerned about the Y2K problem and sought low-cost assistance in preparing their IT systems. These developments created an environment where major corporations were willing to experiment with overseas vendors, and a sufficient number of these experiments were satisfactory. The result was that offshore vendors, particularly Indian firms, were validated as candidates for software projects. These projects also allowed offshore vendors, again particularly Indian firms, to grow rapidly in headcount, experience and financial resources so that they could undertake ever larger and more complicated projects.

Tata Consultancy Services[16] Tata Consultancy Services (TCS), the largest and oldest Indian software services provider, is an excellent example of the growth of Indian vendors (see Table 10.1). TCS was established in 1968 to service the in-house data processing requirements of the Tata Group and, in 1969, offered electronic data processing (EDP) services to outside clients. In 1970, it became the exclusive Indian licensee to sell and maintain mainframe computers built by the American firm, Burroughs. In an effort to encourage the development of an Indian computer industry, the government enacted the Foreign Exchange Regulation Act of 1973, forbidding foreign firms from operating fully-owned subsidiaries. A number of foreign firms established JVs, and the Indian industry grew gradually. During this period, all of these firms including TCS sold and maintained computers and software systems made overseas by their JV partner and offered electronic data processing services to local clients.

TCS's overseas experience in providing software-related services began in 1974, when TCS was asked by Burroughs to install systems at USA-based clients. Burroughs was attracted by the combination of software engineering talent and the English language skills that it had found in the TCS workforce. This was the beginning of the body-shopping business, which entailed the dispatch of Indian programmers to the sites of overseas clients. Typically, these assignments lasted for a few months at a time. During this period, Indian firms were basically labour recruiters.

As the software industry changed and Burroughs continued to lose market share, TCS developed a growing competence in conversion work, that is, converting clients' existing Burroughs' systems to work on IBM hardware. To further its growth, in 1979, TCS opened an office in New York, the first overseas office by an Indian software firm. Entering the 1980s, TCS remained the largest Indian software services firm. In 1980, the Indian software industry exports were $4 million, shared by 21 firms of which TCS and a sister firm, Tata Infotech,

16 This case study draws heavily upon Dossani and Kenney (2004).

accounted for 63 per cent. By 1984, the number of firms increased to 35 and export revenues reached $25.3 million.

Table 10.1 TCS revenues, number of employees and per cent of revenues derived from outside to India, 1991–2005

Year	Annual revenues (in $ million)	Total employees	Revenue derived from abroad (per cent)
1990	28.1	2,300	70.8
1991	45	2,600	75.6
1992	52.3	4,761	80
1993	55.9	6,450	80.7
1994	64	5,589	79.2
1995	90.1	6,071	80.5
1996	123.9	7,864	81.9
1997	169	9,929	84.2
1998	241.8	11,176	88.2
1999	357.8	12,770	89.8
2000	417.9	15,044	86.1
2001	616.2	17,607	91.3
2002	792.1	20,459	92.7
2003	1,000	24,168	
2004	1,560	30,100	
2005	2,240	45,700	

Source: Compilation by Rafiq Dossani and Martin Kenney.

When, in 1985, TI persuaded the government to supply it with scarce satellite bandwidth, Indian firms such as TCS also demanded telecommunications access. But it was the acceptance of UNIX as a programming standard in the 1980s that made offshore work for clients feasible. Again, TCS pioneered the remote project management model as it came to be called. In 1988, only 10 per cent of TCS's work was done in India, but this rose to 37 per cent in 2005. The industry shift to UNIX and workstations also benefited Indian firms since they could secure work converting installed applications into UNIX-compatible programs. Again TCS was a leader, but soon other Indian competitors such as HCL, Infosys, Satyam and Wipro emerged.

The type of work that TCS performed changed substantially in the 1990s. Conversion work tapered off once most corporations completed the adoption of the common UNIX platform. This work was replaced with writing applications programs, a more profitable activity. TCS eagerly sought higher value-added work such as systems integration, and focused considerable effort by the end of the decade into bidding for larger projects, that is, those that required from 20 to 150 people-years. The largest industry serviced by TCS continues to be financial services. Today, 72 per cent of its revenues continue to be in application development and maintenance, while body shopping still provides over 60 per cent of its revenue (Mahalingam 2005).

By 1991, TCS had grown to 2300 employees and had revenues of $28 million. During this period the company pioneered the establishment of India-based, client-specific, offshore development centres (ODCs) which enabled firms such as TCS to undertake large turnkey projects that combined Indian-based and overseas staff (the latter often supplying critical industry expertise otherwise unavailable in India). Y2K was a bonanza for TCS and the other Indian firms. At the end of the fiscal year 2000, TCS had 15,000 employees and revenues of $428 million. To accelerate its growth, in 2001 TCS acquired CMC, an Indian government-owned firm with 2,500 employees. Rather than slow down after 2000, the rapid improvement in telecommunications capabilities combined with serious pressure on the bottom lines of firms in the LDCs expanded the opportunities for TCS which grew to over 20,000 employees in 2003. TCS began offering new services such as real time database management, quality assurance and web services.

By 2005 TCS had grown to over 45,000 employees and was continuing to grow at approximately 25 per cent per year. As TCS continues its efforts to overtake firms such as IBM and Accenture, it is establishing a global network of operations facilities, not only marketing, customer liaison or concentrations of dispatched personnel. In 2005 the company had development centres in Europe, Latin America and Japan, although most of its employees continued to be located in India.

TCS and its major Indian competitors have had a significant cost advantage over their DC rivals. Until very recently, however, they did not have either scale or a sufficiently global footprint to compete against the IBMs and Accentures. This situation is changing as the Indian firms experience annual growth rates in excess of 25 per cent and have significantly better profitability than their USA-based competitors (Hira and Hira 2005). The marketplace dynamic may change as the rivals from DCs increase the percentage of their workforce located in lower-cost environments. Regardless of the outcome, firms such as TCS have successfully forced firms from DCs to increase dramatically the portion of their global workforce located in LDCs, thereby shifting the geography of software service provision.

Softtek Indian firms, due to their size and sophistication, have rightfully received the bulk of the attention from those considering offshoring. However, there are firms in other LDCs that are also providing software services to DCs. One noteworthy example is Softtek, a privately-owned Mexican firm based in Monterrey with development centres in Monterrey, Aguascalientes and Mexico City, two others in Brazil and one in Spain. Like the large Indian firms, Softtek operates certified Six Sigma programs and has reached a CMM 5 rating (Softtek 2005b). The company was established in 1982 to employ graduates of Mexico's best technical university, the Tecnologico de Monterrey, to provide IT consulting services to Mexican firms and later to firms in other parts of Latin America. It entered the USA market in 1997 with the business strategy of providing a nearshore alternative. In recent years, Softtek has grown from 2,000 employees in 2000 to approximately 3,400 in 2005 (Lopez 2005; Softtek 2005a). With about $135 million in revenue it is growing at 30 per cent per year, although it still is only about one-tenth the size of providers in the large DCs or India.

Softtek's value proposition is based on the fact that its software development centres are nearshore, and thus operate synchronically with its customers. Because its employees are more highly paid than those in the Asian LDCs, Softtek had to develop a somewhat different model than Indian vendors.[17] Their business strategy is not to displace offshore vendors, but rather to capture a portion of the total offshore spending. What Mexico offers is an opportunity to diversify risk which is important for highly interactive processes that could benefit from running at the same time. To further their advantage, Softtek even adopted the USA vacation calendar for their USA-focused operations. In addition, the USA and Mexico share similar cultural and commercial environments. Proximity facilitates the logistics of arranging face-to-face meetings. This limits the need for Softtek engineers to be stationed onsite, thus lowering costs and helping to make Softtek competitive with the lower-cost Indian or Chinese competitors (Lopez 2005). Travel is simplified, because, when it is necessary to visit, as a Mexican firm, employees can use NAFTA visas. In general, Softtek works on fixed-price contracts, not the time-and-materials contracting that is typical of body-shopping.

Despite the opportunities, Mexico's growth in the IT area has been limited. Softtek is the largest independent Mexican software services offshoring firm serving the global market, although there are other smaller firms. Only recently has the Mexican government recognized the opportunity in software services offshoring and formed an organization (Prosoft) to improve Mexico's position by funding training projects. Even five years ago, few Mexican universities outside of the Technologico de Monterrey were providing well-trained graduates for this industry. This has changed as Mexican universities and students have recognized the career potential in IT. To improve the preparation of Mexican IT workers, Softtek and the other Mexican IT vendors are interacting with a number of Mexican universities to improve IT training (Lopez 2005).

Softtek's experience demonstrates that it is not only the Indian majors that are finding opportunities to provide software services to DCs. Yet, its status as one of the largest Latin American software services firms indicates the lead the Indian firms have built. This case study also shows that high-level CMM qualification is not confined to Indian firms. Most importantly, it demonstrates the entrepreneurial opportunities available in any LDC that has a reservoir of technically trained personnel.

17 An IT graduate from a Mexican university starts at between $15,000–18,000 per year as opposed to an Indian graduate who starts at $6,000 per year.

Conclusion Software services firms from a number of the LDCs are players in the global economy. They have not yet become significant players in the packaged software industry, and given the propensity for the large international players to buy promising software startups wherever they may be located, it could be difficult for packaged software firms from LDCs to capture significant global market share. The large Indian firms such as TCS, Infosys, Wipro, Satyam and HCL, are at this time the global leaders. However, in China, Mexico and Russia, there are smaller but also rapidly growing software service firms that employ between 1,000 and 5,000 workers. Currently, the firms from other nations are not large enough to compete with either the DC TNCs or the large Indian firms. These medium-size firms in other geographies can reduce country risk for customers, although it is also possible that some of them will be acquired. The larger TNCs and Indian firms are also establishing facilities in other geographies, particularly Eastern Europe and, more recently, Mexico.

Conclusion

The variety of case studies in this chapter illustrates the breadth of the phenomenon of software and software services offshoring. The reasons for offshoring vary by firm and particular recipient nation, and often decisions are made for a complicated amalgam of reasons. In the case of the elite R&D laboratories, the desire to tap into the most talented individuals, wherever they might be in the world, is clearly the foremost motivation. Particularly in the case of China, but also increasingly India, the growing local markets are attractive and a reason for siting software facilities locally. Labour costs are a primary motivation for much of the offshoring being undertaken by the firms examined.

There can be little doubt that offshoring is still small in comparison to how large it is likely to become. The case studies in this chapter are firms that can be considered early adopters; the followers have only recently begun to investigate the opportunities for offshoring. As the case of ATI showed, particularly in the subsidiaries of Western firms, it is likely that more sophisticated work will be relocated during the coming decade. Firms are becoming increasingly willing to entrust core activities to their offshore subsidiaries.

Whereas some believed that a certain size was necessary prior to offshoring, the case studies of startups showed that this is not true. USA startups are establishing offshore subsidiaries even before their headcount reaches 50 people, and for some firms, their entire business plan is built on the premise of using lower-cost offshore IT professionals. This suggests that employment growth in the USA might be constrained. However, the availability of low-cost technical talent also can lower the barrier to entry for entrepreneurship, and this may encourage greater entrepreneurship and, as a result, wealth and job creation in the USA.

Every firm in this admittedly small sample is pursuing a global strategy for R&D and IT provisioning. It is entirely possible that this will become the norm for nearly every firm in the DCs. The labour-cost arbitrage factor is and will remain significant and all executives, in large and small firms, are considering the most economical footprint for their IT operations.

References

Athreye, S. (2002), *Multinational Firms and the Evolution of the Indian Software Industry* (published online 12 December 2002) <http://ssrn.com/abstract=361680>, accessed 30 May 2008.

Barlas, D. (2004), 'SAP Doubles Indian Commitment', <http://www.line56.com/articles/default.asp?articleID=6168&TopicID=3>, accessed 21 September 2005.

Bishop, J. (2005), 'China's Red Flag Software Sees Strong Growth This Year', *Xinhua Financial News* (updated 14 September 2005) <http://www.linuxinsider.com/story/mmLJtyG9rg5jby/Chinas-Red-FlagSoftware-Sees-Strong-Growth-This-Year.xhtml>, accessed 17 September 2005.

BITSAA International (2004), 'How They Did It', <http://www.bitsaa.org/sand paper/leaders/hukku.htm>, accessed 21 August 2005.

Blau, J. (2005), 'Siemens Business Services Faces Change', *Network World* <http://www.networkworld.com/news/2005/0204siemebusin.html>, accessed 18 September 2005.

Carmel, E. (1999), *Global Software Teams: Collaborating Across Borders and Time Zones* (Upper Saddle River, NJ: Prentice Hall).

Carmel, E. and Tjia, P. (2005), *Offshoring Information Technology* (Cambridge: Cambridge University Press).

CIOL (2005), 'IBM Opens Dedicated IT Services Research Center in Bangalore' (updated 18 August 2005) <http://www.ciol.com/content/news/2005/105081813.asp>, accessed 22 September 2005.

Dossani, R. and Kenney, M. (2004), 'Moving Tata Consultancy Services into the Global Top 10', *Journal of Strategic Management Education* 1:2, 383–402.

Dossani, R. and Manwani, A. (2005), *Agilent's Supply Chain: A Locational Analysis of its Indian Operations*, paper presented at the Stanford University Conference on the 'Globalization of Services' (17 June 2005).

Express Computer (2002), 'SISL Grows Out of Parent's Shadow' (updated 1 July 2002) <http://www.expresscomputeronline.com/20020701/company1.shtml>, accessed 13 August 2002.

Hallez, F. (Senior Manager, Business Improvement SOLutions BeLux, Siemens Business Services) (2004), PowerPoint presentation at EU-US Seminar entitled 'Offshoring of Services in ICT and Related Services', Brussels, Belgium (13–14 December 2004).

Hamm, S. (2005), 'IBM'S Passage to India', *Business Week Online* (updated 8 August 2005) <http://www.businessweek.com/technology/content/aug2005/tc2005088_6314_tc024.htm.>, accessed 14 August 2005.

Hayward, D. (1997), 'How Offshore Programmers Save IBM Millions', *TechWire* (updated 18 February 1997) <http://www.offsiteteam.com/outsourcing1.html>, accessed 23 August 2005.

Hellosoft Inc. (2005), 'Corporate Home Page' [website] <http://www.hellosoft.com/aboutus/> (home page), accessed 21 August 2005.

Hira, R. and Hira, A. (2005), *Outsourcing America: What's Behind our National Crisis and How We Can Reclaim American Jobs* (New York: American Management Assoc).

IBM (2004), 'IBM Increases Hiring Plans' (updated 12 August 2004) <http://www.ibm.com/investor/viewpoint/ircorner/2004/12-08-04-1.phtml>, accessed 14 August 2005.

IBM (2005), 'China Software Development Laboratory', <http://ibmcampus.51job.com/businessunit/businessunit_7.htm.>, accessed 16 August 2005.

Indo-Asian News Service (2005), 'SAP India to Hire 2,000 More Techies', <http://www.hindustantimes.com/news/181_1420811,0003.htm.>, accessed 16 August 2005.

Kenney, M. and Dossani, R. (2005), 'Offshoring and the Future of US Engineering', *The Bridge: Linking Engineering and Society* 35:3, 5–12.

Kumar, V.R. (2004), 'NetScaler to Double Headcount' (updated 16 June 2004), <http://www.thehindubusinessline.com/2004/06/17/stories/2004061702170700.htm>, accessed 18 August 2005.

Lopez, B. (Softtek CEO for US Near Shore Services) (2005), Telephone Interview by Martin Kenney (7 September 2005).

Mahalingam, S. (2005), Presentation at the India Conference: 'Shaping the Future' (published online 12 August 2005), Tata Consultancy Services [website] <http://www.tcs.com/investors/pdf/TCS_Presentation.pdf>, accessed 16 August 2005.

NodeInfoTech (2005) Home Page [website] <http://www.nodeinfotech.com/expertise.html.>, accessed 21 August 2005.

Orian, S. (2004), *The Politics of High Tech Growth: Developmental Network States in the Global Economy* (Cambridge: Cambridge University Press).

Palmisano, S. (2005), '2005 Investor's Briefing', <http://www.ibm.com/investor/events/inv_day0505/presentation/inv_day0505prepared.pdf.>, accessed 21 August 2005.

Parthasarathy, A. (2005), 'Global Sourcing – The New Reality', paper presented at *The Globalization of Services*, Stanford University (17 June).

People's Daily Online (2004), 'SAP Vows to Enhance R&D Capacity in China' (updated 16 December 2004) <http://english.people.com.cn/200412/15/eng20041215_167405.html>, accessed 21 August 2005.

Rediff (2005), 'Rediff Interview with Naresh Gupta' (updated 21 April 2005) <http://www.rediff.com/money/2005/apr/21inter.htm>, accessed 22 August 2005.

SAP (2005), SAP Bulgaria Home Page, <http://www.50.sap.com/company/saplabs/bulgaria/>, accessed 21 August 2005.

Shah, R. (2005), 'Ketera India Case Study', *Conference on the Globalization of Services*, Stanford University (17 June).

Siemens Business Services (2004), Corporate Website [website] <http://www.siemens.com/>, accessed 14 March 2005.

Sim, S. (2002), 'Adobe Says it's Committed to China Despite Piracy', *ITworld.com* (updated 15 January 2002) <http://www.itworld.com/Tech/2418/IDG020115adobe/>, accessed 21 August 2005.

Smith, R. (2005), *Partnerships are Important*, Rediff (published online 24 August 2005) <http://us.rediff.com/money/2005/aug/24inter.htm>, accessed 31 August 2005.

Softtek, S.A. (2005a), Corporate Home Page, <http://www.softtek.com>, accessed 21 August 2005.

Softtek, S.A. (2005b), 'Our People … in Their Own Words', <http://www.softtek.com/nearshore/html/about_people.htm>, accessed 25 August 2005.

Tillman, J.I. and Blasgen, N.W. (2005), 'Case Study of Netscaler', paper written for CRD 199 Special Study Course, University of California Davis.

Time News Network (2004), 'Another IT Firm Deserts US for India' (updated 8 December 2004) <http://economictimes.indiatimes.com/articleshow/950969.cms.>, accessed 22 August 2005.

Tsai, M. (2004), 'Inside DSP on Automotive Signal Processing: Feeling the Heat', *Inside DSP* (updated 13 September 2005) <http://insidedsp.eetimes.com/features/printableArticle.jhtml?articleID=30900019>, accessed 21 August 2005.

Udani, D. (2001), 'The Quest for HTMLParser', <http://htmlparser.sourceforge.net/articles/quest.html> (updated 17 September 2006), accessed 22 August 2005.

Verma, P. (2005), 'Adobe to Buy Macromedia's B'lore Development Centre', *The Economic Times* (updated 16 May 2005) <http://economictimes.indiatimes.com/articleshow/1111152.cms>, accessed 21 August 2005.

ZDNet (2005), 'IBM Expands in China', *ZDNet News* (updated 26 May 2005) <http://news.zdnet.com/2100-9589_22-5721779.html>, accessed 23 August 2005.

Chapter 11

Newly Emerging Paradigms in the World Economy: Global Buyers, Value Chain Governance and Local Suppliers' Performance in Thailand

Carlo Pietrobelli, Federica Saliola[1]

Introduction

The fragmentation of production processes and the evolution of internationally-dispersed but functionally-integrated economic activities have shaped the recent economic setting faced by industries and individual firms in LDCs, as it has in the industrialized world. Remarkable trade integration and cross-border investments have been the result of these transformations, and their consequences have been widely studied.

In particular, the international economics literature has had a lasting interest in the static and dynamic effects of this new market setting and in the different forms that the international involvement of countries, industries and firms is taking. The term 'international involvement' was first used by Lall (1980) with reference to the choice of USA TNCs between exports and FDI. The same notion was extended by others (Oman 1984; Markusen 1995), and now comprises a wider set of strategies that firms can jointly or individually use to serve foreign markets and/or gain access to assets available abroad, including licensing and other agreements with foreign partners, the creation of networks of sales agents and the setting up of commercialization affiliates abroad (Helpman et al. 2004; Castellani and Zanfei 2006).

1 The authors would like to thank Giuseppe Iarossi and Giovanni Tanzillo for making the data available. We also wish to thank Davide Castellani, Dave Kaplan, Pierre Mohen, Henny Romijn, Beata Smarzynska Javorcik, Adam Szirmai, Antonello Zanfei and two anonymous referees for helpful comments. We are also indebted to seminar participants in the CNR workshop in Milan, Globelics India in Trivandrum, Eindhoven Technical University and UNU-MERIT Maastricht for many helpful comments. Financial contribution from the MIUR-PRIN project on 'Capabilities dinamiche tra organizzazione di impresa e sistemi locali di produzione' is gratefully acknowledged. The findings, interpretations and conclusions expressed in this text are entirely those of the authors and do not involve their respective organizations.

However, many branches of this literature appear to disregard an important part of the story, that is, the form and the organization of the relationships among the various actors involved in these channels, and their implications for development. We develop this analysis in this chapter, and explore it empirically regarding Thailand. Indeed, following this approach, the forms and patterns of coordination and the level of hierarchy in such relationships may indeed matter for growth and learning processes, especially in LDCs. This trend may influence the benefits and costs of LDCs' integration in global markets, as

> … it is not so much a matter of whether to participate in global process, but how to do so in a way which provides sustainable income growth for poor people and for poor countries (…) Moreover, many of those countries which have suffered from declining income shares have experienced growing trade/GDP ratios. (Kaplinsky 2000, 1, 5–6)

The relevance of coordination or governance of the relationships in international production and trade has been highlighted by the recent GVC literature, which suggests that the governance and the actors of GVCs importantly affect the generation, transfer and diffusion of knowledge (Humphrey and Schmitz 2002a; Schmitz 2004; Gereffi et al. 2005; Altenburg 2006; Pietrobelli and Rabellotti 2007). The GVC literature has tackled these issues by referring primarily to LDCs, and debating the opportunities and threats that GVCs may represent for them. Among the various contributions, a critical view has emerged to argue that by adopting hierarchical forms of coordination, global buyers may operate to confine competences of LDCs' manufacturers simply to the assembly of imported material, making these manufacturers potentially very vulnerable and subject to increasing competition and falling returns (Schmitz 2004).

Global Value Chains: An Overview of Main Concepts and Theory

The concept of a value chain describes the full range of activities that are required to bring a product from its conception, through the different phases of production, to its end use and beyond. This process includes activities such as design, production, marketing, distribution and support to the final consumer.

The 'GVC approach' focuses on the activities and the strategic role of the relationships with other firms and actors. Drawing from the transaction cost literature, Gereffi (1994) developed a framework that ties the concept of the value-added chain directly to the global organization of industries. Later, he introduced the notion of 'governance' of the value chains, defined as authority and power relationships that determine how financial, material, and human resources are allocated and flow within a chain (Gereffi 1994). This concept is now central in the literature. By focusing explicitly on the governance of disintegrated chains and contrasting them to the relationships within vertically integrated chains, the global commodity chains framework draws attention to the role of networks in driving the co-evolution of cross-border forms of industrial organization.[2]

The literature highlights two critical parameters of the value chain governance: what is to be produced, and how it is to be produced. In each case, the level of detail at which the parameters are specified can vary. When the buyer plays this role, we refer to it as the 'lead firm' in the chain (Sturgeon 2002; Sturgeon and Lester 2004).[3]

In studies on the electronics sector, Sturgeon (2002) and Sturgeon and Lee (2001) emphasize the complexity of information exchanged between firms and the degree of asset specificity in production equipment. They highlight three types of supply relationships, based on the degree of standardization of products and processes:

- The 'commodity supplier' that provides standard products through arm's length market relationships.
- The 'captive supplier' that makes non-standard products using machinery dedicated to the buyer's needs.
- The 'turnkey supplier' that produces customized products for buyers, and uses flexible machinery to pool capacity for different customers.

Along similar lines, but more explicitly stressing governance and power relationships, Humphrey and Schmitz (2002a, 2002b) distinguish between suppliers in quasi-hierarchical relationships tied in a 'captive' relationship, and 'network' relationships between firms that cooperate because they possess complementary competences.

Gereffi et al. (2005) acknowledge, as do most other frameworks that seek to explain industry organization (such as transactions costs, global commodity chains, organizational theory) that market-based relationships among firms and vertically integrated firms (hierarchies) make up opposite ends of a spectrum of explicit coordination, and that network relationships comprise an intermediate mode of value chain governance. They identify three key determinants of value chain governance patterns: the complexity of information and knowledge transfer

2 Schmitz and Knorringa 1999; Kaplinsky 2000; Humphrey and Schmitz 2002, 2002a; Gereffi and Memodovic 2003; Gereffi et al. 2005; Giuliani et al. 2005a; Altenburg 2006.

3 In the case of product definition, the buyer can provide different levels of specification. It can set a design problem for the producer, which the producer then solves by providing its technology and design. The buyer might provide a particular design for the producer to work on, or the buyer might even provide detailed drawings for the producer. Buyers can also specify process parameters. Once again, these can be specified at different levels of detail. In some cases, the buyer may merely refer to the process standards to be attained. In other cases, the buyer will specify precisely how particular standards should be attained by requiring and perhaps helping to introduce particular production processes, monitoring procedures, and so on (Sturgeon 2002).

required to sustain a particular transaction, especially with respect to product and process specifications; the extent to which this information and knowledge can be codified and, therefore, transmitted efficiently and without transaction-specific investment between the parties to the transaction; and the capabilities of actual and potential suppliers in relation to the requirements of the transaction.

The concept of governance in the GVC literature is mostly dynamic. Humphrey and Schmitz (2002a) underline three factors which may determine a governance change: (a) power relationships may evolve when existing producers, or their spin-offs, acquire new capabilities; (b) establishing and maintaining quasi-hierarchical governance is costly for the lead firm and leads to inflexibility because of transaction specific investments and (c) firms and clusters often do not operate only in one chain but rather simultaneously in several types of chains, therefore they may apply competences learned in one chain to supply other chains. Gereffi et al. (2005) also explore some possible patterns of evolution of governance, and relate them to the evolution of the complexity of transactions, the codifiability of transactions and the competence of suppliers.

Although the final aim of most of these studies is to understand the reason and determinants of performance within value chains, the link between enterprise *upgrading* and GVC governance has been made explicit only recently. In a GVC context, upgrading is defined as innovating to increase value added (Giuliani et al. 2005b; Pietrobelli and Rabellotti 2007). Enterprises may achieve this in various ways; for example, by entering higher unit value market niches, by entering new sectors or by undertaking new productive (or service) functions; and always deepening technological capabilities.[4] In addition, within this context innovation is clearly not defined only as a breakthrough into a product or a process that is new to the world. It is rather a story of marginal, evolutionary improvements of products and processes that are new to the firm and that allow it to keep up with an international (moving) standard.

The GVC perspective is useful for various reasons: first, because the focus moves from manufacturing only to the other activities involved in the supply of goods and services, including distribution and marketing. These activities account for increasing shares of GDP worldwide. A second new and merit-worthy perspective is that GVC emphasizes the nature of the relationships among the various actors involved in the chain, and their implications for development. Moving beyond firm-specific analysis and concentrating on inter-firm linkages, it allows for an easy uncovering of the dynamic flow of economic and organizational activities between producers within different sectors even on a global scale. For example, even informal-sector, scrap- metal collectors in South Africa are inextricably linked to a global export trade. They bring scrap metal in old trolleys directly to shipping agents who pay them London spot prices and transfer the scrap immediately to ships for export to iron and steel furnaces across the globe

4 On the relationship between technological capability building and global value chains, see Morrison et al. 2007.

(Kaplinsky and Morris 2001). Furthermore, the notion of organizational inter-linkages underpinning value chain analysis may make it easy to analyse the inter-relationship between formal and informal work, with workers, particularly in LDCs, moving often seamlessly from one to the other, rather than viewing them as disconnected spheres of activity. Finally, by focusing on all links and phases in the chain (not just on production) and on all activities in each link, it helps identify which activities are subject to increasing returns in markets characterized by imperfect competition and segmentation.

Addressing these issues however is not straightforward. From an analytical point of view, it implies the study of activities taking place outside firms, and in particular to understanding the strategic role of the relationships with key external actors. Most of this literature is still based on case studies employing an increasingly systematic and rigorous empirical methodology. With this chapter we try to integrate field-based studies with a methodology that may exploit existing databases and lends itself more easily to comparisons and generalizations. This method is explained in the following section.

Objectives and Novelties of the Analysis

In this chapter we develop an analytical framework to evaluate the patterns of governance arising in value chains led by global buyers and their impact on suppliers' performance with specific reference to the Thai manufacturing industry.

We integrate and build on the existing literature in three ways. First, we look at the case of whether the global buyer is represented by a TNC. The GVC literature has the merit of including the *governance* of the relationships and the role played by global buyers in the study of the static and dynamic effects of openness. The concept of governance is central to this analysis. At any point in the chain, some degree of governance and coordination is required. It is often preferable to write of governance rather than only coordination, as the proactive involvement and participation of all the actors within the value chain is crucial. The role of global buyers' characteristics was initially analysed by Gereffi (1994), who introduced a categorization of 'buyer-driven' and 'producer-driven' commodity chains, with their respective forms of governance, that has been widely employed thereafter. Later, the analysis of buyers' forms of governance was further developed, and Gereffi et al. (2005) introduced a useful typology which identifies five different GVC governance patterns and discusses under which conditions these types can be expected to arise. According to the authors, three factors determine the lead firm's choice between one of the different patterns: the complexity of information involved in the transactions, the possibility of codifying information and the competence of suppliers along the chain.

We share their emphasis on GVC governance, and follow up this approach here. Our line of argument is that heterogeneity in global buyers may significantly affect the way cross-border relationships are governed, the extent of the transmission of knowledge and the ensuing learning promoted in LDCs' firms. More specifically, we distinguish between TNCs and other chain leaders. TNCs are increasingly operating as global buyers, with their role not yet confined to production but progressively extending to planning and management of global networks of suppliers and firms. Moreover, the literature has traditionally considered TNCs as possessing some technological lead and exploiting this proprietary advantage in international markets (Dunning 1993), and thereby potentially creating opportunities for knowledge diffusion and learning for their local suppliers (Turok 1993; Albio et al. 1999; Hewitt-Dundas et al. 2002).

A second original contribution of this study is to define a quantitative measure of the value chains' governance that we use in our econometric tests. To the best of our knowledge, most existing studies are based on case studies and surveys that offer the advantage of providing extremely detailed analyses of specific cases, capturing the complexity of the relationships involved in the chain and the role played by each actor. In addition, our effort to build measures of governance and use them across a large sample of firms presents the advantage of allowing interesting generalizations and comparisons that may usefully complement evidence based on present and future field studies.

Finally, we attempt a comparison between global and domestic value chains and then between GVCs led by TNCs and those led by other global actors, in terms of governance patterns and the effects on suppliers' performance and learning. We expect that the forms of governance may differ, and be more or less binding and severe for local suppliers in terms of product specification and standards' enforcement, with the parameters set by the TNC – typically oriented to international markets and more subject to open competition – being more complex, requiring greater assistance and possibly creating opportunities for improving performances. At the same time, however, we are aware of 'cherry picking' followed by TNCs when they carefully select their suppliers: local suppliers would be performing better ex-ante and not as a result of the assistance offered by the chain leader.

For similar reasons the intensity and the extent of buyers' influence on suppliers' performance is likely to vary between firms which are part of a TNC's network, and firms in value chains led by national buyers. The former are likely to be exposed to a larger flow of knowledge and learning opportunities than the latter, which usually operate on a smaller market. Consistently, we expect domestic firms, ex-ante less efficient than domestic firms working for TNCs, to heavily depend on the way the buyer assists them in improving products' features and production processes. In other words, efficiency improvements are expected to be powerfully linked to the governance of the value chain.

Relationships and Governance along Global Value Chain: Characteristics of Thai Firms

Asian countries offer some of the most interesting case studies for analysis of value chains. Thailand represents an attractive case of study in this context, due to the significant increase in value chain networks, and the important challenges the country is presently facing. Thus, several studies have provided evidence that Thailand is 'technologically challenged' (World Bank 2004) and therefore needs to move beyond its traditional role in GVC as a low-cost manufacturing location. Furthermore throughout the past decades, especially since 1986, Thailand has experienced a rapid increase in merchandise exports, growing from around one-fifth of GDP in the early 1980s to almost two-thirds today.

The rapid export growth has also been accompanied by rapid growth in private investment, both local and foreign. Indeed, Thailand has been one of the major FDI recipients in Southeast Asia over the past two decades (Brimble and Sherman 1999; Mephokee 2002). In 2004, the Global Investment Prospects Assessment (GIPA) of UNCTAD, designed to analyse 'future patterns of FDI flows at global, regional, national, and industry levels', ranked Thailand as one of the four 'top hot spots for FDI' in the world over the next four years, preceded only by China, India, and the USA.

The data come from the 'Productivity and the Investment Climate Private Enterprise Survey' (PICS), conducted by the World Bank on a representative (stratified) sample of 1,385 Thai firms from 2001 to 2003.[5] For each firm the information is plant-based. The survey focuses on manufacturing firms (sectors 15–36 in the International Standard of Industrial Classification, ISIC). The industries considered are the following: Food Processing, Textile and Clothing, Wooden Furniture and Product, Auto Parts, Electronics, Rubber and Plastic, and Machinery and Equipment. Finally, the database contains comparable qualitative and quantitative information on foreign ownership, sales, technology, value chains, workforce education and exports and productivity.

We define TNCs' value chains as those where Thai firms sell most of their products to TNCs, but are not owned by them. Then, in order to exploit the information in the PICS database, we broke down the sample distinguishing between Thai firms serving only the domestic market (DOM), Thai firms which are large suppliers of TNCs but are not foreign owned (MNS) and Thai firms that export more than a threshold value of 5 per cent of their output abroad but are not suppliers of TNCs (EXP) and not foreign owned. To avoid ambiguity in the analysis, we do not consider firms with more than 50 per cent of equity owned by foreigners,[6] (13 per cent of firms in the sample). Therefore, we use a reduced

---

5 We performed various tests to control for missing values, zero sales, zero employment, and observations failing to satisfy other basic error checks.

6 OECD and UNCTAD use a benchmark of 10 per cent as threshold ownership level. Other benchmarks taken by other researches include Sjoholm (1997), who had a benchmark

sample of 1197 Thai firms in the analysis, of which about 49 per cent meet the definition of MNS, 14 per cent that of EXP and 35 per cent DOM.

Table 11.1 Distribution of groups of firms across industries

	MNS		DOM		EXP	
	No	%	No	%	No	%
Textiles and clothing	171	28.6	127	29.9	46	26.1
Food processing	115	19.2	16	3.7	43	24.4
Machinery	79	13.2	58	13.6	10	5.6
Electronics	36	6.0	36	8.4	10	5.6
Wood	53	8.8	55	12.9	15	8.5
Rubber and plastics	81	13.5	96	22.6	44	2
Automotive parts	62	10.3	36	8.4	8	4.5
Total	597	100	424	100	176	100

Source: The World Bank – Enterprise Surveys (http://www.enterprisesurveys.org/).

The distribution of firms is spread fairly equally across industries in our sample. The presence of EXP and MNS is concentrated in industries such as Food Processing and Textiles and Clothing, while domestic firms are mainly concentrated in Textiles and Clothing and Rubber and Plastics (Table 11.1).

Table 11.2 Size and sales of firms in the sample

	MNS	DOM	EXP
Permanent workers			
median value	197	51	135
mean	432	95	283
Total sales			
Average sales (current US$)	11,898,767	1,657,910	10,478,366

Source: The World Bank – Enterprise Surveys (http://www.enterprisesurveys.org/).

15 of equity owned by foreigners; Haddad and Harrison (1993) considered foreign firms as those with at least 5 per cent equity owned by foreigners; Djankov and Hoekman (1998) had a benchmark of 20 per cent, while Castellani and Zanfei (2002) considered foreign firms as those with at least 50 per cent equity owned by foreigners.

A comparison of firm size (Table 11.2), computed in terms of total workers, shows that EXP and MNS are generally larger than domestic market-oriented firms. On the basis of the value of sales, domestic firms sell on average less than one-fifth of what MNS and EXP sell. Thus, EXP and MNS appear rather similar according to these statistics.

The next step in our analysis was to define a measure of value chain governance on the basis of selected and available variables. This method takes into account different levels and types of buyers' involvement in the suppliers' specification of product and process standards, R&D activities and dissemination of technology. Following the literature, and considering some characteristics of the Thai economy, we choose the following variables:

- Percentage of sales made exclusively to (suit) buyer's unique specification (Cl. spec).
- Whether the buyer provided information on design/quality (product characteristics) (Prod inf. by client) and imposed product quality standards (Client enforcement).
- Whether the buyer engaged the firm in process or product R&D type of activities (R&D activities).
- Whether the buyer sent employees (personnel exchanges) to disseminate and diffuse new technologies into firms' production facility (Empl. for tech diff.).

Table 11.3 documents some descriptive evidence about these factors. Overall, the degree of buyers' involvement in product definition is high in the sample, but much smaller for R&D and technology dissemination. On average, a larger share of firms which are part of value chains led by TNCs receive specifications of products and design by buyers, and the TNC is also involved in R&D activities and in technology dissemination. DOM firms seem to receive the lowest requirements and product and technology info by clients, while EXP are in between. On the contrary, in terms of sales made according to clients' unique specification, EXP reveal the highest value.

In order to capture different types of governance, we allow different combinations of the key variables above. It is important to recall that our index does not intend to reflect merely a growing involvement of buyers with their suppliers in *all* aspects of production, but rather focuses on crucial elements of the buyer-supplier relationship as setting product standards and quality requirements, and disseminating technology.

Table 11.3 Thai firms' relationships with buyers

	MNS	DOM	EXP
Cl. spec (mean) (1)	44.40	43.31	51.53
Prod inf. by client (2)	78.97	68.87	75.15
Client enforcement (3)	83.3	72.6	77.05
R&D activities (4)	42.68	31.21	32.6
Empl. for tech diff. (5)	39.79	24.42	28.36

Notes: (1) per cent of sales made exclusively to buyers' unique specification; (2) Information on design/quality provided by the buyer; (3) Product quality standards enforced by the buyer; (4) Engagement of the buyer in process or product R&D type of activities; (5) Employees from the buyer to work to disseminate and diffuse new technologies into suppliers' production facility.

Source: Authors' own computation on The World Bank – Enterprise Surveys (http://www.enterprisesurveys.org/).

Table 11.4 Classification of value chains' governance

Types of value chains' governance	Per cent of sales made according to buyers' unique specification	Design/quality and product quality standards	Technology dissemination and process and product R&D
G1 – Low requirements	Less than 20	No	No
G2 – Higher requirements	More than 20	No	No
G3 – Higher requirements and DQ	More than 20	Yes	No
G4 – Higher requirements and Tech_RD	More than 20	No	Yes
G5 – Higher requirements and DQ and Tech_RD	More than 20	Yes	Yes

Source: Authors' own elaboration based on The World Bank – Enterprise Surveys (http://www.enterprisesurveys.org/).

Our typology identifies five basic types of value chain governance (Table 11.4). G1 reflects a situation where less than 20 per cent of total sales are made according to their clients' unique specification, and suppliers do not receive substantial inputs from buyers; G2 type occurs when the percentage of sales made according to buyer's specification is higher, but still suppliers do not receive information or involvement from the buyer; G3 type reflects a situation where there is a relevant share of sales made according to the client's specification and buyers intervene to specify quality and design definition; in G4 a relevant share of sales are made according to the client's specification and buyers are involved in technology dissemination and R&D activities, but without intervening in product design and quality; finally with G5 all forms of buyers' involvement occur.

Before proceeding to the econometric analysis, it is instructive to look at the distribution of governance among the three groups of firms and across different industries. We also computed Chi-squared distribution tests to assess whether differences between MNS, EXP and DOM and across industries were significantly different from zero.

Table 11.5 Distribution of governance across Thai firms and industries

Governance by firms					
	G1	**G2**	**G3**	**G4**	**G5**
MNS	8.5	8.9	29.9	11.1	41.2
DOM	15.4	13.5	31.7	12.1	26.5
EXP	7.0	17.1	35.5	9.2	31.1
Pearson chi2=	45.058	32.408	6.810	3.799	66.862
Pr=	0.000	0.000	0.033	0.15	0.000
Governance by industries					
	G1	**G2**	**G3**	**G4**	**G5**
Textile and clothing	9.0	9.0	39.6	8.5	33.9
Food processing	10.6	12.9	21.8	17.9	36.9
Machinery	11.3	13.0	27.1	13.0	35.0
Electronics	10.8	8.4	24.7	7.2	46.4
Wood	15.2	14.4	34.4	7.2	28.0
Rubber and plastics	9.6	20.1	21.8	12.1	35.6
Automotive parts	7.6	8.3	32.4	16.6	34.5
Pearson chi2=	16.2409	66.204	100.4292	59.1907	36.6493
Pr=	0.000	0.000	0.000	0.000	0.000

All the governance types vary according to firms' status in a statistically significant way (Table 11.5). Importantly, firms selling their products to TNCs are more likely to be involved in governance type G5 than firms selling to other buyers, both global and domestic. This means that TNCs become engaged in their suppliers' R&D and send their experts to work to disseminate and diffuse new technologies more often than do other buyers. In contrast G3 is more frequent for firms which sell only to the domestic market, and for those which export through other channels.

The distribution across industries generally reflects this picture, with higher concentration in both types G3 and G5. However, some industrial specificities emerge, for example, with electronics value chains mainly following a G5 governance, or wood and textiles/clothing with less encompassing forms of governance. What is remarkable and perhaps unexpected is that value chains with forms of governance G1 and G2 do not occur frequently, and not even in firms selling only to the domestic market.

This fact appears to confirm the widespread and growing evidence of various sorts of networks and forms of intense coordination among firms, with stand-alone strategies hardly occurring. Firms are always embedded in multiple linkages, and these linkages appear to be taking forms of increasing complexity.

However, we need to go back to the main question of this chapter: after showing, with quantitative evidence, that governance forms vary across GVCs, how does this matter for local firms' efficiency and performance?

Global Value Chain Governance and Productivity

We explore here the relationship between firms' productivity and governance, focusing on the three groups of firms above, namely MNS, EXP and DOM. As a performance measure we employ total factor productivity (TFP). This measure is typically considered a growth rate and consists of the wedge between the average growth of outputs and the corresponding average growth of inputs (Navaretti and Castellani 2004). Moreover, this estimation technique has become increasingly popular in recent studies on FDI (for example Schoors and var der Tol 2002; Javorcik-Smarzynska 2004; Blalock and Gertler 2003) setting the standard for the current literature.[7]

Our measure of TFP is defined as the residual of a Cobb-Douglas production function. In order to take into account the problem of potential correlation between input levels and the unobserved firm-specific productivity shocks in the estimation of production coefficients, we carried out a panel data analysis using a semi-parametric technique to estimate TFP. The estimator used is that proposed by Levinsohn and Petrin (2003) with intermediate input use serving as a proxy for productivity shocks.[8] More specifically, we utilized the information on the

7 Other authors use 'upgrading' as a multidimensional measure of performance to encompass not only productivity improvements but also product improvements and firms' growing involvement in new functions and sectors (Giuliani et al. 2005b; Humphrey and Schmitz 2002a; Kaplinsky 2000). Regrettably, this was not possible with the presently available dataset.

8 Olayy and Pakes (1996) develop an estimator that uses investments as a proxy for unobservable shocks. Levinsohn and Petrin (2003) suggest that investments are subject to

amount of electricity consumed by each plant. As electricity cannot be stored, its consumption is likely to follow changes in production activity more closely than the use of materials.

The production function considered is the following:

$$Y_{it} = \delta_1 l_{it} + \delta_2 k_{it} + \delta_3 m_{it} + \omega_{it} + \varepsilon_{it} \qquad (1.1)$$

where m_{it} is the intermediate input (electricity). The error term has two components, the transmitted productivity component ω_{it} (or the state variable), and an error term which is uncorrelated with input choices ε_{it}. The state variable is not observed by the econometrician and affects firms' choice of inputs, potentially leading to the simultaneity problem in production function estimation, first mentioned by Marschak and Andrews (1944).

Thus, we construct our TFP measure as:

$$\hat{w} = \exp(y - \hat{\beta}l - \hat{\beta}k - \hat{\beta}m) \qquad (1.2)$$

A comparison of our TFP estimates between the three groups reveals important exporter premia in terms of productivity (average value 5.4, not reported here); MNS show quite similar values (average value 5.2), while DOM firms have lower values (average value 4.6).

We then test the relationship between governance and firms' efficiency through the following specification, and using G1 as baseline category:

$$TFP_i = \delta_0 + \delta_1 G2 + \delta_2 G3 + \delta_3 G4 + \delta_4 G5 + X + e_i \qquad (1.3)$$

where X captures firms' specific characteristics, including size, region and industry. We estimate two different specifications of the above equation: first, with our entire sample, then with the three sub-samples of firms.

As for similar research (for example 'learning by exporting' literature), however, we are aware of the difficulties in defining the direction of causality between buyer-supplier relationships and suppliers' performance. More precisely, do such relationships cause suppliers' performance improvements or rather do buyers select more efficient firms as their suppliers? Unluckily, the limited number of years for which data are available cannot help us to establish the direction of causality with sufficient confidence in this chapter.

Results for the whole sample suggest a significant and positive relationship between firms' productivity and governance G3 and G5 (Table 11.6). A higher share of sales made to client's unique specification jointly with the buyers' involvement in design and quality and in technology is correlated to higher productivity levels than for type G1. This fact confirms the qualitative results obtained by other authors

adjustment costs, thus not smoothly responding to productivity shocks.

(Schmitz 2004, Giuliani et al. 2005b), although the new and different econometric tests make comparisons difficult.

In a second specification of our model, documented in the last three columns of Table 11.6, we repeat the estimation separately for each group, using interaction terms. Restricting our attention to these sub-samples enables us to investigate the role played by different buyers. The results of this specification are qualitatively dissimilar from the previous ones and deserve careful interpretation. The emerging picture reveals that the way the value chain is organized is very relevant for DOM firms, while it does not appear to matter for firms supplying transnational buyers (MNS), or for EXP. DOM firms with high customization of products to buyers' standards, and that also receive assistance on design and quality definition and R&D and technology dissemination (G5) are more productive than the others. The mode of governance of their value chains is positively related to their productivity.

Can we interpret these results to suggest that firms participating in domestic value chains rely on a greater involvement of the chain leader to foster their process of learning and efficiency improvement? As seen above, EXP and MNS have relatively higher level of TFP than DOM. Once again the problem of causality forces us to interpret these results very cautiously. On the one hand, TNCs may select their suppliers among the most efficient firms (that is, 'picking the best cherries') – and indeed our data reveal that firms that are suppliers of TNCs are more efficient than DOM firms. We may explain this by observing that firms are often forced to improve their efficiency before starting the relationship with the TNCs in order to qualify as TNCs' suppliers. In this case the form of governance of the value chain would not matter for them initially, and we would need longer time series to test for the existence of possible dynamic learning processes.

The same line of reasoning may apply to firms serving other foreign buyers (EXP), in agreement with the literature on 'learning by exporting':[9] efficient and above average performer firms are likely to be the ones that are able to cope with sunk costs, and exporters have most of the desirable performance characteristics several years before they enter the export market.

On the other hand, if the self-selection hypothesis were not confirmed, the test of the existence of a learning process would require longer (dynamic) observations. Another possible explanation of these results that may deserve future testing might be due to the different nature of the information and knowledge exchanged within global and within domestic chains. Insofar as the gap of competences between TNCs and their suppliers is smaller in GVCs, it is easier to have cooperative relationships. In contrast, hierarchy is more likely to occur in national chains due to the suppliers' poor level of skills and competences relative to the leader.

9 Bernard and Jensen 1999; Clerides et al. 1998; Kraay 1999; Aw et al. 2000; Blalock and Gertler 2003. For a review, see Wagner 2006.

Table 11.6 Firms' TFP and value chains' governance

Dependent variable: Log of TFP				
	ALL FIRMS	**MNS**	**DOM**	**EXP**
G2	0.21	–0.002	0.32	–0.126
	–1.54	–0.01	(2.38)*	–0.33
G3	0.343	–0.139	0.693	0.007
	(2.59)**	–0.48	(3.79)**	–0.02
G4	0.101	–0.264	0.301	0.194
	–0.76	–0.87	(2.04)*	–0.48
G5	0.389	–0.07	0.668	0.52
	(2.80)**	–0.22	(3.18)**	–0.98
Size dummies	included	included	included	included
Industry dummies	included	included	included	included
Region dummies	included	included	included	included
Year dummies	included	included	included	included
Constant	4.578	4.621	4.367	4.146
	(22.08)**	(11.54)**	(11.61)**	(9.69)**
Observations	4022	4022	4022	4022
R-squared	0.17	0.11	0.15	0.12

Note: Robust t statistics in parentheses: * significant at 5 per cent; ** significant at 1 per cent.
Source: Authors' calculation on The World Bank – Enterprise Surveys (http://www.enterprisesurveys.org/).

As a robustness check of the results, we used an alternative measure of firms' performance: the net value added per worker. Value added is defined as sales minus intermediate input purchases. In Table 11.7 we report results from regressing the governance types on the log of net value added per worker. These findings are generally consistent with the picture we obtained from regressions based on TFP measure, except for G3. Looking at the whole sample, we find that only G5 matters for firms' performance, meaning that only firms receiving assistance on design and quality definition and R&D and technology dissemination are more productive than the others.

Results for the three sub-samples reveal again that governance seems to be very relevant for DOM firms, but not for MNS and EXP, in accordance with the results obtained with TFP.

Table 11.7 Firms' value added and value chains' governance

Dependent variable: Log of value added per worker				
	ALL FIRMS	MNS	DOM	EXP
G2	–0.04	–0.26	0.28	0.35
	–0.43	–1.45	(2.27)*	1.47
G3	–0.02	–0.08	0.25	–0.13
	–0.22	–0.59	(2.35)**	–0.56
G4	0.04	0.10	0.08	–0.45
	0.43	0.61	(0.55)*	–1.22
G5	0.19	0.14	0.42	–0.07
	(2.38)**	1.09	(3.82)**	–0.28
Size dummies	included	included	included	included
Industry dummies	included	included	included	included
Region dummies	included	included	included	included
Year dummies	included	included	included	included
Constant	4.578	4.621	4.367	4.146
	(22.08)**	(11.54)**	(11.61)**	(9.69)**
Observations	4022	4022	4022	4022
R-squared	0.55	0.43	0.48	54

Note: Robust t statistics in parentheses: * significant at 5 per cent; ** significant at 1 per cent.
Source: Authors' calculation on The World Bank – Enterprise Surveys (http://www.enterprisesurveys.org/).

Conclusion

In this chapter we have explored the patterns of governance arising in value chains led by global buyers and their impact on suppliers' performance with specific reference to the Thai manufacturing industry.

In order to address this issue, we have developed a quantitative measure of GVC governance which takes into account different levels and types of buyers' involvement in the suppliers' specification of product and process standards, R&D activities and dissemination of technology. Our typology identifies five basic types of value chain governance. We applied this typology to Thailand and compared the governance patterns and suppliers' performance of GVCs led by TNCs, of domestic value chains and of firms exporting through other channels.

An important descriptive finding is that the relationships that TNCs have with their suppliers is multi-fold, and they seem to become engaged in their suppliers' process or product R&D and send their experts to work to disseminate and diffuse new technologies more often than other buyers. In contrast, firms which are part of domestic value chains and those that sell to other global buyers follow modes of governance that imply only involvement in defining design and products' characteristics.

How do these different modes of governance impact on local firms' efficiency and performance? Our estimates show that more intense buyers' involvement with local suppliers, not only in the definition of products' characteristics, design and quality but also in technology dissemination and R&D, is associated with higher productivity. This observation appears to support evidence from different sources, obtained through different methods (for example, Schmitz 2004; Giuliani et al. 2005b; Pietrobelli and Rabellotti 2007).

As we further focus our attention on the three sub-samples of firms, we find that the way the value chain is organized is very relevant for domestic-led value chains, and affects these firms' productivity. In turn, the mode of governance does not appear to matter for firms supplying TNCs or for exporters. High customization of products to national buyers' standards coupled with assistance on design and quality definition and technology dissemination is associated with higher local firms' productivity. In order to explain this apparent paradox, contrary to the other sources of evidence quoted above, we suggest that it may be explained as a consequence of the different nature of the information and knowledge being exchanged, and of the larger gaps in knowledge and capabilities between the domestic leader and its suppliers. Future research will shed further light on this.

References

Albio, V., Garavelli, A.C. and Schiuma, G. (1999), 'Knowledge Transfer and Inter-firm Relationships in Industrial Districts: The Role of the Leader Firm', *Technovation* 19, 53–63.

Altenburg, T. (2006), 'Governance Patterns in Value Chains and their Development Impact', *The European Journal of Development Research* 18:4, 498–521.

Aw, B., Chung, S. and Roberts, M. (2000), 'Productivity and the Decision to Export: Micro Evidence from Taiwan and South Korea', *World Bank Economic Review* 14:1, 65–90.

Bernard, A. and Jensen, B. (1999), 'Exceptional Exporter Performance: Cause, Effect or Both?', *Journal of International Economics* 47, 1–25.

Blalock, G. and Gertler, P. (2003), 'Technology Diffusion from Foreign Direct Investment through Supply Chain', Working Paper (New York: Cornell University).

Brimble, P. and Sherman, J. (1999), 'Mergers and Acquisitions in Thailand: The Changing Face of Foreign Direct Investment', paper prepared for the United Nations Conference on Trade and Development (UNCTAD).

Castellani, D. and Zanfei, A. (2002), 'Multinational Experience and the Creation of Linkages with Local Firms. Evidence from the Electronics Industry', *Cambridge Journal of Economics* 26:1, 1–15.

Castellani, D., Zanfei, A. (2006), *Multinational Firms, Innovation and Productivity* (Cheltenham, UK: E. Elgar).

Clerides, S.K., Lach, S. and Tybout, J.R. (1998), 'Is Learning by Exporting Important? Micro-Dynamic Evidence from Colombia, Mexico, and Morocco', *Quarterly Journal of Economics* August, 903–48.

Djankov, S. and Hoekman, B. (1998), 'Trade Reorientation and Productivity Growth in Bulgarian Enterprises', *Journal of Policy Reform* 2:2, 151–68.

Dunning, J.H. (1993), *Multinational Enterprises and the Global Economy* (Wokingham: Addison-Wesley).

Gereffi, G. (1994), 'The Organization of Buyer-driven Global Commodity Chains: How US Retailers Shape Overseas Production Networks', in G. Gereffi and M. Korzeniewicz (eds), *Commodity Chains and Global Capitalism* (Westport: Praeger), 95–122.

Gereffi, G. and Memodovic, O. (2003), 'The Global Apparel Value Chain: What Prospects for Upgrading by Developing Countries?', United Nations Industrial Development Organization (UNIDO), Sectoral Studies Series, <www.unido.org/doc/12218>, accessed 22 February 2004.

Gereffi, G., Humphrey, J. and Sturgeon, T. (2005), 'The Governance of Global Value Chains', *Review of International Political Economy* 12:1, 78–104.

Giuliani, E., Pietrobelli, C. and Rabellotti, R. (2005), 'Upgrading in Global Value Chains: Lessons from Latin American Clusters', *World Development* 33:4, 549–73.

Giuliani, E., Rabellotti, R. and van Dijk, M.P. (eds) (2005), *Clusters Facing Competition: The Importance of External Linkages* (Aldershot: Ashgate).

Haddad, M. and Harrison, A. (1993), 'Are there Positive Spillovers from Direct Foreign Investment? Evidence from Panel Data for Morocco', *Journal of Development Economics* 42:1, 51–74.

Helpman, E., Meliz, M. and Yeaple, S. (2004), 'Export versus FDI with Heterogenous Firms', *American Economic Review* 94:1, 300–16.

Hewitt-Dundas, N., Andreosso-O'Callaghan, B., Crone, M., Murray, J. and Roper, S. (2002), *Learning from the Best – Knowledge Transfers from Multinational Plants in Ireland – A North-South Comparison* (Belfast: NIERC/EAC).

Humphrey, J. and Schmitz, H. (2002a), 'Developing Country Firms in the World Economy: Governance and Upgrading in Global Value Chains', INEF Report, No. 61 (Duisburg: University of Duisburg).

Humphrey, J. and Schmitz, H. (2002b), 'How does Insertion in Global Value Chains affect Upgrading in Industrial Clusters?', *Regional Studies* 36:9, 1017–27.

Javorcik-Smarzynska, B.S. (2004), 'Does Foreign Direct Investment Increase the Productivity of Domestic Firms? In Search of Spillovers through Backward', *American Economic Review* 94, 605–27.

Kaplinsky, R. (2000), 'Spreading the Gains from Globalisation: What can be Learned from Value Chain Analysis?', IDS Working Paper No. 110, <http://www.ids.ac.uk/ids/bookshop/wp/wp110.pdf>, accessed 12 January 2001.

Kaplinsky, R. and Morris, M. (2001), *A Handbook for Value Chain Research* (International Development Research Centre: Ottawa).

Kraay, A. (1999), 'Exports and Economic Performance: Evidence from a Panel of Chinese Enterprises', *Revue d'Economie Du Developpement* 1–2, 183–207.

Lall, S. (1980), 'Vertical Inter-firm Linkages in LDCs: An Empirical Study', *Oxford Bulletin of Economics and Statistics* 42, 203–6.

Levinsohn, J. and Petrin, A. (2003), 'Estimating Production Functions Using Inputs to Control for Unobservables', *Review of Economic Studies* 70:2, 317–41.

Marschak, J. and Andrews, W. (1944), 'Random Simultaneous Equations and the Theory of Production', *Econometrica* 12:3–4, 143–205.

Markusen, J. (1995), 'The Boundaries of Multinational Firms and the Theory of International Trade', *Journal of Economic Perspectives* 92, 169–89.

Mephokee, C. (2002), *Japanese Companies in Thailand's IT-related Industry*, Visiting Research Fellow Monograph Series (IDE-JETRO).

Morrison A., Pietrobelli C., Rabellotti R. (2008), 'Global Value Chains and Technological Capabilities: A Framework to Study Learning and Innovation in Developing Countries', *Oxford Development Studies* 36:1, 39–58.

Navaretti, B.G. and Castellani, D. (2004), 'Investments Abroad and Performance at Home: Evidence from Italian Multinationals', CEPR Discussion Paper No. 4284 (published online March 2004) <http://papers.ssrn.com/sol3/papers.cfm?abstract_id=527562>, accessed 13 April 2004.

Olay, G.S. and Pakes, A. (1996), 'The Dynamics of Productivity in the Telecommunications Equipment Industry', *Econometrica* 64:6, 1263–97.

Oman, C. (1984), *New Forms of International Investment in Developing Countries* (Paris: OECD).

Pietrobelli, C. and Rabellotti, R. (2007), *Upgrading and Governance in Clusters and Value Chains in Latin America* (Cambridge, MA: Harvard University Press).

Pietrobelli, C. and Saliola, F. (2008), 'Power Relationships along the Value Chain: Multinational Firms, Global Buyers, and Local Suppliers' Performance', *Cambridge Journal of Economics*, forthcoming.

Schmitz, H. (2004), 'Chain Governance and Upgrading: Taking Stock', in H. Schmitz (ed.), *Local Enterprises in the Global Economy* (Cheltenham: Edward Elgar).

Schmitz, H. and Knorringa, P. (1999), 'Learning from Global Buyers', Working Paper No. 100, November 1999, Institute of Development Studies.

Schoors, K. and van der Tol, B. (2002), 'Foreign Direct Investment Spillovers Within and Between Sectors: Evidence from Hungarian Data', Working Papers of Faculty of Economics and Business Administration, Ghent University, Belgium 02/157.

Sjoholm, F. (1997), 'Technology Gap, Competition and Spillovers from Direct Foreign Investment: Evidence from Establishment Data', Working Paper Series in Economics and Finance No. 211 (published online December 1997) <http://swopec.hhs.se/hastef/papers/hastef0212.pdf>, accessed 23 January 1998.

Sturgeon, T. and Lee, J.-R. (2001), 'Industry Co-evolution and the Rise of a Shared Supply-base for Electronics Manufacturing', Paper presented at Nelson and Winter Conference, DRUID, Aalborg.

Sturgeon, T. (2002), 'Modular Production Networks: A New American Model of Industrial Organization', *Industrial and Corporate Change* 11:3, 451–96.

Sturgeon, T. and Lester, R. (2004), 'The New Global Supply-base: Challenges for Local Suppliers in East Asia', in S. Yusuf, A. Altaf, and K. Nabeshima (eds), *Global Production Networking and Technological Change in East Asia* (New York: Oxford University Press), 35–87.

Turok, I. (1993), 'Inward Investment and Local Linkages: How Deeply Embedded is "Silicon Glen"?', *Regional Studies* 27, 401–17.

Wagner, J. (2006), 'Exports and Productivity: A Survey of the Evidence from Firm Level Data', *The World Economy*, Special Issue on Export and Growth.

World Bank (2004), *Thailand Economic Monitor* (Bangkok: World Bank).

Chapter 12
African Cloth, Export Production and Second-hand Clothing in Kenya

Tina Mangieri

Introduction

Textiles and apparel are currently undergoing dramatic upheavals in Kenya. Domestic production of 'traditional' African print textiles for local consumption, once the backbone of post-independence manufacturing strategies, ceased in the 1990s. Significant changes in the global regulatory environment affecting Kenya, including preferential trade arrangements with the EU (the Cotonou Convention of 2000) and the USA (the African Growth and Opportunity Act or AGOA, passed by the US Congress in 2000), resulted in tremendous, albeit truncated, growth in nascent export apparel production prior to the end of the MFA on 1 January 2005. Gains experienced by the export garment sector in the post-AGOA environment, prior to the completion of the MFA, included a tripling of Kenyan apparel exports from $45 million in 2001 to $150 million in 2003 (Flint 2004). The increasing import of second-hand clothing, once banned in Kenya during a period of import substitution industrialization (ISI), has likewise profoundly affected production, trade and apparel consumption.

This chapter focuses on the parallel and intersecting development of three international clothing systems in Kenya: 'African' print cloth with its own transnational histories of design, production and trade; more recent export-oriented mass-produced clothing primarily for large Western markets; and the import and sale of second-hand clothing sourced from those same markets. By taking as its focus these systems, and by stressing their interrelatedness, this chapter seeks to expand discussions of production in clothing and textile research beyond an 'assembly for export' emphasis, by integrating analyses of export production with both production for domestic consumption and the recent 'production' of local markets for second-hand clothing. Further, by focusing on these systems as they integrate in Kenya, this discussion aims to locate Africa centrally with respect to the international trade in textiles and apparel, and, more broadly, contestations over the processes of globalization.

South-South linkages in the production, trade and consumption of textiles and apparel in Kenya, particularly as Kenya links to India, China and the Arabian Gulf states, are both indicative of the current structure of the global textiles and apparel industry and ignored by research that focuses exclusively on the economic aspects of these ties. Occurring 'beneath' Western-dominated analyses of the global production and trade in cloth and clothing, these geographies are illustrative of long-standing and profound economic and cultural engagements with processes of globalization throughout the Indian Ocean littoral. Like globalization itself, these South-South links are not new, but they are constantly renegotiated. Attention must be paid, I argue, to the critical role played by these histories, in tandem with current cultural and economic links, to gain a better understanding of the contemporary context and future challenges facing not only Kenya, but also other countries for which present analyses of globalization continue to elide truly global integrations.

While this transnational focus forms the basis of a larger study from which this chapter emerged (Mangieri 2007), its specifics lie beyond this text. Rather, here I explore articulations of these multiple textile and apparel systems in Kenya within and through which these transnational linkages are forged. Following an introduction to textile and apparel production in Kenya, the chapter presents discussions of African print cloth, export apparel in the context of changing international trade regimes and second-hand clothing as each system links with the other. I conclude by considering globalization as a reflection of changing global interdependencies as suggested by current practices in textiles and apparel. The information presented is based on field research conducted in Kenya in 2004 and 2006, including 65 semi-structured interviews,[1] participant-observation, archival research and analyses of a variety of print materials including technical reports, newspaper articles and trade data. Throughout, I aim to explore the myriad integrations of economic policies with cultural practices in ways elided by state- and firm-centred approaches to the role of textiles and apparel in economic development.

Kenyan Textiles and Apparel: An Overview

Since independence in 1963, the textile and apparel sector in Kenya has been a centrepiece of national economic development strategies. The 'Directory of Industries 1986 Edition' complied by Kenya's Central Bureau of Statistics indicates considerable growth in the number of textile and garment manufacturers and the number of people employed, following the colonial period through the 1970s. Included in the survey are knitting mills (numbering 19), makers of 'made up textile goods except wearing apparel' (27), producers of 'spinning,

1 These interviews were conducted in Kenya, Dubai (United Arab Emirates) and the Sultanate of Oman as part of the dissertation project 'Refashioning South-South Spaces'. Interviewees included government representatives, business owners and managers, employees in the apparel/textile trades, transnational garment entrepreneurs, shipping and transport staff, consumers and developers and others associated with the new consumption spaces linking East Africa and Arabia for the sale of clothing.

weaving, and finishing textiles' (19) and 240 manufacturers of 'wearing apparel (except footwear)' (Republic of Kenya 1988).

Import substitution industrialization (ISI) policies in the early post-independence years were a boon to both cotton-growers and to textile and apparel industries, due to the imposition of a 100 per cent duty on imported goods. Under ISI, cloth and apparel manufacturing in Kenya was pursued as an aggressive means to industrialization, increased domestic control of the economy and employment generation. Print cloth for domestic consumption was pursued as both an economic policy generating jobs and revenue and as a cultural strategy predicated on a post-independence African pride visibly expressed by wearing 'African' garments. By the early 1980s, the textile and apparel industry emerged as a leading manufacturing activity in Kenya, in terms of size and employment (involving over 200,000 households and engaging approximately 30 per cent of the manufacturing sector's labour force) (EPZA 2005; Omolo 2006, 148). By the 1990s, however, Kenyan textile production was showing signs of impending collapse.

In this atmosphere, characterized by an effective lack of a viable industrial policy (Coughlin and Ikiara 1988), Kenya's Fourth Development Policy (1979–1984) advocated a more open strategy for the industrial sector, including cloth and clothing production, relying on market-based incentives and fewer regulatory structures, coupled with the strengthening of export promotion schemes (Bienen 1990; Ikiara et al. 2004). Manufacturing Under Bond (MUB) was initiated from 1985–1990, together with export compensation schemes, import duty and VAT remission, and, in 1990, the passing of the Export Processing Zones Act (CAP 517 of the Laws of Kenya) (Ikiara et al. 2004). Neither MUB schemes nor EPZs exhibited notable expansion in production, however, until the passage of the African Growth and Opportunity Act in 2000, discussed below.

Together with the founding of export-oriented production policies, protectionist measures implemented after independence to foster domestic industrial growth were eliminated under trade liberalization policies begun in Kenya in 1991. A ban on second-hand clothes enacted in 1984 (McCormick 1998) was overturned in 1991 in an atmosphere of widespread liberalization. Second-hand used clothing was again legally allowed into the country for resale and at prices far below that of new, domestically manufactured apparel. McCormick et al. (2002, 2) describe these transformations in the clothing sector:

> Kenya, after some hesitation, embarked on a liberalisation course in the early 1990s. The country removed foreign exchange controls, dropped quantitative restrictions on imports, reduced tariffs, and promoted exports. The impact on imports was dramatic. Between 1996 and 2000, the value of Kenya's imports rose by nearly half. Exports, on the other hand, increased by a much more modest 12%. Instead of boosting local industry, liberalisation caused an imbalance between imports and exports that created serious problems in a number of sectors. The garment industry was one of those most noticeably affected by the surge in imports. In Nairobi, second-hand clothes, commonly called *mitumba*, were everywhere.

The relationship between imports (of second-hand clothing) and exports (of new garments) is a pivotal one for both industries, exemplifying the irony of an economic landscape in which workers, clothed in second-hand garments, manufacture new clothing to be exported out of the country. I will return to this discussion of the articulations of *mitumba* and export apparel below, following the presentation of a third system contributing to these broader transformations in the geographies of production, and one often elided from discussions of manufacture focusing on both 'garments' and 'exports' – Africa print cloth for domestic consumption.

African Print Cloth: Rise and Decline

African print cloth and the clothing derived from it provides a productive entrée to discussions of global apparel industries, given their transnational nature. For the purposes of this paper, I will focus specifically on *khanga* (known also as *leso*), a cloth worn as a garment by women throughout East Africa. Despite their geographically dispersed development linking Africa with Asia, Europe and North America,[2] *khanga* are paradigmatically East African. *Khanga* developed in Zanzibar and the East African coast as, initially, a clothing style of Muslim women, many of whom were slaves prior to emancipation (see Fair 2001). While Muslim women continue to wear *khanga*, though to a lesser degree, their popularity has spread to other groups in East Africa and beyond.

Khanga are sold as conjoined rectangular pairs of cloth that must be cut and hemmed to form two pieces, most commonly worn in tandem as a skirt and head covering. They are characterized by a border on each side of the cloth and the inclusion of text printed in a narrow box in the bottom third of the fabric. The text, which may include a proverb, insult, flirtation or political slogan, was printed in Arabic on the earliest *khanga* of the late nineteenth century and appears most often in Swahili today. Regional language variations include Arabic (for the Omani market), Somali, Malagasy and other East African languages as appropriate.

While initially manufactured locally on a small scale, *khanga* first appear in British trade reports as an import into East Africa in 1897 (Fair 2001). They were subsequently produced in India, primarily, although a series of trade conflicts between the British colonial government and Indian cloth manufacturers resulted in a wholesale shift of *leso* production to Japan and the Netherlands in the 1950s.

Following independence, the textile industrial sub-sector was identified by the government as a 'core industry' with the potential for inducing rapid economic development in Kenya under national ISI strategies. Local cloth manufacturing expanded from six weaving mills in operation by 1963 to 52 weaving mills by 1983 (Republic of Kenya 2001, 3). The economic and political centrepiece of this

2 The complicated, globalized histories of khanga and other cloth in East Africa are beyond the scope of this text (see Linnebuhr [n.d.]; Fair 2001, 2004; and Prestholdt 1998, 2004 for additional information).

expansion was the domestic manufacture of African print cloth as both a foundation for industrial development and a powerful aesthetic symbol of independent African identity. Indeed, post-independence governments throughout sub-Saharan Africa invested heavily in domestic textile and apparel manufacture, with a particular focus on 'African' prints. These local print-cloth industries, once the core of post-independence manufacturing sectors and celebrated as thriving, tangible symbols of African independence, are today in steep decline throughout sub-Saharan Africa. By 2000, domestic manufacture of *khanga* had ended in Kenya. Currently *khanga* sold in Kenya are again imported from India, although a small number of shops in Mombasa and Nairobi carry *leso* manufactured in Tanzania, where local production persists.

Local woven and print cloth production in Kenya was undermined, in part, by crises in domestic cotton production, management debacles, the import of second-hand clothing from the USA and Europe, a diminished market following the collapse of the East African Community in 1977 and an emerging movement toward the local manufacture of western-style clothing for export. An additional factor, and one absent from analyses focusing on one or another economistic scenario, remains the declining popularity of *khanga* amongst Muslim women in particular, who have widely rejected the garment in urban areas, in favour of *buibui*, a black cloak akin to *abaya* worn throughout the Arabian Gulf states. This fashion shift is linked to wider geopolitical and identity transformations, suggesting the paucity of assessments of domestic print cloth production decline focused exclusively on economic indicators. The recent availability of *khanga* produced in India and China, and selling for 150 Kenyan shillings per pair, as opposed to 180–300 Ksh for *leso* made in Tanzania, has all but ensured that a revival of *khanga* manufacture is unlikely in Kenya.

The primary *khanga* production facilities previously extant in Kenya, including Rift Valley Textiles (RIVATEX) in Eldoret, which closed in 1998, and Nanyuki's Mount Kenya Textile (MOUNTEX), which operated intermittently through the late 1990s before closing in 2000, remain empty and for sale, attracting the sporadic interest of prospective investors from India, the USA and Thailand, amongst others (*East African Standard* 2003). In July 2007, local news outlets reported comments by then-Minister of Trade and Industry, Dr. Mukhisa Kituyi, indicating the defunct RIVATEX factory in Eldoret would reopen as part of the government's efforts to fast track revival of textile factories (Oyuke 2007). The professed aim of this revival is, however, the local production of textiles to supply export-oriented apparel manufacturers (thus addressing the needs of continuing AGOA participation, discussed below) and not a reinvigoration of an African print cloth industry for domestic consumption.

Although *khanga* are no longer manufactured in Kenya, *khanga* popularity persists in diminished form. To meet this demand, *khanga* are imported primarily from India and Tanzania, as noted above, though not always through official channels. The loss of government revenue attributed to cross-border *khanga* smuggling is estimated by the Kenyan government to exceed 105 million Kenyan shillings annually or over $1.3 million in October 2004 rates (Republic of Kenya 2001, 4). The persistent popularity of African print cloth, the resonance of its 'African' aesthetic and its fluctuating economic importance in Kenya are nevertheless transformed by articulations with second-hand garments and an emergent export-clothing manufacturing strategy.

International Trade Regimes and Kenya

Amidst the global restructuring associated with trade liberalization, parallel agreements have been enacted between regional and continental trading blocks, including Africa, resulting in a recent flourish of export apparel-productions industries in Kenya. These transformations are best illustrated by a focus on changes effected in Kenya by the African Growth and Opportunity Act (AGOA), a US response to the brokered agreement between the EU and nations of Africa, the Caribbean and the Pacific, known in its current form as the Cotonou Convention. These agreements, particularly AGOA, resulted in a (re)surge(nce) of investor interest in the garment sector in Africa, including Kenya, following a period of decline and factory closure. While apparel comprised a mere 9 per cent of the products imported into the USA under AGOA in 2003 (second only to energy-related products, which account for 79 per cent of trade by cost), the impact on manufacturing for participating African countries has been substantial. This resurgence is now threatened by global reconfigurations following the end of the MFA.

AGOA was first passed by the US Congress in May 2000 and most recently amended in December 2006.[3] Under its provisions, AGOA eliminates duties and quotas on apparel and certain textiles exported to the USA by (currently 38) sub-Saharan African signatories. AGOA beneficiaries must meet a series of political and economic conditions for inclusion within the agreement. These provisions are broadly defined to permit those countries 'that work to open their economies and build free markets' to participate under AGOA (AGOA 2007). Further distinctions

3 George W. Bush signed the African Investment Incentive Act of 2006, or AGOA IV, on 20 December 2006. As the number suggests, this latest instalment is the third amendment to the original AGOA. AGOA IV 'provides duty free and quota free treatment for eligible apparel articles made in qualifying Sub-Saharan African countries through 2015. Qualifying articles include: 1) apparel made of US yarns or fabric; 2) apparel made of Sub-Saharan (regional) yarns and fabric, subject to a cap until 2015; 3) apparel made in a designated lesser developed country of third-country yarns and fabrics, subject to a cap until 2012; 4) apparel made of yarns and fabrics not produced in commercial quantities in the United States; 5) certain cashmere and merino wool sweaters; 6) eligible hand loomed, handmade, or folklore articles and ethnic printed fabrics; and 7) textiles and textile articles produced entirely in a lesser-developed beneficiary country' (http://www.agoa.gov/eligibility/apparel_eligibility.html).

demarcate LDCs[4] and non-LDCs. In this framework, LDCs are permitted to utilize third-country fabrics in apparel production for export to the US, a key component enabling current garment manufacture throughout sub-Saharan Africa.[5]

AGOA is credited with a resurgence and revitalization of investor interest in garments in Kenya, following a period of decline and deindustrialization (Omolo 2006). Indeed, as Figure 12.1 indicates, the export apparel trade to the EU, in comparison to the US, is minimal. Kenyan garment production for the US market is concentrated in two garment categories: women and girl's slacks and the like (category 348)[6] and cotton men's and boy's trousers (category 347). Taken together in 2006, these two categories accounted for over 60 per cent of Kenyan apparel exports to the USA (OTEXA). Cotton garments similarly accounted for the majority of Kenyan apparel to the USA, or 77 per cent in 2006 (OTEXA). While Kenyan cotton production[7] once supplied the domestic garment industry, and cotton apparel continues to dominate Kenyan manufacturing, local capacities have contracted sharply since the 1980s and the abandonment of import substitution industrialization.

4 LDCs are defined as those countries with a per capita GNP of less than $1500 year in 1998 as measured by the World Bank. While many countries in sub-Saharan Africa meet this designation, South Africa and Mauritius are important exceptions.

5 Kenya is currently estimated to require 225 million square metres of fabric for apparel production. Prior to the decline of the textile industry in the 1990s, there were 52 textile mills producing fabric and yarn with an installed capacity of 115 million SME. With a 50 per cent decline in the number of textile mills in Kenya, coupled with reduced cotton production, Kenya's current fabric demands greatly outstrip current supply (Republic of Kenya 2001; Kimani 2002; RATES 2003; Omolo 2006).

6 The figures '347' and '348' refer to the garment classifications used in the US textile and apparel category system. This three-digit system, correlated on the OTEXA website to 10–digit Harmonized Tariff Schedule Codes used more commonly in the EU and sub-Saharan Africa, are grouped as follows: 200 (cotton and or manmade fiber), 300 (cotton), 400 (wool), 600 (manmade fiber), 800 (silk blends/non cotton vegetable fiber) (OTEXA).

7 Cotton in Kenya is grown by small farmers and is confined, predominantly, to arid and semi-arid areas. The Export Processing Zone Authority (EPZA) estimates that Kenya has 140,000 small-scale cotton farmers, down from over 200,000 in the early 1980s when the industry was at its peak (EPZA 2005). As of 2003, 25,000 hectares were under cotton with a total lint production of 20,000 bales (EPZA 2005). There were 24 ginneries in Kenya in 2005 with an estimated installed capacity of 140,000 bales, thus actual output is considerably less than the industry potential. At these rates, Kenya's present ginneries could process cotton were production increased by as much as 600 per cent (Omolo 2006; see also RATES 2003).

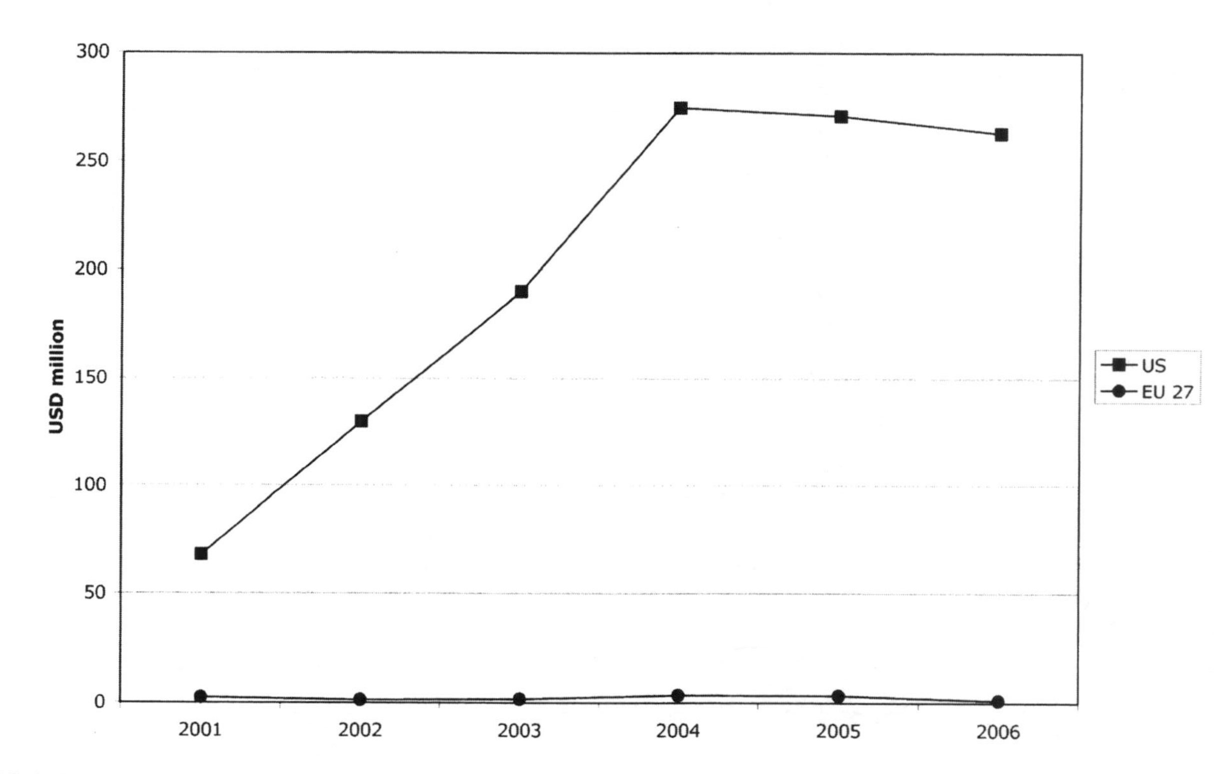

Figure 12.1 Kenya apparel exports to US and EU27, 2001–2006
Source: OTEXA 2007 and Comext 2007.

At the close of 2007, there were 22 garment factories operating within export processing zones (EPZ) in Kenya, down from a high of 34 prior to the MFA phase-out (Table 12.1). EPZ garment production in Kenya follows a 'cut, make, trim' model (CMT), using fabric (and other inputs) sourced in Asia. AGOA supporters tout the benefits of investment and employment creation throughout sub-Saharan Africa. It is, arguably, 'the most far-reaching initiative in both the history of US-African economic relations, and more generally in relation to the claim that concessions in the area of trade provide better long-term prospects for developing countries' economic development than do ones in aid' (Gibbon 2003, 1809). Detractors, while not denying the creation of tens of thousands of export apparel jobs under AGOA, nevertheless question its sustainability, its meaningful contributions to Kenya's economy (aside from the low wages paid its employees) and the downward pressure placed on Kenyan workers (Kamungi and Ouma 2004).

Table 12.1 Kenya EPZ apparel factories and employment, 2000–2006

Year	Number of Kenyan EPZ garment factories	Total number of employees
2000	6	6,487
2001	17	12,002
2002	30	25,288
2003	34	32,095
2004	34	–
2005	–	–
2006	24	30,800
2007	22	27–30,000

Source: EPZA and Republic of Kenya Statistical Abstract 2006.

For Kenya, trade agreements have perceptibly altered the manufacturing landscape in terms of the production of textiles, apparel and the growth of a clothing export sector. Factories previously devoted to the domestic manufacture of African print cloth, as noted above, have either ceased to produce or are being reopened to support export apparel facilities. To support an expansion of export-oriented production prior to the end of the MFA, Kenya increased the number and scope of Export Processing Zones (EPZs) – a brief period of expansion made possible by AGOA (RATES 2003, 79). While prior to the end of the MFA garment factories located in the EPZs had spawned a $163 million apparel manufacturing industry and created over 30,000 jobs (Mireri 2000; The Nation 2004; Omolo 2006), this growth was not without controversy. Labour conditions in these zones have been likened to 'sweatshops of South East Asia' (The Nation 2004; Kamungi and Ouma 2004), resulting in labour unrest and manufacturing shutdowns.

Despite these and other serious human rights contentions, Kenya continued to expand its export market in cloth and clothing to the USA through 2004. As a testament to this expansion, Kenya surpassed Mauritius in 2003 and South Africa in 2004 to place second behind Lesotho for apparel exports to the USA under AGOA measured as SME (AGOA 2007). Since the end of the MFA, factory closures have contributed to the reduction of apparel manufacturers from 34 to 22, reversing the few years of AGOA gains – a devastation Kenyans refer to as the 'Chinese tsunami' (Mulama 2005). To illustrate the speed at which these changes were wrought, by 1 May 2005 – International Labour Day and four months beyond the end of the MFA – Kenya had lost 6,000 jobs in apparel-for-export manufacturing with the closure of four major, seemingly well-established, factories (Mulama 2005). As the ten-year 'tax holidays' offered by Kenya's EPZs begin to conclude for some earlier manufacturers, there is the further threat of closure as producers shift their establishment to capitalize on concessions offered by neighbouring Uganda[8] and Tanzania.

Gibbon's work on clothing manufacture in South Africa and Mauritius indicates that the supply response by African nations to the provisions offered by AGOA have been 'substantially and reasonably broad in geographical terms' (Gibbon 2003, 1821), but, on closer investigation, more narrowly confined to certain types of industrial enterprise. Of the firms that have increased production for the US market under AGOA, the majority are east Asian-owned, specializing in the assembly and finishing of basic garments (Gibbon 2003). The reasons for the greater perceived expansion in East Asian-owned enterprises in South Africa and Mauritius are numerous and speculative (including greater access to capital and longer experience with the type of production pursued under AGOA.

In Kenya a view of the current ownership and management of apparel manufacturing indicates a majority Indian and East Asian investment (RATES 2003, 98–122). Given the deeply implicated histories of South Asians in Kenya, as elsewhere in East Africa, attention must be paid to their role in establishing and indeed expanding textile production under the current AGOA regime. India is currently lobbying East African nations, particularly Kenya, to form partnerships under AGOA by which India would provide manufacturers in Kenya with the raw materials for textile production. For now, however, manufacturers continue to source inputs from East Asia rather than South Asia.

Despite this trend, one possible post-MFA impact will be a decreased investment by Asian firms in African manufacturing. China and Taiwan, in particular, established offshore production facilities throughout sub-Saharan Africa as a

8 This occurred with the departure from Kenya of Tri-Star Apparel, a Sri Lankan firm that was among the first to establish garment production in the Athi River EPZ in 1994. In 2004, at the conclusion of its tax holiday, the firm moved to Uganda. It has since ceased operations (Udalagama, personal interview, 26 April 2006, Nairobi).

way of circumventing country-based quota restrictions placed on garments under the MFA. As these restrictions ended on 1 January 2005, Africa may prove less attractive to Chinese garment manufacturers, in particular, given that the cost of producing apparel (including such variables as raw material inputs, electricity and worker wages) remains higher throughout sub-Saharan Africa than in China. Despite this possible scenario, overall investment by China in sub-Saharan Africa, including Kenya, is increasing as China's economic and political clout rises (Otieno 2005; Morris 2006; Broadman 2007).

The collapse of domestic *khanga* production, including the complete closure of the factories involved, has thus put Kenya in a precarious position under AGOA provisions to cease the use of imported third-country fabrics for apparel exports by 2012. Alongside the expanding export sector, African print cloth struggles to subsist. Government agencies, manufacturing associations and local designers in Kenya are advocating a return to post-independence movements advocating a consciousness of African identity by 'wearing Kenya, buying Kenya' in an attempt to further encourage pride in and popularity of African prints. In 2004 these efforts included a nationwide contest to select Kenya's National Dress. The competition results reflected re-emerging ethnic differences, as Kenya's myriad social groups found little common ground in selecting garments that would adequately represent and appeal to Kenya's diverse populations – with one exception. Despite the inability to choose one form of representative clothing, wry commentators agreed that Kenya does, arguably, have a national dress – *mitumba*, or second-hand clothes.

Global and Local Second-hand

Recent international agreements on export manufacture and trade have prompted considerable research on African production, particularly that of Gibbon (2000a, 2000b, 2002, 2003). Second-hand clothing for consumption in these same markets is mentioned, however, only tangentially. An increasing interest in second-hand clothing research in Africa has appeared in the anthropological literature (Hansen 1994, 1995, 1999, 2000), although by Hansen's own admission 'little research has been done [in anthropology] on clothing production issues' (2004a, 369). These two contrasting approaches indicate the present gulf in studies integrating economic analyses within a cultural framework. In the remainder of this chapter, I will focus on the rise in second-hand clothing in Kenya. How did the second-hand clothing trade in parts of sub-Saharan Africa achieve such prominence, in a relatively short period of explosive growth (since the 1980s)?

Cloth and clothing were among the primary historical commodities linking Africa with an increasingly globalizing economy and gradually altering the subjectivities of the population, creating consumers. The nineteenth- and early twentieth-century presence of colonial administrations and missionaries prompted a profound shift in the clothing worn in communities throughout the continent

– particularly in areas where the predominant coverage came from skin, hides and barkcloth (Comaroff and Comaroff 1997). Both voluntary and forced migration brought men, in particular, together in urban and other concentrated settlements, where new clothing and the wearing of western fashions became increasingly the norm, particularly in southern and East Africa (Hansen 2000).

The second-hand clothing trade in colonial Africa began in the immediate post-WWI period with an influx of surplus military uniforms shipped by used clothing dealers in Europe and from production areas in the USA (Hansen 2000, 66). The amount of clothing appearing as cast-off articles increased throughout the industrial states of the North in the twentieth century with the initial success of Fordist policies and concomitant increases in disposable income. The countries of sub-Saharan Africa now form the world's largest second-hand clothing destination, receiving 30 per cent of total world exports in 2001 with a value of $405 million, up from $117 million in 1990 (United Nations 1996, 2003).

The post-independence period throughout much of sub-Saharan Africa was characterized by an increased emphasis on domestic production of textiles and finished apparel for local consumption, as noted above. African print cloth was the focus of campaigns encouraging or otherwise imposing 'dress codes' or national dress, using domestic print cloth from these new industries. In light of these policies, the importation of second-hand clothing was banned in Kenya in 1984, and elsewhere throughout much of sub-Saharan Africa, as import substitution policies were adopted to strengthen domestic manufacture. Second-hand clothing nevertheless entered Kenya in the 1970s and 1980s as a consequence of regional political crises.

The first post-independence shipments of second-hand clothes into Kenya were linked to geopolitical upheavals in eastern Africa. Conflict in neighbouring Uganda, Sudan and Ethiopia resulted in increasing populations of refugees in Kenyan camps. Charitable organizations working with displaced persons were able to import used clothing to serve the needs of these impoverished communities during this period, with some of the donations reportedly finding their way into surrounding villages and later urban areas as commodities for resale (Kimani 2002, 4). Protectionist measures implemented after independence to foster domestic industrial growth were eliminated under trade liberalization policies begun in Kenya in 1991. Previously banned imports of used clothing were now legally allowed into the country for resale, and at prices far below that of new, domestically manufactured apparel.

The absence of foreign exchange controls, the elimination of restrictions, and reduced tariffs resulted in a rise in value of Kenya's imports by nearly half between 1996 and 2000 (McCormick et al. 2002). The garment industry was the most noticeably affected by the surge in imports, particularly the widespread availability of second-hand clothing, or *mitumba* (McCormick et al. 2001). *Mitumba*, arriving in compressed plastic-wrapped 45 kg bales in East African ports, were cheaper than Kenyan products and rapidly gained favour with consumers (Ongile and

McCormick 1996). This popularity was achieved despite the stigma of the garments being associated with 'dead Europeans' or '*kafa Ulaya*' – an earlier term for *mitumba*. The changing market was reflected in the clothing production index, which plummeted from a high of 378.6 in 1989 to only 154.8 in 1999, a drop of 40.9 per cent (McCormick et al. 2001).

The increasing volume of used clothes entering Kenya throughout the 1990s corresponded with the closure of the vast majority of clothing and textile factories creating products for local consumption in those sub-Saharan African countries that legalized imports. By 2001 in Kenya, as noted above, all domestic manufacturing of African print cloth for the local market ended with the closure of the last remaining producer, MOUNTEX. While multiple factors are necessarily involved, including problems with outmoded equipment, lack of investment, corruption and the unavailability of consistent inputs including electricity, second-hand clothing is popularly portrayed as the primary causal factor for textile and garment factory closure. While the import of *mitumba* is viewed as a fundamental contributor to the cessation of domestic textile and apparel manufacture throughout the 1990s, the rise in the current export-apparel market complicates this scenario. Many of the world's largest importers of second-hand clothing are also among the top exporters of textiles and garments, including Pakistan and Hong Kong (Hansen 2004b, 4). Under the current changes in AGOA, Kenya has likewise shifted to high imports of second-hand clothing for domestic consumption and an increasingly large volume of apparel for export.

Articulations, Globalizations and the Outlook Ahead

The recent past has been dominated by discourses of 'globalization'. Generally accepted as indicative of a degree of global economic integration that is new in intensity, if not altogether novel in practice, economic globalization has succeeded in supplanting the rhetorical strategies of the past few decades of economic policy formulation in East Africa, as elsewhere. Rather than import substitution, the dominant discussion in Kenya is now one of Export Processing Zones. Instead of paring back regulation (in the form of state power) to a bare minimum, as was the tactic under structural adjustment programs at their height, there is increasing discussion of the ways in which regulatory policies might support increased global integration as expressed in the shorthand of globalization. This regulation may take the form of regional or more often 'global regulatory meta-narratives' (Amin 2004, 219).

Economic globalization is perhaps less meaningful as a development strategy than as a spatial ontology (Amin 2004) or a disabling discourse (Hart 2002). As such, globalization may be best viewed as a product rather than a condition, such that limits are placed upon 'the explanatory power of these processes in the absence of other necessary preconditions and contingent factors' (Yeung 2002, 301). In other words, 'globalization' as a term is itself rather meaningless in the absence of critical historical and contemporary contextualizations detailing its specificities. Rather than a paralyzing global phenomenon, globalization itself must be broken down into its component parts to have any coherence or explanatory power. In the last ten years, during which globalization has generated such discursive momentum, interdependencies between places are being reshaped, as is evidenced by the articulation of the three international clothing systems discussed herein. In East Africa, this is reflected not only in changing state relations with places as far-flung as Gujarati *khanga* factories, *buibui* markets in Dubai, Wal-Mart production negotiations in Bentonville and *mitumba* from Rome, but at scales as intimate as the clothed body and as overarching as the WTO.

While local production of cloth and clothing for domestic consumption in Kenya declined (or altogether ceased in the form of *khanga*), these three systems of textiles and apparel bind together producers and the fast-changing fashion whims of consumers in the Global North; Asian and Middle Eastern entrepôt profiting from the resale of imported second-hand to Africa; national tax revenues generated from the income produced by the import of *mitumba*; the declining popularity of indigenous *khanga* and concomitant rise in international Islamic fashions among East African Muslims; the establishment of large second-hand markets; street sellers; and the unexpected and evolving fashions of East African communities making *mitumba* their own.

Are these three systems compatible or contradictory? How do the three intermingle, negate, or build on one another? Building on this discussion, I view each of these systems as offering unique bases for rethinking regional and global systems relationally. Their compatibility or contradictory characteristics cannot be measured on the basis of the brief treatment they have received here. Rather, they suggest ways to view the processes associated with each form of production as necessarily articulated.

Although Kenya has 'risen' to second place behind Lesotho for apparel imports to the USA under AGOA, and is set to further expand this market share in the short-term, the global effects of a quota-free China remain a major threat to the thousands of jobs created in the export apparel sector. Under AGOA, there is nevertheless room to grow. Kenya now relies on the import of inputs from Asia to meet its export apparel manufacturing needs – a third-county fabrics allowance that has been extended to 2012. Current cloth production in Kenya cannot satisfy the demands for export apparel inputs, nor are other AGOA countries poised to fill this need. The collapse of this domestic cloth industry producing African prints in the 1990s was related, in part, to the influx of *mitumba*.

The closure of African cloth manufacturing centres has resulted in a physical landscape of abandoned factories. While devastating to local community economies, these same sites are now attractive to foreign investment interested in restarting these plants to produce export apparel, and textile inputs for export garments, rather than African fashions. While *mitumba* is often blamed for problems in manufacturing in Kenya, its appeal and cost-effective popularity likely preclude the reinstatement of bans on second-hand clothing, although this strategy remains in effect in Ethiopia, and elsewhere. The creation of large wholesale/retail markets for *mitumba* sales and the thousands of legal and illegal kiosks whose vendors rely on the trade reveal both the concrete changes to Kenya's urban landscape (poised to increase with even more *mitumba*) and the cultural effects of the shift from 'traditional' clothing styles, represented in part by *khanga*, to an embrace of *mitumba*.

In the midst of these challenges and transformations, one new trend in Kenya is actually among the oldest – a strengthening of trade networks and relations throughout the Indian Ocean. The reliance on trade between India and Kenya, in particular, linked to shared colonial histories, export-apparel production, *mitumba*, *khanga* and changing domestic fashions in Kenya, may be poised to expand in a post-quota environment as attention turns to China and its relations with the major markets of North America and the EU. The discussion above may be seen as an effort to draw attention to this possibility, as changes in industrial production are linked to national consumption patterns and regional transformations occurring 'beneath' northern-oriented analyses of globalization.

References

AGOA (2007), 'Apparel Trade under AGOA' [AGOA website] <http://www.agoa.info/index.php?view=trade_stats&story=apparel_trade>, accessed 8 November 2007.

Amin, A. (2004), 'Regulating Economic Globalization', *Transactions of the Institute of British Geographers* 29, 217–33.

Bienen, H. (1990), 'The Politics of Trade Liberalization in Africa', *Economic Development and Cultural Change* 38, 713–32.

Broadman, H. (2007), *Africa's Silk Road. China and India's New Economic Frontier* (Washington, DC: The World Bank).

Comaroff, J.L. and Comaroff J. (1997), 'Fashioning the Colonial Subject: The Empire's Old Clothes', in Comaroff and Comaroff (eds), *Of Revelation and Revolution, Vol. 2, The Dialectics of Modernity on a South African Frontier* (Chicago: University of Chicago Press), 218–73.

COMEXT (2007) <http://fd.comext.eurostat.cec.eu.int/xtweb/> (home page), accessed 8 November 2007.

Coughlin, P. and Ikiara, G.K. (eds) (1988), *Industrialization in Kenya* (Nairobi: Heinemann).

East African Standard (2003), 'Indian Investors Eyeing Kenyan Textile Plant', 30 October 2003.

EPZA (Export Processing Zones Authority) (2005), 'Kenya's Apparel and Textile Industry 2005', <http://www.epzakenya.com/news.php?type=press&itemno=45> (home page), accessed 12 July 2007.

Fair, L. (2001), *Pastimes and Politics: Culture, Community, and Identity in Post-Abolition Urban Zanzibar, 1890–1945* (Athens, OH: Ohio University Press).

Fair, L. (2004), 'Remaking Fashion in the Paris of the Indian Ocean: Dress, Performance, and the Cultural Construction of a Cosmopolitan Zanzibari Identity', in J. Allman (ed.), *Fashioning Africa: Power and the Politics of Dress* (Bloomington, IN: Indiana University Press), 13–30.

Flint, H. (2004), 'AGOA Update: Why AGOA III is Vital for Africa?', *Standard Bank African Research*, Economics Division [website] <http://www.tralac.org/scripts/content.php?id=2382>, accessed 8 November 2007.

Gibbon, P. (2000a), 'Global Commodity Chains and Economic Upgrading in Less Developed Countries', CDR Working Paper Subseries No. viii, 00.2.

Gibbon, P. (2000b), '"Back to the Basics" Through Delocalisation: The Mauritian Garment Industry at the End of the twentieth Century', CDR Working Paper 00.7, October 2000, Working Paper Subseries on Globalisation and Economic Restructuring in Africa No. x.

Gibbon, P. (2002), 'South Africa and the Global Commodity Chain for Clothing: Export Performance and Constraints', CDR Working Paper 02.7, April 2002, Working Paper Subseries on Globalisation and Economic Restructuring in Africa No. xix.

Gibbon, P. (2003), 'The African Growth and Opportunity Act and the Global Commodity Chain for Clothing', *World Development* 31:11, 1809–27.

Hansen, K.T. (1994), 'Dealing with Used Clothing: Salaula and the Construction of Identity in Zambia's Third Republic', *Public Culture* 6:3, 503–23.

Hansen, K.T. (1995), 'Transnational Biographies and Local Meanings: Used Clothing Practices in Lusaka', *Journal of Southern African Studies* 21:1, 131–45.

Hansen, K.T. (1999), 'Second-hand Clothing Encounters in Zambia: Global Discourses, Western Commodities, and Local Histories', *Africa* 69:3, 343–65.

Hansen, K.T. (2000), *Salaula: The World of Second-hand Clothing and Zambia*. (Chicago: University of Chicago Press).

Hansen, K.T. (2004a), 'The World in Dress: Anthropological Perspectives on Clothing, Fashion, and Culture', *Annual Review of Anthropology* 33, 369–92.

Hansen, K.T. (2004b), 'Helping or Hindering? Controversies around the International Second-hand Clothing Trade', *Anthropology Today* 20:4, 3–9.

Hart, G. (2002), *Disabling Globalization. Places of Power in Post-apartheid South Africa* (Berkeley, CA: University of California Press).

Ikiara, G.K., Olewe-Nyunya, J. and Odhiambo, W. (2004), 'Kenya: Formulation and Implementation of Strategic Trade and Industrial Policies', in C. Soludo, O. Ogbu, and H.-J. Chang (eds), *The Politics of Trade and Industrial Policy in Africa. Forced Consensus?* (Trenton, NJ and Ottawa: Africa World Press/IDRC), 205–24.

Kamungi, P. and Ouma, S. (2004), *The Manufacture of Poverty: The Untold Story of EPZs in Kenya* (Nairobi: Kenya Human Rights Commission).

Kimani, M. (2002), 'The Textile Industry in Kenya', Nairobi [Kenya] Ministry of Trade and Industry Working Paper, <http://www.intracen.org/worldtradenet/docs/whatsnew/atc_lesotho_november2002/country_paper_kenya.pdf>, accessed 10 June 2007.

Linnebuhr, E. [n.d.], 'Zawadi ni tunda la moyo' [A gift is a fruit from the heart], *Form+Zweck* [website] <http://www.formundzweck.com/eng/themen.php?D+Kleidersprache>, accessed 10 November 2007.

Mangieri, T. (2007), *Refashioning South-South Spaces: Cloth, Clothing, and Kenyan Cultures of Economies*, PhD dissertation (Chapel Hill, NC: University of North Carolina).

McCormick, D. (1998). 'Policies Affecting Kenyan Industrialization: 1964–94', in N. Ngethe and W. Owino (eds), *From Sessional Paper No. 10 to Structural Adjustment: Towards Indigenizing the Policy* (Nairobi: Institute for Policy Analysis and Research).

McCormick, D., Kimuyu, P. and Kinyanjui, M. (2001), 'Kenya's Garment Industry: An Institutional View of Medium and Large Firms', IDS Working Paper No. 531 (Nairobi: University of Nairobi Institute for Development Studies).

McCormick, D., Kimuyu, P. and Kinyanjui, M. (2002), 'Weaving through Reforms: Nairobi's Small Garment Producers in a Liberalised Economy', Paper presented at the East African Workshop on Business Systems in Africa, *Institute for Development Studies, University of Nairobi and the Centre for Development Research*, Copenhagen [website] <http://bij.hosting.kun.nl/iaup/esap/publications/nairobi/Nairobi_garment_paper_%204.pdf>, accessed 10 June 2007.

Mireri, C. (2000), 'The Impact of Export Processing Zone Development on Employment Creation in Kenya', *Singapore Journal of Tropical Geography* 21:2, 149–65.

Morris, M. (2006), 'Globalization, China, and Clothing Industrialization Strategies in Sub-Saharan Africa', in H. Jauch and R. Traub-Merz (eds), *The Future of the Textile and Clothing Industry in Sub-Saharan Africa* (Bonn: Friedrich-Ebert-Stiftung), 36–53.

Mulama, J. (2005), 'International Labour Day – Kenya: A Murky Future for Textile Workers', *Inter Press Service News Agency* [website] <http://www.ipsnews.net/interna.asp?idnews=28520>, accessed 8 November 2007.

Omolo, J. (2006), 'The Textiles and Clothing Industry in Kenya', in H. Jauch and R. Traub-Merz (eds), *The Future of the Textile and Clothing Industry in Sub-Saharan Africa* (Bonn: Friedrich-Ebert-Stiftung), 147–64.

Ongile, G. and McCormick, D. (1996), 'Barriers to Small Firm Growth: Evidence from Nairobi's Garment Industry', in D. McCormick and P.O. Pedersen (eds), *Small Enterprises: Flexibility and Networking in an African Context* (Nairobi: Longhorn), 40–62.

OTEXA (Office of Textiles and Apparel) (2007) <http://otexa.ita.doc.gov/> (home page), accessed 18 November 2007.

Otieno, J. (2005), 'Move Over Uncle Sam, the Chinese are Now in Town', *The Nation [Kenya]*, 23 August 2005.

Oyuke, J. (2007), 'Kenya: Rivatex to Open Soon says Trade Minister', *East African Standard* [website] (updated 25 July 2007) <http://allafrica.com/stories/200707241173.html>, accessed 12 August 2007.

Prestholdt, J. (1998), 'As Artistry Permits and Custom May Ordain: The Social Fabric of Material Consumption in the Swahili World, circa 1450 to 1600', PAS Working Paper No. 3, Program of African Studies, Northwestern University, [website] <http://www.northwestern.edu/african-studies/working%20papers/wp3prestholdt.pdf>, accessed 12 June 2007.

Prestholdt, J. (2004), 'On the Global Repercussions of East African Consumerism', *The American Historical Review* 109:3 [website] <http://www.historycooperative.org/journals/ahr/109.3/prestholdt.html>, accessed 12 June 2007.

RATES (2003), *Cotton-textile and Apparel Value Chain Report for Kenya* (Nairobi: Regional Agricultural Trade Expansion Support Program).

Republic of Kenya (1988), *Directory of Industries 1986* (Nairobi: Government Printers).

Republic of Kenya (2001), 'Position Paper on the Textile Industry', <http://www.tradeandindustry.go.ke/documents/di_report_textile.pdf>, accessed 12 November 2007.

The Nation [Kenya] (2004), 'Kenya: Let Locals Reap AGOA Gains', 14 April 2004.

Udalagama, B. (2006), *Personal Interview*, Nairobi, 26 April 2006.

United Nations (1996), *1995 International Trade Statistics Yearbook. Vol. 2: Trade by Commodity* (New York: United Nations).

United Nations (2003), *2001 International Trade Statistics Yearbook. Vol. 2: Trade by Country* (New York: United Nations).

Yeung, H. (2002), 'The Limits to Globalization Theory: A Geographic Perspective on Global Economic Change', *Economic Geography* 78:33, 285–305.

Chapter 13
Conclusion

Lois Labrianidis

The increased mobility of production – regionally, nationally and internationally – constitutes one of the key challenges confronting social scientists and policy makers. The main dimensions of the challenge could be usefully discussed in relation to four questions: 1) *why* are these changes taking place, 2) *how* are they taking place, 3) *so what* and how significant are these changes (impact), and 4) is there scope for action? This book attempts to provide new insights and rigorous arguments about a wide range of issues that are captured within these questions. It does so by bringing together academics from different disciplinary backgrounds, located in diverse national settings, and bringing together expertise from a variety of sectoral contexts. Interestingly, the evidence presented in the chapters of this book provides a picture of diversity between individual industrial contexts that in some respects questions the rationale behind the emphasis placed upon sectors with considerable labour-intensity.

This chapter attempts to bring together the main arguments regarding the four dimensions of the delocalization challenge. It attempts to decipher similarity and difference, and to advance our understanding of the phenomena under investigation. It places a special emphasis on exploring the possible scope and avenues for action, and aims to inform policy decision-making at different levels.

Why?

Industrial activity is becoming increasingly mobile – within as well as between states. Technological advances in ICT and change towards more liberal governance regimes underpin delocalization. The prevailing wisdom suggests that this process is instigated by 'relative labour scarcity'. This scarcity is manifested in terms of wages, alternative opportunities in the labour market, outmigration, fast rising expectations/cultural aspirations and changing preferences in terms of work and lifestyles (for example, young people tend to prefer working in services rather than in manufacturing). The impact of these changes is context specific: it would depend on the level of economic development and degree of social stability, among other factors, and would vary across countries.

In this sense, the shape and direction of delocalization may be influenced by a broad range of factors, rather than being exclusively dependent on labour cost. While labour cost is certainly a very important factor, this book demonstrates that even in some of the most labour-intensive sectors, it is not always the main consideration. For example, labour cost considerations cannot solely determine issues such as the choice of destination or the means of achieving delocalization. *It is not only the labour cost that matters*, but rather the *overall environment*, since it appears that 'Stable environments save costs'. Within this context, global regulatory frameworks, national institutional settings and proximity (not only geographic but also cultural) may be of considerable importance.[1] The evidence presented in the main body of the book suggests that when firms decide to delocalize, they do not act as perfectly rational, fully informed actors. This is because in some cases the uncertainty of the environment makes such choices impossible or there is lack of sufficient information, but more importantly on many occasions companies seem to ignore or misunderstand difficulties that would be obvious to the 'rational economic agent'. Some such considerations would include issues around logistics, regulation and the impact of cultural factors, to name but a few. There is also the question of stability and of certainty that could lead to substantial additional costs. The decision to delocalize to the EU, NAFTA or some other well regulated regions provides a more stable and less uncertain environment. While one cannot calculate the exact value added by operating in stable and well regulated environments, working in uncertain environments would most certainly lead to higher costs. Moreover, companies tend to 'imitate' others, follow what is 'in the air', without having the whole picture. One company imitates the strategy of the other, or maybe some consultancy companies actually promote the dominant decision of a sector to delocalize to a certain place. Even further, questions about cost more generally, even when fully rationalized, are still open to considerations such as what constitutes the 'core' of the business and what is the appropriate time horizon over which it should be calculated. In this sense 'cost', especially in the case of complex organizations, is also often a rhetorical device through which very different priorities and directions of change could be argued to be appropriate.[2] Thus, we argued that the overemphasis in the literature on labour cost in particular, and cost more generally, vis-à-vis other factors needs some correction as well as further elaboration.

How?

As far as delocalization is concerned, *diversity is the name of the game.* On the *firm level*, there are firms delocalizing that are large TNCs and at the same time

---

1 For example in the automotive industry there are cases of companies that delocalized in China and they realised afterwards that the production costs there are in fact high. Take the example of Honda, it argues that it is more expensive to build the Accord in China than in the USA (Donnelley et al. 2007).

2 This was well illustrated in the cases of two UK TNCs in the electronics and clothing sectors.

there are very small firms (for example, Greek companies delocalizing to Bulgaria taking advantage of spatial proximity – *local delocalization* – or small software start-ups moving from the USA to countries as far away as India). So there is diversity in terms of the firm size. There is also great diversity depending on the sectors and on the types of internationalisation pursued. As far as *countries* are concerned, there is a great diversity there too; there is *a wide mosaic of countries with various levels of development*. There is no need to go into simplistic categories like new and old members of the EU; things are much more diverse. It is really too simplistic to classify countries like UK and Greece in the same category. There are all types of differentiations in that. Of course we must not go to the other extreme and say that everything is different, so there is nothing to connect things and no way to analyse things apart from speaking about particularities.

Our analysis shows that enterprise strategy is *multidimensional*. There are no ideal type models. Enterprises may opt for different strategies when they operate even in the same segment of the market and in the same national context. This is because the choice of strategy is not influenced only by contextual factors (that is, industrial- or locality-specific) but also by the firms' resources and competences. These competences are the result of a trajectory, a pathway of changes that have taken place on both the organizational and individual (in the case of entrepreneurs in SMEs) levels.

Interestingly, success is not exclusive to one (or more) enterprise strategies. There are successful enterprises across the board of strategies adopted in all (four) industries and all (five) countries. This suggests that success depends first of all on their appropriateness for the industrial enterprise and regional context and also on how well they are applied. All strategies, of course, entail an element of risk.

However, there are strategies that are linked with a somewhat stronger performance than others. These strategies involve a lower degree of dependence upon individual customers or singular markets. This may take the form of either developing generic competences or focusing – in part – upon servicing the needs of the domestic market (a factor that is often overlooked in the literature).

In the companies' breakout strategy enterprises develop these competences without necessarily using them. They develop them maybe in order to minimize risks involved in engaging in strong relationships. There are the generic competences that would allow them to look elsewhere if something goes wrong. Something that is quite interesting is that there are competence lock-in strategies, which are not exclusively linked to the price-sensitive segment of the market as one would have expected.

Enterprise strategies also appear to be linked to specific patterns of external linkages. It is in this area that some interesting findings emerge. These stress the importance of the interplay between strong linkages, but also the multiplicity of linkages that will facilitate information flows and diminish the threat of 'ossification'. In these instances even dependence may exist alongside relationships of mutual confidence and trust.

Indeed, proximity is a complex and multi-dimensional concept; it takes various forms and it goes well beyond geographical proximity. For example, apart of geography, proximity between Greece and Bulgaria involves social and cultural proximity as well as issues of trust, all of which are very important. The question of confidence in relations and dependence is also important, but they are not exclusive; sometimes you have dependence or confidence and trust. Of course proximity is not something with universal consequences.

Border areas seem to be the ones most heavily involved in delocalization, at least with regard to small firms. However, there are significant differences between the countries as to the density of delocalization activities along the borders (for example, in our case studies there is, on the one hand, intense delocalization at the Greek-Bulgarian and Estonian-Finnish borders, while on the other hand, there is low level of delocalization of the studied LII at the Polish-German borders), reflecting and at the same time conditioning the development level of these areas.

So What? (Impact)

Delocalization can operate as a key mechanism to spread prosperity to LDCs

The *consequences of delocalization are once again very diverse*. They are diverse in relation to time; time span is really important. There are differences according to whether the focus is on short-term, medium-term or long-term consequences of delocalization. Finally, there are differences according to country, region, sector or firm.

At the regional/national level we may distinguish between strong and weak regions/countries. In the long-run, in weak regions growth in LII may be viewed a 'window of opportunity'. It may not be always bad news at the end. Something that is happening in LII may result in something else emerging that will change the overall outcome, probably from a pessimistic one to one that is either neutral or positive. In strong regions the importance of growth and/or decline of LII may be of limited significance on the whole.

As far as the home countries are concerned there is an over-exaggeration about job losses in terms of delocalization, whether it is FDI, subcontracting or outsourcing. Social consequences are not just about numbers of jobs, something that is often not reflected in the literature, where there is a strong focus on number of jobs. It is the qualitative dimensions that in our view are very important.

There is obviously no direct and immediate impact on relation between decreasing number of jobs and unemployment levels, not at the national level and usually not on the regional one as well. It is mainly at the local level that we may see this.

The *home country effects of delocalization as far as loss of competences is concerned are important.* In some cases there is a shrinking industry and this leads to the shrinking or disappearance of industry-specific skills. This may create problems in the medium- and long-term. For example, in the case of Greek clothing firms in the medium-run, delocalization may be impossible if people with key skills such as technicians are scarce, or unwilling to leave Greece in order to move to the host country (Bulgaria, FYROM, etc). In the long- term, we might have the case if a company of the same sector tries to start afresh, or to go upmarket. Let's say that in ten years' time, a company would like to become the Zara of that period. They might realize that there are no competences around.

To what degree does the irreversible character of losing some competences constitute a problem? This is not necessarily a huge problem in macro-economic terms for the national economy – it is usually not for the regional economies. And in the long-term, it doesn't matter; some activities, some industries disappear and new ones appear. So from the economic point of view, even from the social well-being point of view, it may not matter.

Delocalization from one European country to another might be seen as a relocation from one part of Europe to another (even more so within the context of the EU) which will enhance its integration. When a firm is relocated from one area to another within the same country, this is considered as a regional rather than a national problem. In a similar way one might think that what has been described in the second part of this book is not a delocalization from UK or Greece or Italy or Spain to the CEECs (new Europe), but rather a relocation within the EU–27 and hence a preservation of jobs. To the extent that those jobs can be preserved for the next decade in the new members these countries will be integrated more easily within the EU.

The net employment effects of delocalization within Europe are not indicating a 'race to the bottom'; they are rather positive at least in the mid-term. Moreover, the social effects of delocalization are more limited than is often maintained. This is due to the fact that there are intermediating factors (for example, socio economic and political features of the region/nation) that influence whether the impact is strong or weak. Even in terms of numbers of jobs, the overall effect of delocalization processes at the EU level or European level is positive for these industries. Obviously we cannot go on and generalize this for other industries. This is particularly because new jobs come partly, or in some countries to a large extent, to less developed areas where there are limited physical alternatives.

The importance of the environment, whether national or regional, on the outcome of delocalization is more clearly visible in the case of Africa, where it seems to be leading towards 'a race to the bottom' since TNCs quite often can play one country against another.

Delocalization does cause job losses in the home countries but they are quite moderate; the phenomenon is not as intense as the hype about it is suggesting. In the context of this study, the only possible exception to that is the UK, where the majority of firms surveyed reported a decline in employment. Surprisingly, at the same time, the UK is one of the countries with the lowest unemployment rates among the EU. Even more surprisingly, areas within the UK with historically high concentrations of the sectors under study (for example, footwear in

Northamptonshire) are displaying some of the lowest unemployment rates only a few years after the decline of the sectors in question. Perhaps this is due to the particular European countries that are the focus of Part II of this book while in other countries with a tradition in labour-intensive activities (such as Italy and France) the problems might be much more intense.

The effects of delocalization go far beyond the companies directly involved in the process; they influence the economy as a whole. For example, due to delocalization, trade unions might become more resilient while other companies in the area/country have to face more intense competition. Even the threat that a big company will move influences the policies of the trade unions; they are under threat: 'behave, otherwise we will delocalize'.

Relation of delocalization and economic development

Then, going to the dimension of economic development, there is a convergence of GDP per capita versus convergence of industrial structures. The delocalization process is going to converge the industrial structure and, at the same time, make an impact on convergence in terms of GDP per capita. The *time factor is very important* for this process, since the convergence of GDP per capita is changing the conditions of labour cost. So, *in one point, convergence of GDP per capita especially in cases where this is influenced mainly by the labour cost is going to stop the process of convergence of the industrial structures, because delocalization will not proceed anymore.*

Convergence process within the EU will lead to lagging countries losing their competitiveness in the LII. In the sense that in LII, labour cost plays a much more important role and, from some point onwards, convergence means losing the comparative advantages that led to the delocalization of the LII.

Another thing that might not be so totally new is this *maxi profundis* impact of FDI. Obviously, one of the major conclusions could be showing the extent to which and how delocalization affects the long-term economic growth or competitiveness – using the notion of economic growth at various geographical scales. This is a crucial result of the study. There is a need for a dynamic view of these globalization production networks, which empirically is not so easy. But the interpretation should be made in a dynamic way. The triangle dimension has been emphasized several times.

What really matters is how in the long run delocalization affects competitiveness and economic growth and consequently indirectly affects the number of jobs and the social well-being. So this is what really matters in the long run.

Is there Scope for Action?

Delocalization can be managed

Delocalization, at least within Europe, is definitely not 'a race to the bottom' in terms of its effects on employment, as well as its social effects. One might even argue that they are quite positive for both home and host countries. Whilst we can not use the experience of five European countries to generalize about the reality of delocalization the world over, we argue that the evidence presented here suggests that delocalization can be managed, at least regarding socio-economic and political formations that seem to be broadly similar.

In doing so, governance matters. Indeed, Gereffi and Mayer (2004, 2) argue that today there is crisis of governance in the sense that there is an inadequacy of institutions not only to facilitate market growth and stability but also to regulate markets and market actors and to compensate for undesirable effects of market transactions. The rise of an increasingly global economy, no longer firmly rooted in nation-states, and one that encompasses a large portion of the developing world, is challenging the regulatory and compensatory capacities of both DCs and LDCs. Moreover, at the international level little regulatory capacity has evolved to take up the slack. These developments have led to a governance deficit of considerable magnitude.

Moreover, the *outcome of certain changes is not unidirectional*. For example, the abolition of MFA/or Bulgaria's accession to the EU might have unforeseen repercussions. Trade policy (for example, MFA) shapes the geography of production. Hence, on the one hand, Bulgarian firms might benefit from the breakdown of triangular schemes, since they can be in direct contact with the customer. On the other hand, the very same changes might be a 'kiss of death' for some industries (for example, clothing), which could gradually miss their comparative advantages (for example, low labour cost, more relaxed labour legislation and more tax incentives to companies). Again, the impacts of the global regulatory environment on developments in particular manufacturing sectors can be considerably more important to LDCs (for example, Kenya). Hence what this book looks for is softer policy recommendations and if this is linked to the recognition of the importance of diversity it will lead to clever and flexible policy recommendations, not a one-size-fits-all kind of approach.

However, the key challenge is: what are the dimensions of action? More specifically, is action impacting upon the pace and/or direction of change, or ameliorating the effects of change?

Impacting upon the pace and/or direction of change

Regulations on a trans-national level are important because they set the framework. However, they do not determine the form that delocalization is going to take.

In delocalization there is an element of regulation coming from the consumers in industrialized countries that may avoid ultimately the worst excesses of capitalism. It is becoming increasingly apparent that large groups of consumers do not want to buy goods that are made based on the over-exploitation of human beings (for example, clothes made by a 7-year-old). So it is consumers who also put pressure on the producers in terms of standards and regulations. It is not only regulations that are government driven, but regulations that are changing the attitude of the consumers.

In this sense market mediated pressures certainly have a role to play as mechanisms of governance and our study clearly demonstrated that TNCs operating in CEE are actively involved with enforcing not only quality but also ethical standards in the organization of production of their business partners. Consumer pressure, however, constitutes only one among many mechanisms of regulation; it has both its limits and limitations and thus can add to, but cannot be a substitute for, formal regulation.

With regard to ethical trade and governance, all approaches are very much oriented towards the DCs (primarily EU and USA) and the basic idea is that markets are in the EU and USA. But if we look at what is really going on in these industries that have their headquarters in Europe, then we see that other markets – such as Russia and China – become increasingly important. Furthermore, these are no longer places to produce cheap products, but are gradually becoming important markets. As the example of Nokia and other Finnish electronics firms shows, nowadays, even product development and also markets move to China. This however does not mean that rules within the context of EU are becoming less relevant but rather that their impact is no longer simple and uni-directional, but instead is complexly inter-dependent on changes that are taking place on the global level as well as in the level of individual states. States, for example, are acquiring new powers of coordinating or steering, and thus have the ability to influences other levels of governance (for example, sub-national and supra-national) (or *meta-governance*).

Substantively, given that both governments and markets fail, although in different ways, governments can still play a role in developing correcting mechanisms for the failures of the market, where short and medium-term orientations are predominant. While some market players can also have longer-term temporal horizons as well as being able to tolerate higher degrees of risk, national governments and supra-national organizations such as the EU seem to be best placed in providing longer term vision and support for sustainable economic and social restructuring. Furthermore, EU institutions in particular can be instrumental in extending the scope and depth of governance on the global level, as well as in shaping the global agenda, particularly in countering overly-enthusiastic neo-liberal visions of globalization. Withdrawal, while a possible option, is neither the only nor necessarily the best one. Our analysis suggests that both states and supra-national organizations such as the EU have a *new*, rather than *no*, governance role to play.

Developing a discussion on such a broad level can neither be supported nor rejected by the experiences of individual companies. As it could be expected company experiences and attitudes towards specific forms of regulation and towards regulation in general varies. Nevertheless, understanding the concerns and conflicting interests of different stakeholders is key to informing specific policy mechanisms. Thus, for example, the introduction of several EU regulations is costly for the companies and particularly costly for the small businesses. For the TNCs this is a less significant problem, firstly because of the availability of resources and secondly, because they already have these standards in most of the countries. Even if they are not passed by the national regulations, they introduce them for internal reasons. For small companies that look for every penny, this is a big issue. This is really important: we normally think that regulations are generally good, but they might lead some companies out of business, or put them in a different position. This might be an important point since it is not advanced in the public governance literature.

There are of course also cases where one has to see with caution what the real meaning of this information is. For example, companies consider EU regulations to be very costly, but on the other hand this is a condition in order to gain access to the EU markets and to a certain extend exclude others from these markets. A basic question that one has to respond to is *why do companies decide to delocalize within the EU, in a more or less regulated environment* and certainly more regulated compared to other countries like China, Moldova or Morocco? Do they benefit from the regulations or some sort of regulations, or is it that they delocalize to countries of the EU because *they do not plan to behave differently?* It is not really the regulation itself that pushes them to behave like that; it is possibly because they care for the image of the company since, especially for larger company, this matters. This depends on the company: IKEA, for example, will behave more or less the same in other countries.

Regulations have some positive aspects for the companies. Perhaps the main reason behind FDI or subcontracting to some countries that are not necessarily really rigid in global terms is the relatively stable environment, the question of reliability and stability – economic, political, etc. And this is provided by some sort of regulations. So they have these benefits and it pays to be in a more regulated environment.

The EU can influence the rest of the world if it sets up a European fair trade trademark that will be rigorously implemented, no matter where the product is produced. It would be the responsibility of the company, as it is now for some big companies even in the clothing industry, to make sure that their suppliers comply with certain standards. This is a way in which big buyers of the EU can have a considerable impact on the rest of the world. In this respect, and coming back to our earlier argument, the development of active, though not necessarily only and always directly intervening regulative mechanisms (on state, the EU, etc levels) is crucial.

Ameliorating the effects of change

There is a big difference between forms of management of an economy. For example, management of the UK economy, which is one of the most liberal economies in Europe, is done with the right touch (for example, there is a minimum wage in the country) and this management has not led to the collapse of enterprise activities. The UK economy has been managed, and is still highly managed, without disturbing the conditions of supply but through soft and subtle measures that are collectively evaluated as positive (for example, the role of embassies in creating a facilitating environment for firms to delocalize).

This study does not support policy recommendations that change the conditions of supply (for example, that artificially reduce the price of labour), because this is not sustainable in the long-term. Even more so it does not support policies that restrict firms to take decisions on how to handle their operations (for example, 'forbid' them to delocalize).

Policies by DCs and international bodies (for example, EU, USA, WTO, IMF) are often focused to protect the DCs against the interests of the LDCs. That is, the EU and USA, in those cases in which their economy is competitive, argue that there must not be any trade barriers; while in the cases in which their economies are less competitive (for example, agricultural products and the clothing industry) they argue for protection. In order to protect the production within EU and within the DCs in general, the EU acted against the interests of the LDCs. However, as seen from the exploration of social consequences, these are often modest.

It is almost imperative to give LDCs the opportunity to find a niche (for example, agricultural products, part of products that are related to the LII). What the EU can do with the *globalization fund* is to manage the effects of delocalization and not the delocalization per se. Certainly, it is not only the unemployment issue; it is also the effects of delocalization in terms of firms that still operate in the same sectors and how they face their own competitiveness increasing or decreasing, because some other firms have delocalized themselves. Trying to understand the effect of delocalization is really trying to find out not only why firms are delocalizing, but also what the effects on the firms, the regions, the people, that stay behind have been. Hence policies should be oriented not only to unemployment but to all the issues related to delocalization.

Regional blocks that include countries with different levels of development often lead to shocks for all the parties involved. Total trade among the NAFTA countries has more than doubled between 1993 and 2002; however, as Anderson and Cavanagh (2004) argue, there are problems in all three countries. That is, on the one hand Mexico did indeed attract a significant number of jobs in export processing factories. However, despite substantial productivity growth, real wages in manufacturing dropped between 1994 and 2000 and this is due in part to the fact that NAFTA has failed to protect the rights of workers to fight for their fair share of economic benefits. Mexico in the 2000–2003 period lost more than 230,000 export assembly jobs, 35per cent of which were due to shifts in production to China.

This job flight has raised fears that *Mexico's strategy of attracting investment by offering low wages is short-sighted.* NAFTA forbids governments from placing requirements on foreign investors that would ensure benefits for the broader economy (for example, to require that investors use a set amount of local content in manufacturing).

Final Thoughts

The delocalization process, explored by social scientists and addressed by policy-makers at different levels, is a complex and continuously evolving phenomenon. In fact, one may argue that delocalization may be best captured as a multitude of often converging but sometimes diverging phenomena that have industrial, locational and enterprise specificities. Research to date has focused primarily upon the commonality and tended to diminish the importance of diversity. We believe that this book offers the point of departure for the introduction of a corrective.

This book also highlights the importance of distinguishing between analytical units: for example the reality of delocalization processes may differ between firms, networks, regions and nations. The importance of this rests with its impact upon policy. Indeed, specific actions and/or initiatives may have significantly differential results. This necessitates an explicit statement of intent.

Action can be taken in order to influence the processes as well as the consequences of delocalization. However, it is apparent from this book that what is needed from the outset is a clear identification of the ultimate aims of policy action. Indeed, a clear distinction between competitiveness and 'employment' social actions may enable us to identify initiatives that are more focused and sustainable in the long-term.

Moreover, it is important to recognize that there are clear boundaries regarding the potential impact of such action. Indeed, policy initiatives that attempt to alter the conditions of supply by artificially reducing costs in one location in relation to all others may be hard to sustain in the long-term. Rather perversely, the more successful such actions are (thus impacting positively on growth and subsequently incomes) the less sustainable the policy becomes.

References

Anderson, S. and Cavanagh, J. (2004), 'Fact Sheet on the NAFTA Record: A 10th Anniversary Assessment, <http://www.zmag.org/content/showarticle.cfm?ItemID=4865> (accessed 27 May 2007).

Donnelley, T., Collis, C., Gomes, E. and Morris, D. (2007), 'The Chinese Car Industry: Opportunity of Threat?', Paper presented at the International Conference, *The New International Division of Labour? The Changing Role of Emerging Markets in Automotive Industry*, Krakow, February 2007.

Gereffi, G. and Mayer, F. (2004), 'The Demand for Global Governance', Working Paper Series SAN04-02 (Duke: Terry Sanford Institute of Public Policy).

Index

Name Index